Medical Intelligence Unit

Immunosenescence

Graham Pawelec, MA, Ph.D.
Professor of Experimental Immunology
University of Tübingen Center for Medical Research (ZMF)
Tübingen, Germany

Landes Bioscience
Austin, Texas
U.S.A.

Springer Science+Business Media
New York, New York
U.S.A.

Immunosenescence

Medical Intelligence Unit

Landes Bioscience
Springer Science+Business Media, LLC

ISBN: 978-0-387-76840-3 Printed on acid-free paper.

Springer Science+Business Media, LLC, 233 Spring Street, New York, New York 10013, U.S.A.
http://www.springer.com

Please address all inquiries to the publishers:
Landes Bioscience, 1002 West Avenue, 2nd Floor, Austin, Texas 78701, U.S.A.
Phone: 512/ 637 6050; FAX: 512/ 637 6079
http://www.landesbioscience.com

Printed in the United States of America.

9 8 7 6 5 4 3 2 1

Library of Congress Cataloging-in-Publication Data

Immunosenescence / [edited by] Graham Pawelec.
p. ; cm. -- (Medical intelligence unit)
ISBN 978-0-387-76840-3 (alk. paper)
1. Immune system--Aging. 2. Aging--Immunological aspects. I. Pawelec, G. (Graham) II. Series: Medical intelligence unit (Unnumbered : 2003)
[DNLM: 1. Aging--immunology. 2. Immunity--physiology. 3. Age Factors. 4. Disease Susceptibility--immunology. 5. Longevity--immunology. 6. Longitudinal Studies. 7. T-Lymphocytes--immunology. QW 540 I336 2007]
QR184.5.I48 2007
616.07'9--dc22

2007042878

About the Editor...

GRAHAM PAWELEC received a BA, MA and Ph.D. from the University of Cambridge, England, and Dr. habil and venia legendi from the University of Tübingen, Germany, where he became Professor of Experimental Immunology in 1997. He is currently a visiting professor at Nottingham Trent University in the U.K. He is a member of the British Society for Immunology, Deutsche Gesellschaft für Immunologie, Deutsche Gesellschaft für Gerontologie und Geriatrie, European Association for Cancer Research, Association for Immunotherapy of Cancer and the American Association for Cancer Research. He received the Sandoz Award for Ageing Research in 1996. His research interests are centered on alterations to T-cell immunity in ageing and cancer in humans and the impact these have on vaccination. He is on the Editorial Boards of *Mechanisms of Ageing and Development, Experimental Gerontology, Biogerontology* and *Immunity and Ageing*. He is Co-Editor-in-Chief of *Cancer Immunology Immunotherapy* (2006 Impact Factor 4.313). He has authored nearly 100 peer-reviewed original articles out of a total of more than 250 publications and has edited 3 books.

CONTENTS

EDITOR

Graham Pawelec
Professor of Experimental Immunology
University of Tübingen Center for Medical Research (ZMF)
Tübingen, Germany
Email: graham.pawelec@uni-tuebingen.de
Chapters 1, 6 and 15

CONTRIBUTORS

Note: Email addresses are provided for the corresponding authors of each chapter.

Corona Alonso
Department of Immunology
"Reina Sofia" University Hospital
University of Córdoba
Córdoba, Spain
Chapter 3

Richard Aspinall
Imperial College of Science, Technology and Medicine
Department of Immunology
Chelsea and Westminster Campus
London, U.K.
Chapter 7

Carmela Rita Balistreri
Gruppo di Studio sull'Immunosenescenza
Dipartimento di Biopatologia e Metodologie Biomediche
Università di Palermo
Palermo, Italy
Chapter 14

Giuseppina Candore
Gruppo di Studio sull'Immunosenescenza
Dipartimento di Biopatologia e Metodologie Biomediche
Università di Palermo
Palermo, Italy
Chapter 14

Calogero Caruso
Gruppo di Studio sull'Immunosenescenza
Dipartimento di Biopatologia e Metodologie Biomediche
Università di Palermo
Palermo, Italy
Email: marcoc@unipa.it
Chapter 14

Javier G. Casado
Immunology Unit
Department of Physiology
University of Extremadura
Cáceres, Spain
Chapter 3

Antonino Crivello
Gruppo di Studio sull'Immunosenescenza
Dipartimento di Biopatologia e Metodologie Biomediche
Università di Palermo
Palermo, Italy
Chapter 14

Olga DelaRosa
Department of Immunology
"Reina Sofia" University Hospital
University of Córdoba
Córdoba, Spain
Chapter 3

Julian de la Torre-Cisneros
Department of Immunology
and Infectious Diseases Unit
"Reina Sofia" University Hospital
University of Córdoba
Córdoba, Spain
Chapter 3

Rita B. Effros
Department of Pathology and Laboratory Medicine
David Geffen School of Medicine at UCLA
Los Angeles, California, U.S.A.
Email: reffros@mednet.ucla.edu
Chapter 4

Andrea Facchini
Laboratorio di Immunologia e Genetica
Dipartimento di Medicina Interna
e Gastroenterologia
Istituto di Ricerca Codivilla-Putti
University of Bologna
Bologna, Italy
Chapter 9

Steven R. Fauce
Department of Pathology and Laboratory Medicine
David Geffen School of Medicine at UCLA
Los Angeles, California, U.S.A.
Chapter 4

Frederick Ferguson
Department of Veterinary Science
College of Agricultural Sciences
Pennsylvania State University
University Park, Pennsylvania, U.S.A.
Chapter 1

Rosalyn J. Forsey
LCG Bioscience
Bourn, Cambridge, U.K.
Chapters 1 and 15

Giusi Irma Forte
Gruppo di Studio sull'Immunosenescenza
Dipartimento di Biopatologia
e Metodologie Biomediche
Università di Palermo
Palermo, Italy
Chapter 14

Carl Fortin
Research Center on Aging
Immunology Graduate Program
Geriatric Division
University of Sherbrooke
Sherbrooke, Quebec, Canada
Chapter 6

Tamas Fulop
Research Center on Aging
Immunology Graduate Program
Geriatric Division
University of Sherbrooke
Sherbrooke, Quebec, Canada
Email: tamas.fulop@usherbrooke.ca
Chapters 1 and 6

Inmaculada Gayoso
Department of Immunology
"Reina Sofia" University Hospital
University of Córdoba
Córdoba, Spain
Chapter 3

Maria Paola Grimaldi
Gruppo di Studio sull'Immunosenescenza
Dipartimento di Biopatologia
e Metodologie Biomediche
Università di Palermo
Palermo, Italy
Chapter 14

Beatrix Grubeck-Loebenstein
Institute for Biomedical Aging Research
Austrian Academy of Sciences
Innsbruck, Austria
Email: beatrix.grubeck-loebenstein@oeaw.ac.at
Chapter 10

Sudhir Gupta
Medical Sciences I
University of California
Irvine, California, U.S.A.
Email: sgupta@uci.edu
Chapter 5

Hajo Haase
Institute of Immunology
University Hospital
RWTH Aachen University
Aachen, Germany
Chapter 11

Sine Reker Hadrup
Department of Natural Science and Biomedicine
School of Health Sciences
Jönköping University
Jönköping
Sweden
and
Tumor Immunology Group
Institute of Cancer Biology
Danish Cancer Society
Copenhagen, Denmark
Chapter 1

Dietmar Herndler-Brandstetter
Institute for Biomedical Aging Research
Austrian Academy of Sciences
Innsbruck, Austria
Chapter 10

Katsuiku Hirokawa
Department of Comprehensive Pathology
Tokyo Medical and Dental University
and
Nakano General Hospital
and
Institute for Health and Life Science
Tokyo, Japan
Email: hiro-hiroz@tiara.ocn.ne.jp
Chapter 2

Milena I. Ivanova
Central Laboratory of Clinical Immunology
University Hospital Alexandrovska
Sofia, Bulgaria
Chapter 13

Boo Johansson
Institute of Gerontology
School of Health Sciences
Jönköping University
Jönköping, Sweden
and
Department of Psychology
Göteborg University
Göteborg, Sweden
Chapter 1

Yuko Kikuchi
Department of Comprehensive Pathology
Tokyo Medical and Dental University
Tokyo, Japan
Chapter 2

Masanobu Kitagawa
Department of Comprehensive Pathology
Tokyo Medical and Dental University
Tokyo, Japan
Chapter 2

Anis Larbi
Tübingen Ageing and Tumour Immunology Group
Center for Medical Research
University of Tübingen Medical School
Tübingen, Germany
Chapter 6

Florinda Listì
Gruppo di Studio sull'Immunosenescenza
Dipartimento di Biopatologia e Metodologie Biomediche
Università di Palermo
Palermo, Italy
Chapter 14

Robin Longdin
LCG Bioscience
Bourn, Cambridge, U.K.
Chapter 15

Marco Malavolta
Immunology Center
Nutrition, Immunity and Ageing Section
Research Department
I.N.R.C.A
Ancona, Italy
Chapter 12

Adriana Rita Mariani
Dipartimento di Medicina Interna
e Gastroenterologia
University of Bologna
Bologna, Italy
Chapter 9

Erminia Mariani
Laboratorio di Immunologia e Genetica
Dipartimento di Medicina Interna
e Gastroenterologia
Istituto di Ricerca Codivilla-Putti
University of Bologna
Bologna, Italy
Email: marianie@alma.unibo.it
Chapter 9

Dawn J. Mazzatti
Unilever Corporate Research
Sharnbrook, Bedfordshire, U.K.
Chapter 15

Wayne A Mitchell
Imperial College of Science, Technology
and Medicine
Department of Immunology
Chelsea and Westminster Campus
London, U.K.
Email: w.mitchell@imperial.ac.uk
Chapter 7

Eugenio Mocchegiani
Immunology Center
Nutrition, Immunity and Ageing Section
Research Department
I.N.R.C.A
Ancona, Italy
Email: e.mocchegiani.inrca.it
Chapter 12

Elissaveta J. Naumova
Central Laboratory of Clinical
Immunology
University Hospital Alexandrovska
Sofia, Bulgaria
Email: immun@alexandrovska-hospital.bg
Chapter 13

M. Luisa Pita
Department of Immunology
"Reina Sofia" University Hospital
University of Córdoba
Córdoba, Spain
Chapter 3

Esther Peralbo
Department of Immunology
"Reina Sofia" University Hospital
University of Córdoba
Córdoba, Spain
Chapter 3

Jonathan R. Powell
Unilever Corporate Research
Sharnbrook, Bedfordshire, U.K.
Chapter 15

Lothar Rink
Institute of Immunology
University Hospital
RWTH Aachen University
Aachen, Germany
Email: lrink@ukaachen.de
Chapter 11

Letizia Scola
Gruppo di Studio sull'Immunosenescenza
Dipartimento di Biopatologia
e Metodologie Biomediche
Università di Palermo
Palermo, Italy
Chapter 14

Rafael Solana
Department of Immunology
"Reina Sofia" University Hospital
University of Córdoba
Córdoba, Spain
Email: rsolana@uco.es
Chapter 3

Per thor Straten
Tumor Immunology Group
Institute of Cancer Biology
Danish Cancer Society
Copenhagen, Denmark
Chapter 1

Jan Strindhall
Department of Natural Science
and Biomedicine
School of Health Sciences
Jönköping University
Jönköping, Sweden
Chapter 1

Raquel Tarazona
Immunology Unit
Department of Physiology
University of Extremadura
Cáceres, Spain
Chapter 3

Masanori Utsuyama
Department of Comprehensive Pathology
Tokyo Medical and Dental University
Tokyo, Japan
and
Sanritsu Medical Laboratory
China
Chapter 2

Sonya Vasto
Gruppo di Studio sull'Immunosenescenza
Dipartimento di Biopatologia
e Metodologie Biomediche
Università di Palermo
Palermo, Italy
Chapter 14

Anders Wikby
Department of Natural Science
and Biomedicine
School of Health Sciences
Jönköping University
Jönköping, Sweden
Email: anders.wikby@hhj.hj.se
Chapter 1

Jacek M. Witkowski
Department of Pathophysiology
Medical University of Gdansk
Gdansk, Poland
Email: jawit@sanus.amg.gd.apl
Chapter 8

PREFACE

Human immunosenescence contributes to morbidity and mortality in later life. The age-associated increasing incidence of cancer and cardiovascular disease plateaus at around 80 years of age in industrialised countries, but death due to infectious disease continues to increase up to 100 years of age and beyond. Understanding the reasons for age-associated alterations to protective immunity in the elderly would facilitate the development of interventions to reconstitute appropriate immune function, increase responsiveness to vaccination and extend lifespan. The majority of the papers collected in this volume therefore address not only the mechanisms responsible for immune ageing in humans but consider what might be accomplished to redress the erosion of immune competence with age.

The first problem facing the gerontologist investigating human ageing is their longevity: most studies are conducted in a cross-sectional manner, in which parameters of interest in elderly cohorts are compared to young controls. However, the ageing trajectories of people now 80 years old, born at the beginning of the 20th century, will have been very different in mostly unidentifiable ways from those born towards the end of that century. These differences include population genetics, nutrition, stress, disease and medical treatment—all of which make these two populations hardly comparable. An approach to overcoming some of these problems is presented in the first chapter, in which Wikby et al review findings from pioneering longitudinal studies of very elderly people from one location in Sweden. These decades-long studies have revealed parameters of immune function which predict mortality and which may provide a rationale for immune interventions. Although the subjects studied were exceptional in that they survived to beyond the average lifespan, and one could therefore argue that this is a non-representative selected population, the "immune risk profile" (IRP) being identified in this work seems likely to be of value in other circumstances as well, especially in patients with cancer. The issue of how to assess immune status in middle-aged individuals in order to predict risk categories and to intervene is discussed in Chapter 2 by Hirokawa et al. They propose a scoring system taking some characteristics of the IRP, as well as other parameters, into account in both healthy subject and cancer patients, and discuss how immunological interventions could be targeted.

The mechanisms responsible for such measurable changes to the immune system in elderly people are discussed in the next several chapters. Gayoso et al focus on the immune cell type which appears most affected by ageing, the CD8 cell. Longitudinal studies reviewed in Chapter 1 had defined clonal expansions and contractions of CD8 cells specific for persistent herpesviruses, especially cytomegalovirus (CMV) as being an important part of the IRP. Gayoso et al investigate the surface markers and functions of these cells in detail, and in the next chapter, Fauce and Effros consider whether telomere shortening in such cells is the driving force resulting in immune dysfunction not only in ageing but also in HIV disease. The hallmark accumulation of these "senescent" or "exhausted" CD8 cells in ageing, HIV and possibly many other diseases of chronic antigenic exposure, including cancer, may be affected by the balance of pro- and anti-apoptotic influences, as discussed in Chapter 5 by Gupta. An underlying mechanism affecting all these outcomes is altered T-cell stimulation status in old cells, due to changes to membrane characteristics as signalling cascade, as discussed next in the chapter by Fulop et al. A consideration of why dysfunctional T-cells are not simply replaced by newly-generated naïve cells is contributed by Mitchell and Aspinall in Chapter 7, who link immunosenescence and thymic involution with the dangers of autoimmunity. Witkowski then considers the age-associated prevalence of different autoimmune diseases in the context of altered T-cell function as a consequence of age. The final chapter in this section on mechanisms of immunosenescence is contributed by Mariani et al and describes alterations in chemokines and their receptors which may maintain the inflammatory state commonly associated with frailty in the elderly.

The clinical impact of immunosenescence is considered in Chapter 10 by Herndler-Brandstetter and Grubeck-Loebenstein in the very important context of the responses of the elderly to vaccination. Improving the outcome of vaccination in old people would make a great impact on health and well-being in later life. Whether dietary supplementation might contribute to improving immune function is discussed in the next two chapters by Haase and Rink and Mocchegiani et al, taking zinc as the example, and considering how the individual's genetic background is likely to influence the outcome. The issue of the impact of genetic differences is discussed in the chapter by Naumova and Ivanova, illustrated with reference to the highly polymorphic HLA system and cytokine gene polymorphisms. This is explored further in the penultimate chapter in the context of atherosclerosis and Alzheimer's disease by Caruso and colleagues.

Finally, one chapter describing a model experimental system, by Mazzatti et al, paves the way for utilising cutting-edge technical approaches involving T-cell cloning and sophisticated mass spectrometric analysis in an attempt to seek unsuspected biomarkers of immunosenescence which may shed light on underlying mechanisms and offer novel avenues for intervention.

This book arose from collaborations and conferences initiated under the aegis of European Commission-supported projects in human development and ageing. I am grateful to the foresight of the Commission in supporting these initiatives and to all the contributors to this book, together with whom I trust much more fruitful work remains to be carried out.

Graham Pawelec, MA, Ph.D.

Acknowledgements

This book arose from collaborations established with the support of a series of European Commission-sponsored projects (EUCAMBIS, ImAginE, and most recently T-CIA and LifeSpan [FP6 036894]).

Chapter 1

Immune Risk Phenotypes and Associated Parameters in Very Old Humans:

A Review of Findings in the Swedish NONA Immune Longitudinal Study

Anders Wikby,* Frederick Ferguson, Jan Strindhall, Rosalyn J. Forsey, Tamas Fulop, Sine Reker Hadrup, Per thor Straten, Graham Pawelec and Boo Johansson

Abstract

In the previous OCTO immune longitudinal study of free-living subjects >85 yr. selected for good health, we identified an Immune Risk Phenotype (IRP) associated with increased mortality. The IRP was characterised by high $CD8^+$, low $CD4^+$ T-cell counts and a poor T-cell proliferative response, inversion of the CD4/CD8 ratio and evidence of persistent cytomegalovirus infection. In the NONA immune longitudinal study the aim was to examine whether the same IRP parameters applied to a population-based sample aged 86-94 years who were not selected for very good health. More sophisticated analytical parameters were studied, as well as the role of inflammatory processes in relation to longevity. The immune panel included the analysis of T-cell subsets, inflammatory markers, virus serology, cytokines, TCR clonotype mapping and functional and phenotypic analysis of virus-specific $CD8^+$ cells by HLA/peptide multimers, in collaborations between participants of the EU funded T-CIA project.

The present review of findings from a 6 year study of Swedish nonagenarians focuses on the IRP and its associations with persistent virus infection, $CD8^+$ T-cell differentiation, cytokines, cognitive functioning, inflammatory activity, virus-specific $CD8^+$ cells and $CD8^+$ T-cell clonal expansions. It also reports on low grade inflammation processes of importance in predicting longevity in very late life.

Introduction

The very old constitute the fastest growing age group in most developed nations and often present with compromised health and significant requirements for service and health care. Much of this late-life health care is necessary for treating infectious disease, which is more frequent and more severe in the very eldery. Clinical interventions to improve health and quality of life in the elderly are therefore likely to focus on the immune system and its age-associated alterations correlating with dysfunction at old age. Cross-sectional studies have revealed many changes in both the

*Corresponding Author: Anders Wikby—Department of Natural Science and Biomedicine, School of Health Sciences, Jönköping University, Box 1026, 551 11 Jönköping, Sweden. Email: anders.wikby@hhj.hj.se

Immunosenescence, edited by Graham Pawelec. ©2007 Landes Bioscience and Springer Science+Business Media.

adaptive and innate immune systems but assessing their causative association with morbidity and mortality is problematic. In this review we summarise results from six years of the Swedish NONA Immune Longitudinal Study.[1] The longitudinal design allows direct association of parameters at baseline with events at subsequent measurement periods, but very few longitudinal immune studies have been performed in humans.[2] One obvious reason is probably that they require sustained effort, extensive financial support and careful control of investigated panels.[2] Particularly rare are longitudinal studies on samples of people over 80 years of age, the group deliberately focused on in the Swedish Immune Longitudinal Studies. The inclusion of very old individuals in these studies is justified by the fact that oldest-old samples provide a model to detect intra-individual changes in a period in life with high probability for changes in immune parameters, health conditions and mortality.[3]

The increased frequency of disease is one of the primary problems in the selection and definition of a sample in population studies of ageing. To handle this problem, most studies have used various selection schemes to exclude individuals with underlying disease from participation in studies of the immune system. The *SENIEUR* Protocol represents an application of a set of exclusion criteria to select individuals in optimal health. However, this selective sampling results in exclusion of all but a minority of individuals aged 80 and older which is nonrepresentative with regard to the entire population.[4] Another way to diminish factors confounding between ageing and disease has been to employ exclusion criteria tailored to the study situation.[5] This was used in the previous Swedish OCTO Immune Longitudinal Study, an integrated component of the Swedish OCTO Longitudinal Study, focusing on psychosocial and functional parameters of importance in late life.[6]

Participants in the OCTO Immune Study were included if they were aged 88-92 years, were not institutionalised, had normal or only mild cognitive dysfunction according to neuropsychological tests and were not on a drug regimen that might influence the immune system. Of the 213 potential subjects available at baseline in 1989, 110 met these inclusion criteria and of these 102 individuals participated in the study. Twenty-three could participate at all four measurement times in 1989, 1990, 1991 and 1997. Lack of participation at various measurement occasions was mainly due to mortality in the sample.

The analysis of immune data at baseline of the OCTO Immune Longitudinal Study revealed a cluster of parameters predictive of subsequent 2-year mortality,[7] later designated the Immune Risk Profile (IRP).[4] This cluster consisted of high levels of $CD8^+$ T-cells, low levels of $CD4^+$ T-cells and poor proliferative responses to mitogens, as well as low numbers of B-cells. The longitudinal nature of the study allowed the demonstration that additional individuals developed the IRP as they aged, caused by increases in the number of peripheral $CD8^+$ cells, decreases in $CD4^+$ cells and altered CD4/CD8 ratios.[8] These new IRP^+ individuals again were found to have increased mortality over the next 2-year study period. In addition, it was found that the IRP could be defined solely using the inverted CD4/CD8 ratio as a surrogate.[8] The OCTO Immune Longitudinal Study documented that 31% individuals of the 102 participating subjects were either IRP^+ at baseline (16%) or developed the IRP (15%) over the 8 years of the longitudinal study.[9] It was noteworthy that individuals in the IRP category at baseline or those who moved into it during the 8 year longitudinal period, were never observed to move out of this elevated mortality risk category.[9]

In 2000 it was shown that the IRP is associated with persistent CMV infection, prevalent (90%) in this very old Swedish population, as well as with significant increases in the level of $CD8^+CD28^-$ cells.[9] The former result was unexpected, because CMV infection had been considered to be quite harmless. However, these findings suggested that changes in the T-cell balance of IRP^+ subjects might be caused by generation of $CD8^+$ effector cells specific for the persistent CMV infection, with subsequent homeostatic decreases in the CD4 cell number and CD4/CD8 ratio. This was supported by the application of the recently-introduced tetramer technology, demonstrating significant accumulation of $CD8^+$ T-cells specific for the CMV_{NLV} peptide in $HLA\text{-}A2^+$ individuals in association with both age and the IRP.[10]

Results from the OCTO Immune Longitudinal Study provided the basis for the subsequent NONA Immune Longitudinal Study.[1] The overall aims were to advance and refine our knowledge about various predictive factors for longevity with special focus on the IRP, but in the broader context of functional and disability parameters also studied in the NONA.[1] More specific aims included a focus on chronic viral infection, inflammation, $CD8^+$ T-cell phenotype and differentiation, longitudinal changes, mortality in the context of cognitive impairment and $CD8^+$ T-cell clonal expansion and the role of inflammatory parameters in prediction of longevity in the very old.

NONA Immune Subjects

NONA set out to examine a population-based old sample without excluding individuals due to compromised health and to include a continuous evaluation of various individual health parameters.[2] This strategy was used in the NONA Immune Longitudinal Study with data collection at baseline in 1999 and follow-ups in 2001, 2003 and 2005 (Table 1).[1] The clinical variables needed for the evaluation of health and morbidity status are of great significance in the comparison of immune system findings from individuals categorised into subgroups according to their health status. This allows analyses of the impact of change in health status for various outcomes.

The NONA immune sample was recruited among participants in the Swedish NONA Longitudinal Study, in which a population-based sample of oldest-old individuals is investigated.[1] In this study, the oldest-old are tested and interviewed across the domains of physical and mental health, cognitive functioning, personal coping and control, social networks, provision of service and care as well as everyday functional capacity of importance in late life.[2] The sample was drawn in the municipality of Jönköping, located in South Central Sweden and the sampling frame was based on available census information in September 1999 on which a nonproportional sampling procedure was employed including all individuals permanently residing in the municipality. The goal was to have an equal number of individuals aged 86, 90 and 94 at baseline.[1]

Subjects were examined in their place of residence by trained Registered Nurses with extensive experience of working with the elderly. The tests and interviews took about 3 hours, including breaks, for individuals who were able to participate in all parts. Blood was drawn at baseline in 1999 from 138 individuals (Table 1), of which 42 belonged to the oldest birth cohort, 47 were 90 years old and 49 were 86 years old. The mean age of the sample at baseline was 89.8 years with 70% women (Table 1). About 60% of them lived in ordinary housing and 40% resided in sheltered housing or in an institution. At the second wave, 61% of individuals participated, at the third 40% and at the fourth only 22% (Table 1). Nonparticipation at the various measurement occasions was mainly due to mortality in the sample. A younger group of twenty-two healthy middle-aged men and women working at the Ryhov Hospital in Jönköping volunteered (mean age 44.7, SD = 8.9 at baseline) across measurement occasions to act as controls

Table 1. Characteristics of the subjects participating in the NONA immune longitudinal study

Occasion (Year)	No. of Subjects Investigated	Proportion of Women (%)	Age (Years)	
			Mean	Range
1999	138	70	89.8	86-95
2001	84	69	91.6	88-97
2003	55	69	93.2	90-99
2005	31	81	94.7	92-101

Health Parameters

Health was defined based on medical records and from clinical chemistry data, supplemented with information gathered in a health interview that focused on diagnosed illness, current symptoms and medication.[11] The neuropsychological battery used to identify cognitive impairment included The Mini-Mental State Examination (MMSE) and the Memory-in-reality (MIR) test.[12,13] MMSE is a screening device used in epidemiological studies to identify cognitive impairment. In the present study we used the following three cognitive status categories: 1) cognitively intact, 2) mild cognitive dysfunction or questionable cases (MCD, evidence of compromised memory/cognition, not fully meeting DSM-IV criteria for dementia, APA, 1994) and 3) dementia (according to DSM-IV criteria; APA 1994). These two latter diagnostic categories were pooled under the "cognitive impairment" and compared with those rated as cognitively intact.

Immune System Parameters

Blood samples were drawn in the morning between 9:00 and 10:00 for immediate transport to the laboratory at Ryhov Hospital in Jönköping. At the laboratory, fresh blood samples were subjected to various clinical laboratory analyses, T-cell subset enumeration by flow cytometry and various functional tests. Remaining blood components were prepared and frozen for storage and analyses to be performed later, in several collaborations within the EU-supported T-CIA project. The immune system parameters examined consisted of:

- Plasma proteins, for example albumin, transthyretin, C-reactive protein (CRP)
- Antibody-defined T-cell surface molecules using three-colour flow cytometry.
- IgG and IgM serology for CMV and EBV to detect persistent and recurrent viral infections.
- Cytokine (IL-2, IL-6, IL-10, interferon-γ) production and secretion by PBMC, mainly by enzyme-linked immunosorbent assays (ELISA).
- Virus-specific $CD8^+$ cells quantified using MHC/peptide tetramers or multimers for CMV (HLA-A2/NLVPMVATV) and EBV (HLA-A2/GLCTLVAML)
- TCR clonal expansion by TCR clonotype mapping combining RT-PCR and denaturing gradient gel electrophoresis (DGGE) for rapid detection and characterisation of T-cell clonality using specific primers covering the TCR Vβ 1-24 variable regions.

Immune Parameters and Morbidity

The health examination allowed a comparison of findings from the application of a modified SENIEUR protocol with results using the exclusion criteria of the OCTO Immune Study.[11] The modified SENIEUR protocol excluded 90.6% of the NONA Immune sample at baseline, indicating that only 9.4% were rated as very healthy. The use of the original protocol, suggesting additional laboratory analysis for exclusion, would probably have excluded even more individuals, demonstrating the need for using less stringent criteria in studies of the immune system in later life.[11] Thirty-eight (27.5%) participants, defined as moderately healthy, met the criteria used in the previous OCTO Immune Study of not residing in an institution, not being demented and not using medication known to effect the immune system. The remaining sample (63%) comprised frail individuals not meeting the above health criteria.[11]

Applying the five most common exclusion criteria, cardiac insufficiency, medication, laboratory data, urea and malignancy, the modified SENIEUR protocol excluded 87% of the original sample.[11] When the OCTO Immune protocol was applied, medication was found to be the most common criterion, excluding 43%, institutionalisation the second, excluding 39% and cognitive dysfunction the third, excluding 14%. Among different disease conditions, cardiac insufficiency (51%), malignancy (15%), dementia (14%), chronic obstructive pulmonary disease (12%), diabetes mellitus (11%), rheumatoid arthritis (9%), hypothyroidism (6%) and pernicious anaemia (6%) constituted the eight most prevalent diagnoses. These figures demonstrate the considerable morbidity and comorbidity in this representative sample of very old individuals.[11]

The application of the above health protocols allowed us to define three independent subgroups of overall health: very healthy, moderately healthy and frail individuals. A comparison across these subgroups indicated no differences between them for T-cell subsets characteristic of the immune risk profile, previously identified in octogenarians.[11] Interestingly, the IRP might thus serve as a significant biomarker of ageing, independent of the individuals' health condition. This important finding is compatible with results in non-inbred mice populations, showing that clusters of immune markers can predict longevity in individuals independently of health conditions.[14]

Immune Risk Phenotype, Cognitive Impairment and Mortality

Analysis of mortality in the very old NONA Immune individuals (n = 138) confirmed the previous findings in the OCTO Immune Study of an elevated risk in individuals with the IRP (Fig. 1).[15] Moreover, it showed that these findings were generalisable to the more representative elderly NONA sample.[15] The findings were also in agreement with a UK study, the Healthy Ageing Study, of a large sample of younger elderly people from the area of Nottingham/Cambridge, showing that an inverted CD4/CD8 ratio is predictive of survival.[16]

The results also supported the previous findings from the OCTO Longitudinal Study in which an elevated mortality risk was observed in individuals with cognitive impairment.[15] Moreover, the results showed that these two conditions (IRP and cognitive impairment) independently predicted survival after controlling for age, sex and various kinds of prevalent diseases and comorbidity in the NONA sample.[15] This was in agreement with the previous findings that the IRP constitutes a major predictor of nonsurvival in very late life independently of morbidity. This finding should be interpreted in light of the fact that only 9% of the NONA Immune individuals conformed to the SENIEUR criteria for optimal health.[11]

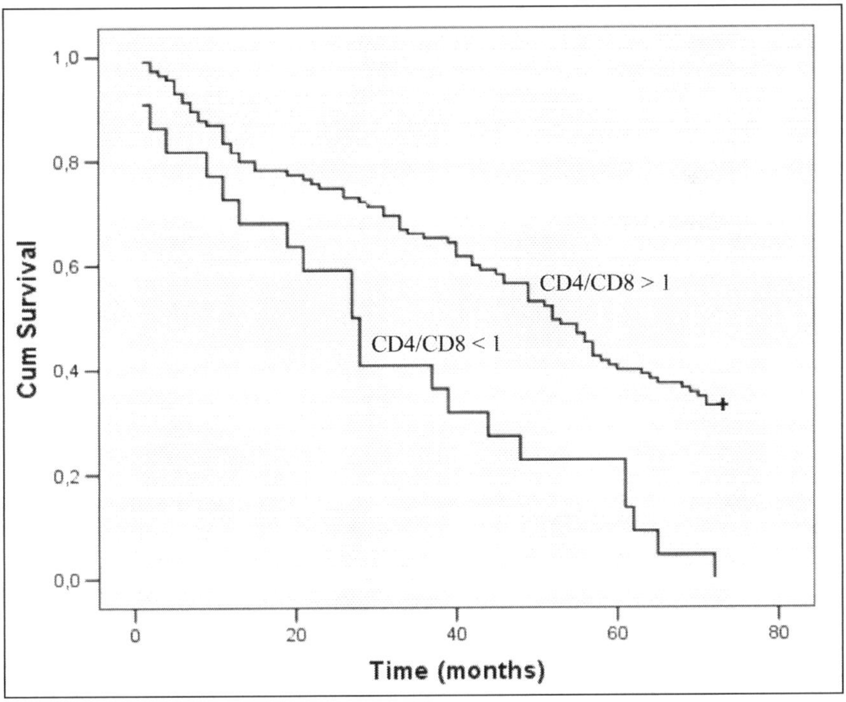

Figure 1. Kaplan Meier survival curves for NONA Immune very old individuals with CD4/CD8 < 1 (IRP) and CD4/CD8 > 1 (non-IRP). Test for equality of survival distribution for the subgroups showed: log rank: 14.20, $p < 0.001$.

Allostatic Load

Only a few individuals (n = 8) with the IRP and compromised cognitive status were identified. A Kaplan-Meier survival analysis revealed that these individuals showed a significantly higher annual mortality rate (42%/year) compared with those with only one or none of these conditions (15%/year and 8.5%/year, respectively), corresponding to relative mortality rates of 5:2:2:1 (Fig. 2).[15]

These observed mortality effects, indicating immune and central nervous system interactions, can be integrated into a general concept of allostatic load, suggesting that cumulative dysfunctions across multiple systems may have more than an additive impact on overall health and survival.[17-19] Allostatic load derives from the concept of allostasis which in turn is derived from homeostasis. Allostasis focuses more specifically on challenges to the specific regulatory nervous, immune and endocrine systems which must adapt in order to maintain balance though changes in various psychosocial or physical situations in life.[18] Although these processes may be adaptive in the short term, they are likely to be damaging when becoming excessive in duration, frequency and magnitude.[18]

The allostatic load in individuals with the IRP and cognitive impairment was associated with changes in the levels of the cytokines IL-2 and IL-6.[15] Cytokines in general are considered to have a central role in the mediation of allostasis by communicating between the nervous, immune and endocrine systems.[18] A significantly lower IL-2 responsiveness on mitogen stimulation, reflecting a state of T-cell anergy, as well as excessive increases in the plasma levels of the proinflammatory cytokine IL-6, were positively associated with cognitive impairment (Fig. 3), represented changes associated with an allostatic load in these individuals.[15]

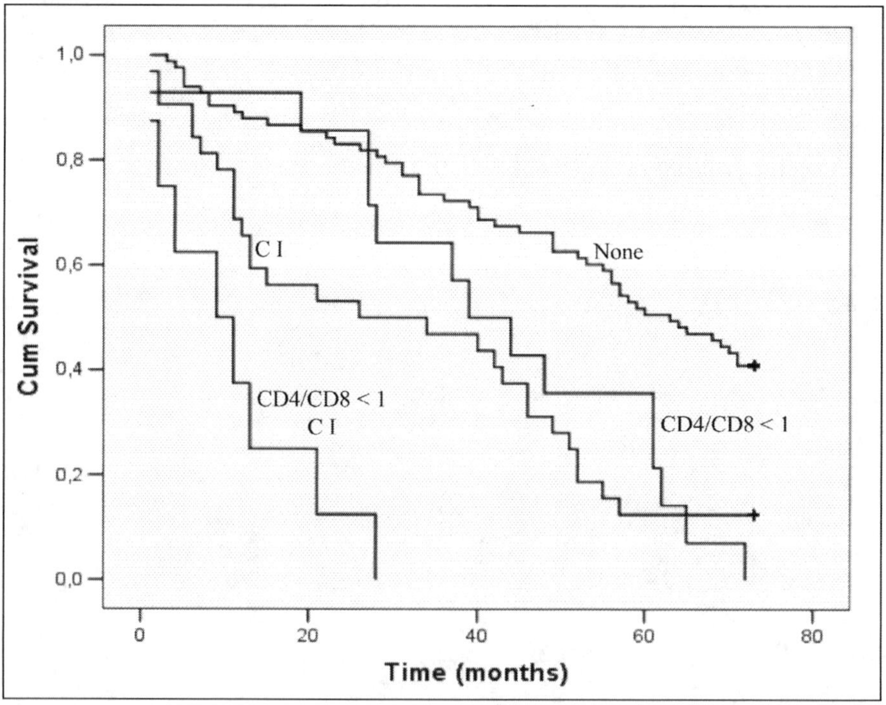

Figure 2. Kaplan Meier survival curves for NONA Immune very old individuals in subgroups created by CD4/CD8 ratio combined with cognitive status. The subgroups were: "CD4/CD8 < 1, CI" (IRP, cognitively impaired); "CI" (cognitively impaired, non-IRP); "CD4/CD8 < 1" (IRP, cognitively intact) and "None" (non-IRP, cognitively intact). Test for equality of survival distribution for the subgroups showed: log rank = 52.11, $p < 0.001$.

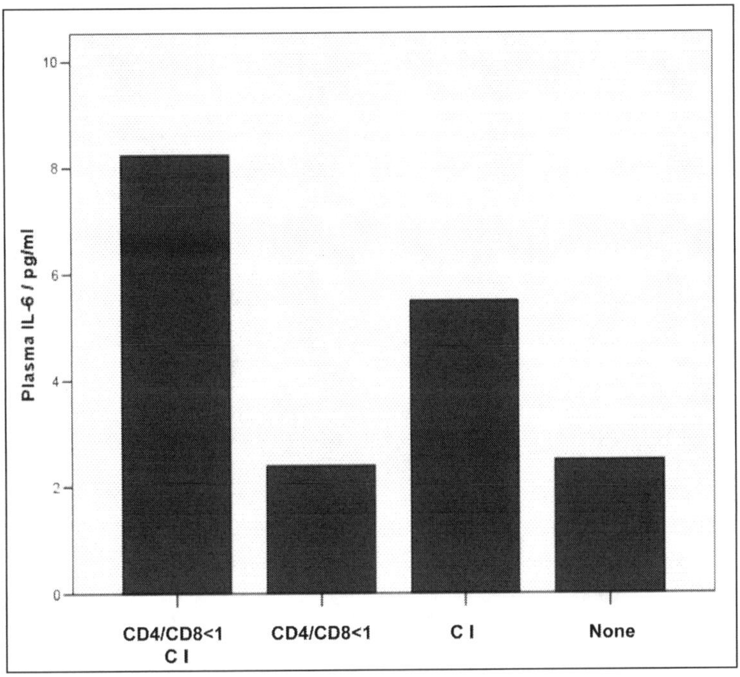

Figure 3. Median plasma Interleukin 6 (IL-6, pg/ml) in subgroups of NONA Immune very old individuals created by CD4/CD8 ratio combined with cognitive status. The subgroups were: "CD4/CD8 < 1, CI" (IRP, cognitively impaired); "CD4/CD8 < 1" (IRP, cognitively intact); "CI" (cognitively impaired, non-IRP); and "None" (non-IRP, cognitively intact). "CD4/CD8 < 1, CI" individuals showed significantly higher plasma IL-6 compared with all other subgroups and "CI" individuals showed significantly higher levels compared with the "CD4/CD8 < 1" and "None" subgroups ($p < 0.001$).

IRP, T-Cell Differentiation and Persistent Viral Infection

Results from baseline measurements of the NONA Immune Study confirmed findings from the OCTO Immune Study that showed an association between the IRP and the prevalence of persistent CMV infection.[1] As for the OCTO study, the NONA study also indicated that a subset with the $CD3^+CD8^+CD28^-$ phenotype was markedly expanded in IRP^+ and CMV-positive individuals.[1] This led us to determine T-cell differentiation in more detail, using the $CD45RA^+$, $CCR7^+$, $CD27^+$ and $CD28^+$ markers in a sequential model suggesting positive expression for naive cells, gradual losses of the markers in the various memory stages and lack of expression for late differentiated effector/memory cells.[20-22] A final differentiation step occurs by reversion of $CD45RO^+$ to $CD45RA^+$ to yield a CD27-CD28-CCR7-$CD45RA^+$ terminally differentiated so-called TEMRA phenotype.[23]

Our results suggested profound decreases in the number of naive cells in the very old, changes that were even more pronounced in IRP^+ individuals (Fig. 4). The results also showed significant increases in the number of $CD8^+$CD27-CD28-CCR7-$perforin^+$ cells (Figs. 4,5) and because a majority of these cells also were $CD45RA^+$, these data confirmed that the IRP is strongly associated with increases in the number of terminally differentiated cells. Recent evidence suggests that increased proportions of terminally differentiated $CD8^+$ cells may possess some of the characteristics of replicative senescence, including short telomeres and apoptosis-resistance (although these may also be qualities of TEMRA cells per se). The inclusion of high proportions of potentially

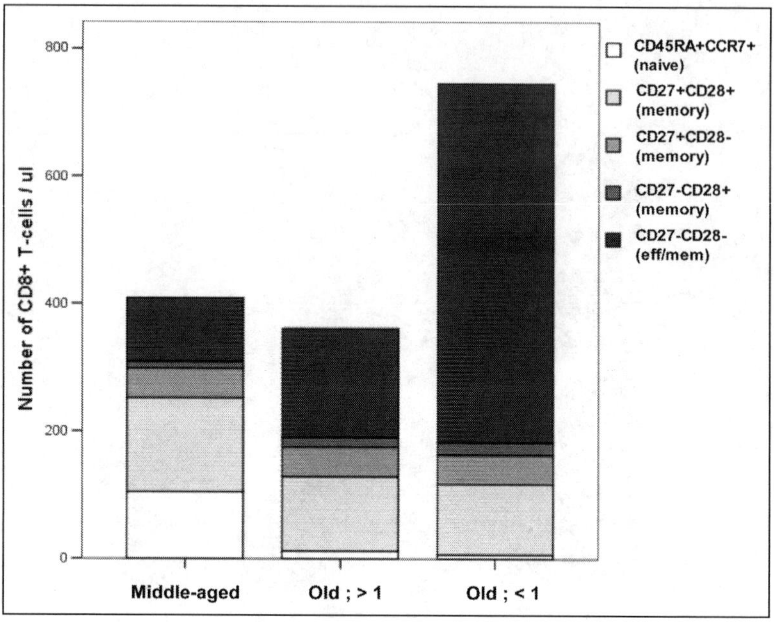

Figure 4. Mean numbers of $CD8^+$ T-cells (per μ1) and subsets of $CD8^+$ differentiation versus subgroups. The subgroups were: "Middle-aged"; "Old, >1" (NONA Immune non-IRP's); "Old, <1" (NONA Immune IRP individuals). The number of naïve cells was significantly ($p < 0.001$) higher in the "middle-aged" compared with other subgroups. The number of effector/memory cells was significantly ($p < 0.001$) higher in "Old, <1" individuals compared with other subgroups.

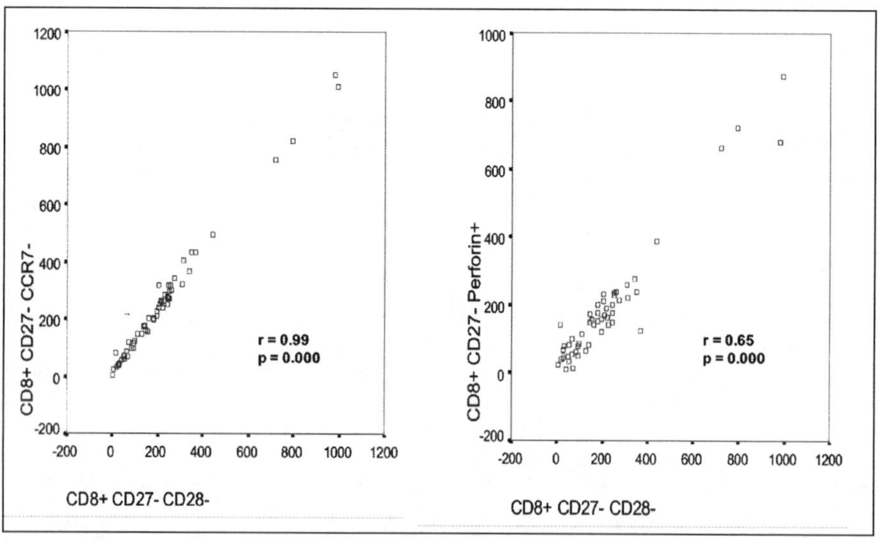

Figure 5. The number of $CD8^+$CD27-CD28- cells plotted versus CCR7- and $perforin^+$ $CD8^+$ subsets. Close associations between the subsets show that $CD8^+$CD27-CD28- cells also are CCR7- and $perforin^+$, giving evidence for a $CD8^+$CD27-CD28-CCR7-$perforin^+$ phenotype characteristic for the IRP.

senescent T-cells in the IPR may therefore represent clinical confirmation of the importance of the Hayflick Limit in ageing of organ systems in humans.[24,25] The results are also in line with findings that individuals with high proportions of associated CD8⁺CD28- T-cells show diminished antibody responses to influenza vaccination.[26,27]

Evidence for a major impact of CMV in generating terminally differentiated $CD8^+$ cells was demonstrated in the OCTO subjects by tetramer technology and was also confirmed in the NONA Immune Study.[28] We found CMV_{NLV}-specific expansions, mainly composed of terminally differentiated cells, in the range 1-20 % of total $CD8^+$ cells, similarly to findings in the OCTO Immune Study.[28] Increases in the percentages of CMV_{NLV}-specific cells were also associated with decreases in antigen-specific IFN-γ responsiveness, suggesting that the accumulation of these T-cells may be a result of compensatory mechanisms to maintain control of CMV by balancing the compromised functionality that occurs with increasing age.[10,28]

The NONA results also support the suggestion that in addition to CMV infection, persistent EBV infection plays a role as a bystander associated with the IRP.[15] IRP individuals were in all cases double sero-positive, suggesting that chronic viral load in the very old might be a prerequisite for developing the IRP. Especially high levels of late differentiated $CD8^+$ cells, characteristic of the IRP, were found in double sero-positive individuals and to a lesser but significant extent in only CMV-infected and to a still lower extent in those individuals infected with EBV only (Fig. 6).[15] In line with these results we found significant expansions of EBV_{GLC}-specific $CD8^+$ cells, although their frequency was tenfold lower than for the CMV-specific cells in double seropositives.[29]

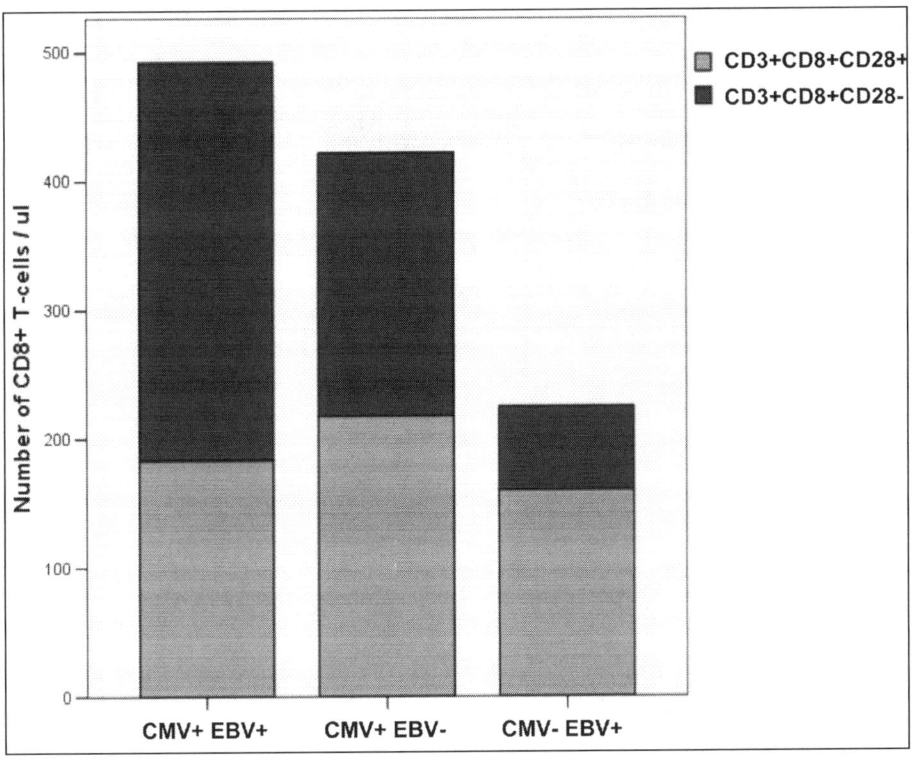

Figure 6. Mean number of $CD8^+$ T-cells (per μl) and $CD3^+CD8^+CD28\pm$ subsets in NONA Immune individuals relative to CMV IgG and EBV IgG antibody status. The "CMV^+EBV^{+}" subgroup showed a significantly ($p < 0.001$) higher number of $CD3^+CD8^+CD28-$ cells compared with the "$CMV-EBV^{+}$" subgroup.

TCR Clonotype Mapping

Clonal expansions have been detected in healthy old individuals and accumulating evidence suggests that these are associated with chronic antigen stimulation caused by persistent viral infections.[30] We analysed the CD8$^+$ T-cell clonal composition in NONA Immune (n = 39) and middle-aged (n = 9) individuals using TCR clonotype mapping.[28] The method combines RT-PCR and denaturing gel electrophoresis (DGGE) for rapid detection and characterisation of T-cell clonal expansions using specific primers covering the majority of TCR-Vβ 1-24 variable regions.[31] With a polyclonal T-cell population, a nondistinct smear in the denaturing gradient gel is seen while, while the TCR of a population of clonally expanded cells present at a level >1% is seen as a distinct band.[31] Thus, when the clonal expansions were quantified by staining with anti-TCR-Vβ mAbs, an individual CMV_{NLV}-specific clone was detected as expanded when it exceeded 1% of the CD8$^+$ repertoire.[28]

The mean number of different expanded clones was significantly higher in nonagenarians compared with the middle-aged (Table 2), suggesting a considerable impact of clonal expansions in the very old.[28] Importantly, most of these clonal expansions in most subjects were also found to be stable across a two-year period of time.[28] The results showed a very strong association between the number of expansions and persistent CMV infection (Table 2), suggesting that the majority of clonal expansions in the elderly are of CMV-specific cells.[28] Direct evidence for this was garnered by sorting CMV_{NLV}-specific cells and mapping their TCR-Vß expansions. This revealed that this specific T-cell population was oligoclonal with a mean number of six CMV-related clone types. These results are comparable with the findings of Khan and coworkers showing that when different CMV epitopes were studied by tetramer technology, the summed percentages of the specific cells were more than 10% and as high as 50% of the total number of CD8$^+$ cells.[32] Such substantial accumulations of CMV-specific cells in a limited number of clones may reduce the available space for T-cells with other specificities, which could be lost through competition and result in reduced clonal diversity and immune protection capability, particularly relevant for IRP's.[33,34] A demonstration that clonal expansions of specific T-cells can compromise the response to other antigens by a mechanism through competition was recently documented in mice.[35] The observations that infection with CMV can reduce prevailing levels of immunity to EBV and that VZV-specific populations are significantly decreased when CMV-specific CD4$^+$ cells expand in old humans also support this hypothesis.[36,37]

We found that among sero-positive individuals, IRP$^+$ individuals possessed a significantly lower number of different expanded clones than the other equally old people (Table 2).[28] Strikingly, we also found that a decrease in the numbers of different clones in IRP individuals was associated

Table 2. The number of clonal expansions in the CD8$^+$ repertoire determined by DGGE in subgroups of individuals

Subgroup	Number of CD8$^+$ T-Cell Clonal Expansions	p<
All individuals		
Middle-aged	10.1 ± 1.8 (9)*	0.01
NONA individuals	19.4 ± 2.4 (39)	
NONA individuals		
CMV positive	22.6 ± 2.7 (31)	0.001
CMV negative	7.4 ± 1.7 (8)	
CMV-positive NONA individuals		
CD4/CD8 < 1	15.0 ± 2.7 (8)	0.05
CD4/CD8 > 1	25.2 ± 3.4 (23)	

* Mean ± SE (n)

with increased inflammatory activity, as reflected by elevated plasma IL-6, as well as with shorter survival times.[28] This suggests that increased numbers of clonal expansions may be sequentially used to maintain essential immune protection capability against CMV. It also supports the hypothesis that when a certain point is reached at a terminal stage in the very elderly, clonal exhaustion leads to shrinkage of the CD8 clonal repertoire, detrimental to immune protection both for unrelated antigens and perhaps even for CMV itself.[28]

Low Grade Inflammation

Results using data from the second and third waves of the NONA Study confirmed results from other studies demonstrating that ageing is associated with low-grade inflammation and that inflammatory markers are significant predictors of mortality in very old humans (Table 3).[38] Logistic regression analysis also revealed that the IRP and low-grade inflammatory activity, defined by the marker IL-6, were independently predictive of 2-year survival, an outcome that remained when CRP and albumin were entered as covariates.[38] The independent main effect predicted 57% of nonsurvival and, impressively, 97% of survival, showing that the IRP and IL-6 are even better predictors of survival than of death and that these parameters are strong candidates as markers of healthy ageing. Remarkably, the IRP and IL-6 were predictive of mortality in a manner not significantly affected by eight prevalent diseases, including AD, cardiovascular disease and type-2 diabetes and independently of gender or exact age. These results are in agreement with findings demonstrating that low-grade inflammation and the IRP can predict mortality independently of disease and comorbidity.[11,39] While the IRP reflects changes in the adaptive T-cell system, which we believe are primarily associated with lifelong persistent CMV infection, the increases in IL-6 seem to reflect innate immune system changes, or non-immune sources, including a wide range of alterations associated with the development of frailty. This is supported by our findings of changes in the plasma levels with decreases in albumin and increases in acute-phase proteins.[38]

The results outlined above may at first seem contradictory to our baseline findings of elevated IL-6 plasma levels specifically associated with cognitive impairment and mortality. This association was not seen at second wave follow-up.[38] However, cognitively-impaired individuals who survived until the follow-up or who became incident cases were more likely to be in the early stages of the disease process compared with those who showed manifest cognitive impairment already at baseline and with subsequent higher mortality rate. Thus, it is likely that sample composition variously reflects reasons for survival or selective mortality in late life.[3]

IRP Movement

In the NONA Immune Longitudinal study only 5 individuals (4%) moved into the category at risk by changes in the CD4/CD8 ratio, which was a significantly lower percentage as compared to the previous OCTO Immune Study. Intriguingly and contrary to anything seen in the OCTO Immune, we have found in the NONA Immune that a few individuals (n = 3) can move out of the IRP category.[38] The changes were associated with increases in plasma IL-6 and IL-10 levels, suggesting that IL-6 may induce an anti-inflammatory rather than a pro-inflammatory effect when

Table 3. ***Inflammatory parameters in plasma at Time 2 in very old individuals surviving (survivors) or not surviving (nonsurvivors) at Time 3 of the NONA Immune Longitudinal Study***

Parameter	Survivor	Nonsurvivor	p<
IL-6 (pg/ml)	4.9 (61)*	9.2 (21)	0.001
CRP (mg/ml)	1.4 (60)*	3.6 (22)	0.05
Albumin (g/l)	36.9 (62)**	33.7 (22)	0.01

* Median (n) ** Mean (n)

associated with enhanced IL-10, neutrocytosis and lymphopenia to limit the potential injurious effects of sustained inflammation in these particular and rare individuals.[40]

Conclusions and Future Directions

In the NONA immune longitudinal study, the IRP and low-grade inflammation were found to be the main predictors of survival in the very old. This outcome was not significantly affected by individuals' health status, suggesting that the physiological ageing processes of T-cell immunosenescence and low-grade inflammation are of primary importance in late life survival.[11,38] The results suggest a sequence of stages that probably begins in early life with CMV infection, followed by the generation of large $CD8^{+}CD28^{-}$ effector cell expansions to control lifelong persistent infection, homeostatic T-cell changes and a gradual change towards an IRP. These individuals show decreased numbers of the $CD8^{+}$ cell clonal expansions associated with increases in levels of plasma IL-6 and shorter survival, suggesting a stage in ageing where clonal exhaustion may lead to shrinkage of the clonal repertoire detrimental to immune capabilities.[28] It ends in a terminal decline stage with a low-grade inflammatory process that occurs in late life.[38] This supports the inflame-ageing hypothesis in human ageing, suggesting that age-associated chronic inflammation causes frailty and that immunosenescence is driven by a chronic antigen load, most often associated with CMV infection, that induces a progressive expansion of compromised poorly functional $CD8^{+}$CD28- effector T-cells.[41,42] The $CD8^{+}$CD28- cells are able to secrete pro-inflammatory cytokines like IL-6 and TNF-α that may compensate for the defective T-cellular function, and/or amplify an ongoing inflammatory process.[43]

In future studies it will be important to more specifically address the question why only a certain fraction of CMV sero-positive individuals reside in or move into the category of risk. It is also urgent to further characterise those exceptional individuals that move out of the category of risk, allowing insight into clinical intervention approaches for those who remain in the IRP category until death. We also need to further study the phenomena of clonal expansion regarding frequencies and specificities of cells for the various clones detected. In particular, it will be important to gain a better understanding of the nature of the link between CMV infection, phenotypic T-cell changes and changes in pro-inflammatory cytokines associated with the IRP.

Acknowledgements

The authors acknowledge the considerable support from the EU project "T-cell immunity and ageing" T-CIA, contract no QLK6-CT-2002-02283, the Research Board in the County Council of Jönköping and the Research Council in the Southeast of Sweden (FORSS) for funding these projects. We also acknowledge Länsjukhuset Ryhov for provision of laboratory resources for the completion of these studies. The authors are also indebted to our coworkers Sture Löfgren, Bengt-Olof Nilsson, Jan Ernerudh, Jadwiga Olsson and Per-Eric Evrin for their important contributions to these studies. We particularly would like to thank the nursing staff including Annica Andersson, Inga Boström, Gerd Martinsson, Agneta Carholt, Lene Ahlbäck, Lena Blom, Monica Janeblad, Gun Karlsson and Lena Svensson for their efforts in obtaining the blood samples used. We are also indebted to Roberta Valeski, Florence Confer, Margaret Kensinger, Penn State University, United States and Andrea Tompa, Gunilla Isaksson, Inger Johansson, Cecilia Ottosson, Helen Olsson, Lisa Stark, Jönköping, Sweden, for secretarial and technical assistance. We finally acknowledge our ImAginE collaborators, Qin Ouyang, University of Tubingen, Germany, Yvonne Barnett, Paul Hyland, Owen Ross and colleagues, University of Ulster, Northern Ireland, Julie Thompson, Unilever, UK and Tania Kollgaard and Tina Seremet, Danish Cancer Society, Copenhagen, Denmark, for fruitful collaboration.

References

1. Wikby A, Johansson B, Olsson J et al. Expansions of peripheral blood CD8 T-lymphocyte subpopulations and an association with cytomegalovirus seropositivity in the elderly: the Swedish NONA immune study. Exp Gerontol 2002; 37:445-453.
2. Wikby A, Johansson B, Ferguson F. The OCTO and NONA immune longitudinal studies: a review of 11 years studies of Swedish very old humans. Adv Cell Aging Gerontol 2003; 13:1-16.
3. Pawelec G, Akbar A, Caruso C et al. Human immunosenescence: is it infectious? Immunol Rev 2005; 205:257-68.
4. Pawelec G, Ferguson F, Wikby A. The SENIEUR protocol after 16 years. Mech Ageing Dev 2001; 122:132-134.
5. Hallgren HM, Berg N, Rodysill KJ et al. Lymphocyte proliferative response to PHA and anti-CD3/Ti monoclonal antibodies, T-cell surface marker expression and serum IL-2 receptor levels as biomarkers of age and health. Mech Ageing Dev 1988; 43:175-185.
6. Wikby A, Johansson B, Ferguson F et al. Age-related changes in immune parameters in a very old population of Swedish people: a longitudinal study. Exp Gerontol 1994; 29:531-541.
7. Ferguson FG, Wikby A, Maxon P et al. Immune parameters in a longitudinal study of a very old population of Swedish people: a comparison of survivors and nonsurvivors. J Gerontol Biol Sci 1995; 50A:B378-B382.
8. Wikby A, Maxson P, Olsson J et al. Changes in CD8 and CD4 lymphocyte subsets, T-cell proliferation responses and nonsurvival in the very old: the Swedish longitudinal OCTO-immune study. Mech Ageing Dev 1998; 102:187-198.
9. Olsson J, Wikby A, Johansson B et al. Age-related change in peripheral blood T-lymphocyte subpopulations and cytomegalovirus infection in the very old: the Swedish longitudinal OCTO immune study. Mech Ageing Dev 2000; 121:187-201.
10. Ouyang Q, Wagner WM, Zheng W et al. Dysfunctional CMV-specific $CD8^+$ T-cells accumulate in the elderly. Exp Gerontol 2004; 39:607-613.
11. Nilsson BO, Ernerudh J, Johansson B et al. Morbidity does not influence the T-cell immune risk phenotype in the elderly: findings in the Swedish NONA Immune Study using sample selection protocols. Mech Ageing Dev 2003; 124:469-76.
12. Folstein MF, Folstein SE, McHugh PR. "Mini-Mental State": A practical method for grading the cognitive state of patients for the clinician. J Psychiatr Res 1975; 12:189-98.
13. Johansson B. The MIR—Memory in Reality Test. Stockholm, Sweden: Psykologiförlaget, AB:1988/89.
14. Miller RA. Biomarkers of aging: prediction of longevity by using age-sensitive T-cell determinations in a middle-aged genetically heterogeneous mouse population. J Gerontol Biol Sci 2001; 56A:B180-B186.
15. Wikby A, Ferguson F, Forsey R et al. An immune risk phenotype, cognitive impairment and survival in very late life: impact of allostatic load in Swedish octogenarian and nonagenarian humans. J Gerontol Biol Sci 2005; 60A:556-565.
16. Huppert FA, Pinto EM, Morgan K et al. Survival in a population sample is predicted by proportions of lymphocyte subsets. Mech Ageing Dev 2003; 124:449-451.
17. McEwen BS, Stellar E. Stress and the individual: mechanism leading to disease. Arch Int Med 1993; 153:2093-2101.
18. McEwen BS. Interacting mediators of allostasis and allostatic load: towards an understanding of resilience in aging. Metabolism 2003; 52:10-16.
19. Karlamangla AS, Singer BH, McEwen et al. Allostatic load as a predictor of functional decline: MacArthur studies of successful aging. J Clin Epidemiol 2002; 55:696-710.
20. Tomiyama H, Matsuda T, Takiguchi M. Differentiation of human $CD8^+$ cells from a memory to memory/effector phenotype. J Immunol 2002; 168:5538-5550.
21. Van Baarle D, Kostense S, van Oers MHJ et al. Failing immune control as a result of impaired $CD8^+$ T-cell maturation: CD27 might provide a clue. Trends Immunol 2002; 23:586-591.
22. Appay V, Dunbar PR, Callin M et al. Memory $CD8^+$ T-cells vary in differentiation phenotype in different persistent virus infections. Nat Med 2002; 8:379-385.
23. Wallace DL, Zhang Y, Ghattas H et al. Direct measurement of T-cell subset kinetics in vivo in elderly men and women. J Immunol 2004; 173:1787-1794.
24. Pawelec G, Akbar A, Caruso C et al. Is immunosenescence infectious? Trends Immunol 2004; 25:406-410.
25. Effros R. From Hayflick to Walford: the role of T-cell replicative senescence in human aging. Exp Gerontol 2004; 39:885-890.
26. Trzonkowski P, Mysliwska J, Szmit E et al. Associations between cytomegalovirus infection, enhanced proinflammatory response and low levels of antihemagglutinins during the anti-influenza vaccination—an impact of immunosenescence. Vaccine 2003; 21:3826-36.

27. Saurwein-Teissl M, Lung TL, Marx F et al. Lack of antibody production following immunization in old age: association with CD8(+)CD28(-) T-cell clonal expansions and an imbalance in the production of Th1 and Th2 cytokines. J Immunol 2002; 168:5893-99.
28. Hadrup SR, Strindhall J, Kollgaard T et al. Longitudinal studies of clonally expanded CD8 T-cells reveal a repertoire shrinkage predicting mortality and an increased number of dysfunctional cytomegalovirus-specific T-cells in the very elderly. J Immunol 2006; 176:2645-2653.
29. Ouyang Q, Wagner WM, Walter S et al. An age-related increase in the number of $CD8^+$ T-cells carrying receptors for an immunodominant Epstein-Barr (EBV) epitope is counteracted by a decreased frequency of their antigen-specific responsiveness. Mech Ageing Dev 2003; 124:477-485.
30. Khan N, Shariff N, Cobbold M et al. Cytomegalovirus seropositivity drives the CD8 T-cell repertoire toward greater clonality in healthy elderly individuals. J Immunol 2002; 169:1984-92.
31. thor Straten P, Barfoed A, Seremet T et al. Detection and characterisation of alpha-beta-T-cell clonality by denaturing gradient gel electrophoresis (DGGE). Biotechniques 1998; 25:244-250.
32. Moss P, Khan N. $CD8^+$ T-cell immunity to cytomegalovirus. Human Immunol 2004; 65:456-464.
33. Akbar AN, Fletcher JM. Memory T-cell homeostasis and senescence during aging. Curr Opin Immunol 2005; 17:480-485.
34. Pawelec G, Koch, Franceschi C et al. Human immunosenescence: does it have an infectious component? Ann N Y Acad Sci 2006; 1067:56-65.
35. Messaoudi I, Lemaoult J, Guevara-Patino JA et al. Age-related CD8 T-cell clonal expansions constrict CD8 T-cell repertoire and have the potential to impair immune defence. J Exp Med 2004; 200:1347-1358.
36. Khan N, Hislop A, Gudgeon N et al. Herpesvirus-specific CD8 T-cell immunity in old age: cytomegalovirus impairs the response to a coresident EBV infection. J Immunol 2004; 173:7481-9.
37. Fletcher JM, Vukmanovic-Stejic M, Dunne PJ et al. Cytomegalovirus-specific $CD4^+$ T-cells in healthy carriers are contineously driven to replicative exhaustion. J Immunol 2005; 175:8218-8225.
38. Wikby A, Nilsson BO, Forsey R et al. The immune risk phenotype is associated with IL-6 in the terminal decline stage: findings from the Swedish NONA immune longitudinal study of very late life functioning. Mech Ageing Dev 2006; 127:695-704.
39. Krabbe KS, Pedersen M, Bruunsgaard H. Inflammatory mediators in the elderly. Exp Gerontol 2004; 39:687-699.
40. Steensberg A, Fischer CP, Keller C et al. IL-6 enhances plasma IL-1ra, IL-10 and cortisol in humans. Am J Physiol Endocrinol Metab 2003; 285:E433-E437.
41. Franceschi C, Bonafe M, Valensin S et al. Inflam-aging: an evolutionary perspective on immunosenescence. Ann NY Acad Sci 2000; 908:244-254.
42. Fulop T, Larbi A, Wikby A et al. Dysregulation of T-cell function in the elderly: scientific basis and clinical implications. Drugs Aging 2005; 22:589-603.
43. Zanni F, Vescovini R, Biasini C et al. Marked increase with age of type 1 cytokines within memory and effector/cytotoxic $CD8^+$ T-cells in humans: a contribution to understand the relationship between inflammation and immunosenescence. Exp Gerontol 2003; 38:981-87.

Chapter 2

Scoring of Immunological Vigor: Trial Assessment of Immunological Status as a Whole for Elderly People and Cancer Patients

Katsuiku Hirokawa,* Masanori Utsuyama, Yuko Kikuchi and Masanobu Kitagawa

Abstract

The three leading causes of death in industrialized countries are cancer, stroke and ischemic heart disease, whereas infections are more prevalent in developing countries. Nonetheless, infections such as pneumonia and urinary tract infection rank 4th and 5th even in Japan. These data are mainly based on death certificates from medical doctors without autopsy. However, infection is found to be a major cause of death in the elderly where autopsy is performed. Also in cancer patients, autopsy reveals that the proximal cause of death is commonly infection. These findings indicate that an immune deficient state is likely to contribute to morbidity and mortality in many elderly people and cancer patients. Thus, developing methods to properly assess immune function as a whole on an individual subject basis is urgently required. The assessment of immune functions should be then followed by immunological restoration, if necessary. This chapter presents one possible method for the proper assessment of global immunological function using comprehensive assessment of various immunological indices, mainly of T-cells. This algorithm is designation "scoring of immunological vigor" (SIV); its usefulness is presented for cancer patients in the first instance in humans. We have also conducted several trials of different immunological interventions in aging animal models. The most recent effective intervention is infusion of activated T-cells from syngeneic young or old mice into old mice. The infusion of autologous activated T-cells has already been applied clinically in humans, mainly for the treatment of cancer and is feasible also in old people. Thus, this approach could be useful for immunological restoration in the elderly, in analogy to the mouse models.

Infection Is a Major Cause of Death in the Elderly

In most developed countries, the three leading causes of death are cancer, stroke and myocardial infarction, with infections such as pneumonia ranking 4th or 5th.[1] In Japan, this information is based on death certificates from physicians throughout the country forwarded to the Ministry of Health and Welfare (MHW). These are written based on the clinical and laboratory findings, but without autopsy in most cases. In fact, approximately 1 million people die of various diseases per year, but less than 3% are autopsied in Japan. Autopsy, however, does reveal that infections are one of the major causes of death.[1]

Table 1a shows that infection was considered the cause of death in nearly 40% of 923 autopsy cases of the elderly in Tokyo Metropolitan Geriatric Hospital.[1] It is interesting to note that cancer ranks the 3rd in this group. In 2005, the number of people over 100 years of age (centenarians) exceeded 25,000 in Japan. Autopsy examination of 44 centenarians revealed as many as 2/3 of

*Corresponding Author: Katsuiku Hirokawa—Nakano General Hospital, 4-59-16, Chuo Nakano-ku, Tokyo 164-8607, Japan. Email: hrokawa@nakanosogo.or.jp

Immunosenescence, edited by Graham Pawelec. ©2007 Landes Bioscience and Springer Science+Business Media.

Table 1a. Major causes of death in elderly people: Tokyo Metropolitan Geriatric Hospital

	People Over 60 Years	People Over 70 Years
Infections	39.2%	27.6%
Vascular diseases of Brain and Heart	29.7%	43.1%
Malignancies	18.7%	22.4%
Others	12.4%	6.9%

Table 1b. Direct cause of death in 44 centenarians

Pneumonia	15 }	
Sepsis	13 }	66.7%
Asphyxia	5	11.9%
Heart failure	4	9.5%
Malnutrition	3	7.1%
Brain stroke	2	4.8%

Malignant neoplasms were observed in 16 cases, but were not the direct cause of death.

them died of infections such as pneumonia and sepsis[2] (Table 1b). In this study, it was again interesting to note that cancer was detected in 1/3 of them, but was not considered as the direct cause of death. Similar findings were also reported by MacGee et al[3] from a geriatric institution in Geneva, Switzerland. The most common cause of death in the elderly was infections such as bronchopneumonia and urinary tract infection, exceeding 50% of the total (Table 2)

The occurrence of severe acute respiratory syndrome (SARS) in the winter of 2003 clearly highlighted the immune deficient status of the elderly. Fatality rates increased with age in SARS in Hong Kong, China, reaching >50% in people aged 65 years and over (Table 3). These data underscore the conclusion that elderly people become more susceptible to infection, most likely due to the age-related decline of immune functions.

A Significant Number of Cancer Patients Die of Infection

The incidence of cancer increases with age, at least for most cancers and up to around 80 years of age.[4] Because of the increasing size of the population of elderly people, cancer is now the most common disease in many industrialized countries. Cancer patients, if untreated, mostly die of their

Table 2. Causes of death in a geriatric hospital in Geneva

Bronchopneumonia	42.9%
Malignancies	28.1%
Pulmonary thrombo-embolism	21.1%
Acute myocardial infarction	19.6%
Urinary tract infection	12.3%
Acute cerebro-vascular disease	6.5%
Internal hemorrhage	5.5%
Congestive cardiac failure	3.3%

Mac Gee W. Causes of death in a hospitalized geriatric population: an autopsy study of 3000 autopsy patients. Virchows Arch. A 424:343. 1993

Table 3. Age-related increase of fatality rate of people suffering from SARS in Hong Kong

	Fatality Rate
24 years and under	less than 1%
25-44 years	6%
45-64 years	15%
65 years and over	52%
Total	14-15%

WHO report consensus document on the Epidemiology of severe acute respiratory syndrome (SARS), 17 October 2003 http://www.who.int/csr/sars/guidelines/en/

cancer; however, most cancer patients receive surgery, anti-cancer drugs, radiation, etc. and many survive for extended periods. A significant number still do die of cancer, or die of other diseases such as stroke or cardiac infarction, but it is of note that a significant number of cancer patients die of infection, for two reasons. One is that cancer treatment suppresses the immune function and the other is that most cancer patients are elderly people with advanced age-related decline of immune functions. Thus, cancer patients become susceptible to infection due to immune deficient status. In fact, immune status of cancer patients is generally lower than that of healthy people.

In the Tokyo Medical and Dental University Hospital, 716 cases were examined by autopsy from 1996 to 2003, of which 514 were cancer patients. 96 of these cases (18.7%) had died of infections (Table 4a). In Nakano General Hospital in Tokyo, 140 autopsies were carried out over the 7 years from 1998 to 2004. From clinical the viewpoint, the most common disease was cancer (51 cases, 36%) and the second commonest infection (44 cases, 31%). However, autopsy revealed that 12 cancer cases (24%) died of infections (Table 4b). Thus, among the 140 autopsy cases, average age 73 yr., the main cause of death was infection, reaching about 40% of total cases. Taken together, about around 20% of cancer patients are likely to die of infection. This is an important aspect of treating cancer patients, but unfortunately does not attract much attention. Cancer is of course the underlying disease in these patients, but one should recognize that a significant percentage of cancer patients die of infections due to their immune deficient status. Therefore, the assessment of immune status is very important not only for elderly people, but also for cancer patients.

Table 4. Infection is one of major causes of death in cancer patients

A: Tokyo Medical and Dental University Hospital	
Autopsied cancer cases	514 (100%)
Cancer cases, died of infection	96 (19%)

Autopsy cases performed in Tokyo Medical and Dental University Hospital were 714 from 1997 to 2004 (average age, 62.4 years).

B: Nakano General Hospital	
Autopsied cancer cases	51 (100%)
Cancer cases died of infections	12 (24%)

Autopsy cases performed in Nakano General Hospital were 140 from 1998 to 2004 (average age, 72.8 years).

The immune deficient status is caused by variable factors such as aging, stress, disease per se and various therapies for disease, especially for cancer. Unfortunately, however, immunological data are not satisfactorily available in most patients who die of cancer, stroke, heart disease or infection. Clinicians are of course aware of the importance of the immune system for protection against infection, but usually merely the number of lymphocytes in peripheral blood is assessed. There are two reasons for this. The first is that it is not clear which immunological parameters and functions should be assessed and the second is that the cost for assessment is generally not covered by insurance.

Assessment of the Immunological State

Their immune status is important information for patients suffering from various diseases as well as frail elderly people.[5] In many cases, however, immune status is not systematically assessed, because the most suitable among many immunological indices for the assessment of "immunological vigor" is not clear. This is mainly due to polymorphism of the immune system, a complicated organ composed of many heterogeneous elements. The first protection against infection is a physical barrier such as the skin or mucosal layers, the second is the innate immunity of granulocytes, macrophages and other immune cells and the third is acquired immunity mainly mediated by lymphocytes. Acquired immunity is most susceptible to aging.[1] This arm of the immune system is itself a complicated organization, composed of various cells, tissues and factors produced by these immune cells. It is difficult to say which cells, tissue or factors are most reflective of the immune vigor of each individual person. We have finally arrived at an interim conclusion that the comprehensive analysis of several immunological indices and functions is reasonable for the assessment of global immunological vigor and have designated this as a scoring system for immunological vigor (SIV).

As noted above and elsewhere in this book, T-cell functions are probably the most vulnerable to aging.[1] This is ascribed partly to thymic involution starting in early puberty. Age-related changes of the T-cell-dependent immune system are characterized by a decrease in T-cell number, changes in T-cell subsets and in qualitative changes such as proliferation and cytokine production. Therefore, T-cells and their characteristics were selected as core indices for the assessment of immune status; we have taken 10 parameters as follows: 1) number of T-cells/μl, 2) capacity for T-cell proliferation, 3) ratio of $CD4^+$:$CD8^+$ T-cells, 4) number of $CD4^+$ naïve T-cells, 5) ratio of $CD4^+$ naïve T-cells:$CD4^+$ memory T-cells, 6) number of B-cells, 7) number of NK- cells, 8) IL-2 production, 9) IFNγ production,10) IL-4 production. Each index was given a score from 1 to 3 (1: low, 2: medium and 3: high) by comparison to values in that population's data base. The summation of total scores was used to generate each individual's SIV, giving a value ranging between 10 and 30. This was then classified into 5 grades. Grade V (SIV: 30-29) means that immunological vigor is well preserved. Grade IV (SIV: 28-26) is fairly good, III (SIV: 25-22) is borderline and in the initial phase of decline, II (SIV: 21-17) is moderate decline and I (SIV: 16-10) is severe decline.

Figure 1 gives examples of 2 cases for which the SIV have been assessed, shown as radar-graphs. The immune status of a healthy 39 yr-old female is well-preserved, SIV 29/30 (grade V/V). In contrast, a 53 yr-old female with an immune system depressed because of chemotherapy for lymphoma manifests an SIV of 20/30 (grade II/V). Figure 2 shows SIV obtained from 60 healthy people and 20 cancer patients before treatment. SIV levels of most healthy people were distributed between 22 and 30, whereas those of cancer patients were distributed between 14 and 28. The distribution of SIV is clearly different in healthy people and cancer patients. The correlation equation is quite different between these two groups. In cancer patients, SIV rapidly declines with age. In Figure 3, the same data are converted from SIV levels to grading, clearly showing that most healthy people belong to grades 4 and 5 (although several were grade 3), but half of all cancer patients are in the range of grades 2 and 3.

In our opinion, people with SIV grade 2 or 3 are candidates for considering interventions to restore immunological vigor in some way. People with grade 1 should be treated with rapidly effective

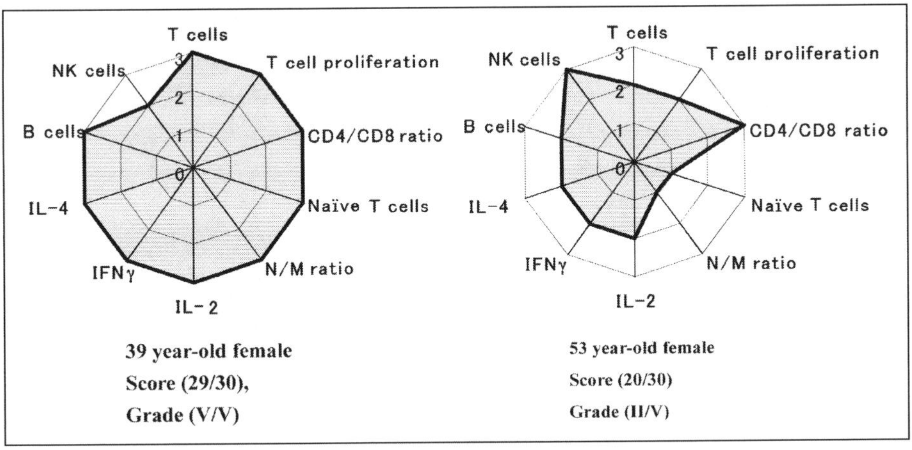

Figure 1. Scoring of immunological vigor (SIV), shown by radar graphs in a healthy 39 year-old female and a 53 year-old cancer patient.

methods for immunological restoration or at least be kept in a clean room. By measuring blood pressure and pulse, one can easily assess the condition of cardiovascular function. From the levels of blood glucose and HbA1c, one can estimate diabetic status. In a similar manner, simple assessments of immune status, would be desirable. Assessment of the immunological status is important for patients suffering from different diseases, because they undergo various treatments that may suppress immune functions. Although the approach presented in this paper is still preliminary and subject to modification, it makes a start in this direction. The next question is how to intervene to reverse the depressed immunological state.

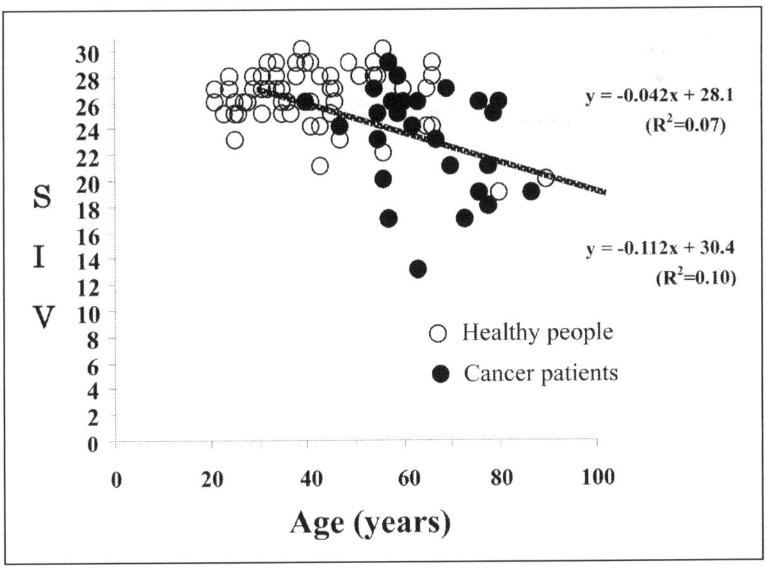

Figure 2. Scoring of immunological vigor (SIV) in 60 healthy people (open circles) and 20 cancer patients (closed circles).

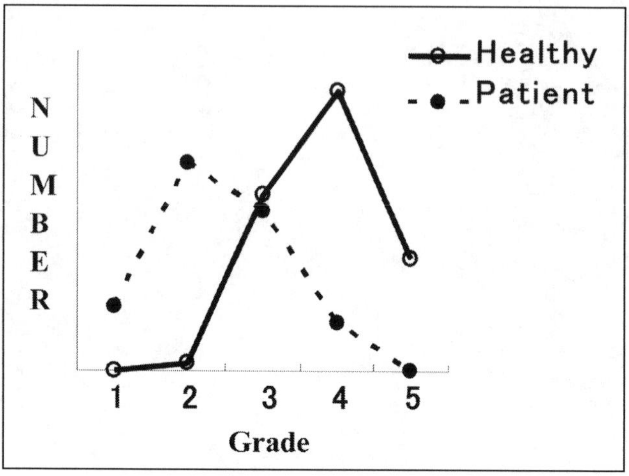

Figure 3. Grading of SIV in 60 healthy people (open circles) and 20 cancer patients (closed circles).

Restoration of Immune Function

Here, we briefly summarize experiments on immunological restoration previously conducted in our laboratory using animal models.[6,7]

a) Activation of existent immune cells.
   1. Anti-oxidants such as VE.
   2. Low dose anti-cancer drugs.
   3. Caloric restriction.
   4. Exercise.
   5. Vaccination.
   6. Japanese herbal medicine
b) Grafting of tissues.
   1. Combined grafting of bone marrow and newborn thymus.
   2. Grafting of newborn thymus.
c) Infusion of activated autologous/syngeneic T-cells.

Methods for activation of existent immune cells are already applied in patients with a certain efficacy. Stem cell transplantation for the treatment of leukemia is routine. For the immunological restoration, grafting of newborn thymus is necessary, but this is, needless to say, not applicable to humans in the present situation. Infusion of activated autologous/syngeneic T-cells may be expected to restore the immune deficient status. T-cells obtained from peripheral blood can be expanded 500- to 1000-folds in vitro in the presence of immobilized anti-CD3 Ab and IL-2 (Fig. 4). These autologous activated T-cells propagated ex vivo could be infused into patients or frail elderly people and a significant immunological restoration could result. This method has been widely used for cancer patients as a form of immunotherapy,[8] but without significant impact on cancer treatment in many cases. However, one can expect that activated T-cells expanded in vitro in a non-antigen-specific manner could improve the immune deficient status of elderly people and cancer patients. Although it might be feared that activated T-cells expanded in vitro could also have adverse effects, such as promotion of autoimmunity, a recent paper reported that immunotherapy with such activated T-cells propagated in this way did not enhance or promote autoimmune phenomena.[9]

T-cells obtained from aged people or immune-deficient patients do not proliferate as efficiently in vitro as T-cells from healthy donors. Therefore, the ideal source of activated T-cells for infusion

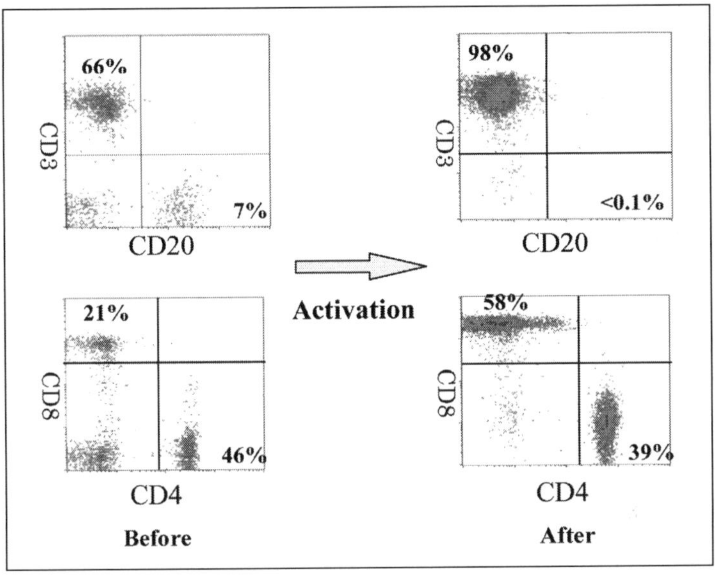

Figure 4. Flow cytometric patterns of peripheral blood lymphocytes before and after activation in the presence of anti-CD3 monoclonal antibody and IL-2. After activation, 98% of lymphocytes are composed of T-cells; 58% are CD8^{+} T-cells and 39% CD4^{+} T-cells.

is healthy young donors. We are now proposing a T-cell banking system for the infusion in later life of autologous activated T-cells taken earlier. Lymphocytes obtained from individuals can be stored in liquid nitrogen for many years and be available for immunological restoration when the SIV for that particular individual indicates such a necessity.

There is a great deal of clinical experience already with infusion of activated autologous T-cells (IAAT), most frequently for the treatment of cancer, as mentioned above. It is reasonable to expect the restoration of immune functions in patients with immune deficiency by IAAT even if this approach has not proven very effective in curing cancer so far. In addition, children with congenital immune deficiency are already treated with IAAT, where the procedure is effective. For example, one study found that 6 of 12 patients showed some recovery in the number of lymphocytes and T-cell proliferative responses after such treatment.[10]

Effect of Infusing Activated T-Cells in the Mouse Model

We tested the effect of infusing activated T-cells using young and old mice.[11] In this study, we employed a congenic combination of B10.Thy-1.1 mice as donors and C57BL/6 Thy-1.2. mice as recipients, to determine how many activated T-cells survived in the recipients. We divided the animals into 6 groups as follows:

a) Young control C57BL/6.Thy1.2 without treatment.
b) Activated T-cells from young Thy-1.1 mice —> young C57BL/6 Thy1.2 recipients.
c) Activated T-cells from aged Thy-1.1 mice—> young C57BL/6 Thy1.2 recipients.
d) Aged control, C57BL/6.Thy1.2 without treatment.
e) Activated T-cells from young Thy-1.1 mice—> aged C57BL/6 Thy1.2 recipients.
f) Activated T-cells from aged Thy-1.1 mice—> aged C57BL/6 Thy1.2 recipients.

The mice was sacrificed 11 days and 25 days after the experiment and used for immunological assessment. For the infusion of activated T-cells, splenic T-cells can be expanded 10- to 15-fold in the presence of immobilized anti-CD3 and IL-2. T-cells activated in this way are composed of approximately 70 to 80% of CD8 T-cells and 7 to 14% of CD4 T-cells. The activated T-cells prepared from aged donors contain many more CD8 T-cells. Although the $CD4^+$ cells are fewer in number, most did have a phenotype of naïve T-cells. After infusion of these activated T-cells, the absolute number of T-cells increased in the spleen of recipient mice, especially in the aged mice. In the peripheral blood and spleen, donor-type Thy-1.1 T-cells were significantly more numerous in aged than young recipients. In addition, many more donor-type T-cells were present in the spleen than in peripheral blood in both young and old recipients.

The magnitude of antibody formation against SRBC did not change significantly in young recipients after this procedure. In about half of the aged recipients, however, a significant enhancement of antibody formation was observed. Importantly, this was seen even in aged recipients infused with activated T-cells from aged donors (Fig. 5).

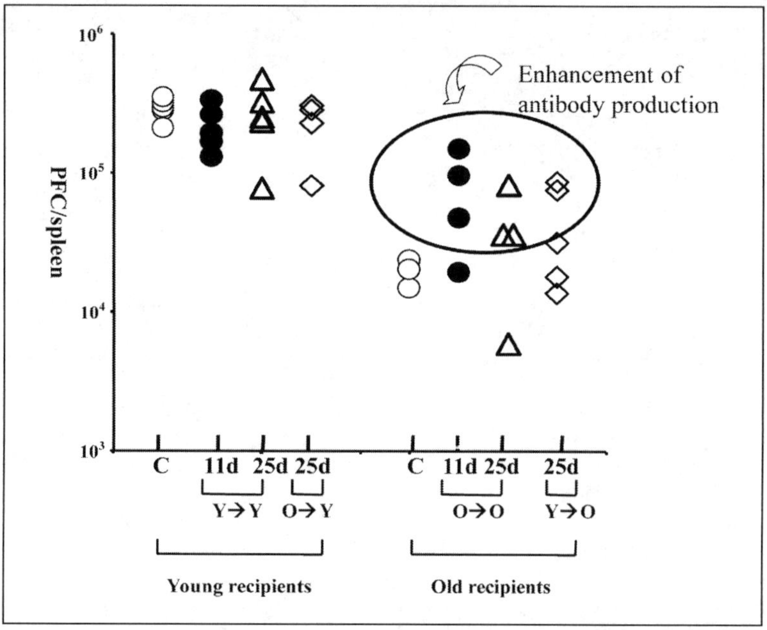

Figure 5. Effect of infusion of activated T-cells on anti-SRBC antibody formation in young and old recipient mice. No positive effect was observed in young recipients compared with controls. In contrast, enhancement of antibody formation was observed in some of the old recipients following infusion of activated T-cells from either young or old donors. Magnitude of antibody formation against SRBC is indicated by the number of PFC (plaque-forming cells) per spleen. The assessment was done 11 days (11d) or 24 days (25d) after the infusion of activated T-cells. Y→Y, activated T-cells from young donors were transferred to young recipients. O→Y, activated T-cells from old donors were transferred to young recipients. O→O, activated T-cells from old donors were transferred to old recipients. Y→O, activated T-cells from young donors were transferred to old recipients.

Conclusions

Autopsy examination has revealed that infection is a major cause of death in elderly people as well as a significant fraction of cancer patients.

Elderly people and cancer patients are both immunodeficient.

Appropriate assessment of immune status is urgently needed, especially for the elderly and patients suffering from various diseases. Here, we have proposed one method to assess the immune status as a whole by scoring immunological vigor.

Assessment of the immune status should be followed by immunological restoration or reconstruction, when necessary.

Infusion of activated previously banked autologous T-cells could be used for such immunological restoration.

References

1. Hirokawa K, Utsuyama M, Makinodan K. Immunity and aging. In: Pathy MSJ, Sinclair AJ, Morley JE, eds. Principles and Practice of Geriatric Medicine. 4th ed. 2006:19-36.
2. Hirokawa K: Aging and immunity. Jpn J Geriat 2003; 40:543-552.
3. MacGee W. Cause of death in a hospitalized geriatric population: an autopsy study of 3000 autopsy patients. Virchow Arch A 1993; 423:343-349.
4. Ershler WB, Longo DL. Aging and cancer: issues of basic and clinical science. J Natl Cancer Inst 1997; 89:1489-1497.
5. Castle SC, Uyemura K, Fulop T et al. A need study the immune status of frail older adults. Immun Ageing Published online 2006; 3:1.
6. Hirokawa K, Utsuyama M. Animal models and possible human application of immunological restoration in the elderly. Mech Ageing Develop 2002; 123:1055-1063.
7. Tsunemi A, Utsuyama M, Seidler BKH et al. Age-related decline of brain monoamines in mice is reversed to young level by Japanese herbal medicine. Neurochem Res 2005; 30:75-81.
8. Rosenberg SA. Progress in human tumour immunology and immunotherapy. Nature 2001; 411:380-384.
9. Yamaguchi T, Bamba K, Kitayama A et al. Long-term intravenous administration of activated autologous lymphocytes for cancer patients does not induce anti-nuclear antibody and rheumatoid factor. Anticancer Res 2004; 24:2423-2429.
10. Morio T. Adoptive immunotherapy by activated T-cells. Tokyo Pediatrician Conference Report 2001; 20:22-25 (In Japanese).
11. Utsuyama M, Hirokawa K. Unpublished data.

Chapter 3

Remodelling of the CD8 T-Cell Compartment in the Elderly:

Expression of NK Associated Receptors on T-Cells Is Associated with the Expansion of the Effector Memory Subset

Inmaculada Gayoso, M. Luisa Pita, Esther Peralbo, Corona Alonso, Olga DelaRosa, Javier G. Casado, Julian de la Torre-Cisneros, Raquel Tarazona and Rafael Solana*

Abstract

Immunosenescence is a complex series of alterations that affect most aspects of immunity, including innate and adaptive immunity and is dependent not only on chronological ageing itself, but also on exogenous factors such as persistent antigenic stress leading to chronic activation of the immune system. The most significant changes occur in the T-cell compartment, with decreasing naïve cells and increasing numbers of cells with a memory/activated phenotype. T-cells from elderly individuals also show altered responsiveness to antigen and display altered profiles of cytokine production when compared to T-cells from young individuals. The alterations in the T-cell compartment observed in the elderly have been implicated in the impaired immune response to viral infections and low response to vaccines. In particular, the most characteristic changes of the CD8 compartment associated with ageing are the accumulation of $CD8^+CD28^{null}$ T-cells, antigen receptor repertoire shrinkage, increased expression of NK-associated receptors and the downregulation of CCR7 and CD45RA, suggesting an expansion of T-cells with an effector-memory 2 phenotype. These changes can be explained by the reduction of naïve $CD8^+$ T-cell-output, due to age-associated thymus involution and the oligoclonal expansion of $CD8^+$ T-cells, likely as a consequence of persistent viral infections. The role of cytomegalovirus in the oligoclonal accumulation of $CD8^+CD28^{null}$ T-cells has been recently stressed.

Introduction

The term "immunosenescence" refers to the decline in the immune response observed in elderly persons and in aged animals. The alterations in the immune response were thought to be due to physiological deterioration associated with chronologic ageing.[1] Clinical and epidemiological studies support the idea that immunosenescence contributes to the increased morbidity and mortality due to infections and the low response to vaccines found in elderly individuals as well as possibly to autoimmune phenomena and cancer.[1-6]

*Corresponding Author: Rafael Solana—Department of Immunology, Faculty of Medicine, "Reina Sofia" University Hospital, University of Córdoba, 14004 Cordoba, Spain. Email: rsolana@uco.es

Immunosenescence, edited by Graham Pawelec.

Studies in the last decade have demonstrated that immunosenescence is a complex series of alterations that affect most aspects of immunity, including innate and adaptive immunity. They are dependent not only on chronological ageing itself, but also on exogenous factors such as persistent antigenic stress leading to chronic activation of the immune system.[1,7-9] Thus, as suggested by Franceschi et al[10,11] immunosenescence should not be considered as a unidirectional deterioration, but on the contrary, this complex phenomenon is much better described by terms such as 'remodelling', 'reshaping' or 'retuning'. Whereas the number and functional capacity of some cell subsets of the immune system is decreased by ageing, other subsets are increased and/or show an increased or aberrant response. Thus, both innate and adaptive components of the immune system undergo significant age-related changes. The T-cell immune response is the most dramatically affected by ageing, although age-associated alterations in the phenotype and function of other cells of the immune system have been demonstrated.[1,2,6,8,9,12-14]

Evidence from studies using T-lymphocytes obtained ex vivo from healthy elderly donors, including centenarians, as well as long term in vitro cultures, support the notion that changes associated with T-cell immunosenescence include not only decreased proliferation and IL-2 production in response to mitogens but also CD28 downregulation and telomere shortening.[1,15,16] These changes can be found both in the CD4 and CD8 T-cell subpopulations, although they do affect the CD8 T-cell subset more. T-lymphocytes from elderly individuals reveal that T-cell senescence is characterized by the decrease of naïve cells and the increase of cells with memory/activated phenotype as well as a decreased in vitro responsiveness to antigen and mitogen stimulation. They also display altered profiles of cytokine secretion when compared to T-cells from young individuals, independently of being naïve or memory T-cells.[3,4,17,18] This decline in T-cell function is in part due to alteration in the signalling pathways, including alterations in the immune synapse and decreased fluidity of lipid rafts with high levels of cholesterol.[19,20]

In this chapter we summarize the phenotypic and functional characteristics of T-cells expressing NK-associated receptors (NKRs) that are increased in human ageing. The expansion of these cells in the elderly supports the idea that ageing is associated with the remodelling of the CD8 T-cell compartment with a dramatic decrease in naïve cells and the expansion of effector-memory type 2 and terminally-differentiated effector cell subsets. The possible relevance of latent virus infection by CMV in this process is also discussed.

Expression of NKR on T-Cells in Ageing: The Expansion of CD8 T-Cells Expressing NK Associated Receptors in the Elderly Is Due to the Expansion of Effector Memory 2 T-Lymphocytes

One of the most characteristic age-associated alterations found in T-cells is the increased expression of differentiation markers that are preferentially expressed in NK cells including CD16, CD56, CD57 or CD94. It has been shown that a very low percentage of T-lymphocytes from newborns express NKR and NKR-positive T-cells represent a minor proportion of circulating T-cells in healthy young individuals.[21] It has been previously shown that a significant increase in the proportion of $CD3^+$ T-cells from elderly individuals, healthy SENIEUR donors or centenarians co-expresses some of these NKRs.[22-26] An increase in the frequency of T-cells expressing NKRs can also be found in several clinical situations involving chronic activation of the immune system including tumours, HIV infection or rheumatoid arthritis.[27,28-32] The expression of other receptors that are widely expressed on NK cells such as CD244, or the killer cell lectin-like receptor G-1 (KLRG-1) is also increased on CD8 T-lymphocytes from elderly individuals. Whereas CD244 and NKG2D can act as costimulatory receptors on T-cell cytotoxicity,[33,34] KLRG-1 is a marker of end-stage differentiation and apoptosis resistance.[35]

Another hallmark of age-associated changes in T-lymphocytes is the decrease in the expression of the CD28 costimulatory molecule on T-cells, in particular in the CD8 T-cell subset, as well as a downregulation of markers of naïve T-cells such as CCR7 and CD45RA (Figs. 1 and 2). Thus, $CD8^+CD28^{null}$ T-cells, that are virtually absent in the newborn, become the majority of circulating $CD8^+$ T-cells in the elderly. $CD28^{null}$ T-cells are derived from the repeated stimulation of naïve

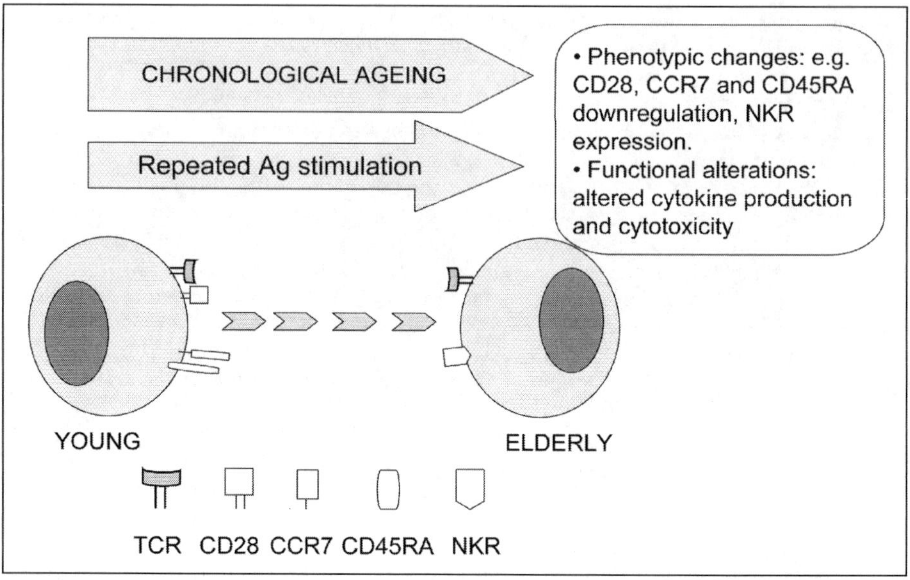

Figure 1. The phenotypic and functional changes observed in CD8 T-cells in the elderly are the consequence not only of chronological ageing but also chronic antigenic stress. Phenotypic alterations include downregulation of costimulatory molecules, modification of the CD45 isoforms, changes in the chemokine receptor pattern and increased expression of NK-associated receptors. Changes in the functional capacity include decreased proliferation in response to antigens and mitogens, altered signal transduction and disbalanced cytokine production.

$CD28^+$ progenitors and the deficiency in the expression of CD28 can be considered a marker of T-cells that have undergone a process of replicative senescence after repeated antigenic stimulation. Thus $CD28^{null}$ T-cells can be considered "senescent" T-cells as they have high levels of expression of mitotic inhibitors, have short telomeres, are highly resistant to apoptosis and show a decreased proliferative capacity probably as a consequence of the reduced levels of CD3-ζ phosphorylation after TCR triggering.[1,9,16,36-41] In particular those $CD8^+CD28^{null}$ cells that also express CD57 have a very low proliferative capacity and their telomere lengths are significantly shorter than in $CD8^+CD28^+$ cells. These results support the notion that CD28-positive and -negative T-cell subsets have distinct replicative histories and that the phenotype $CD28^{null}$ is associated with a state of "replicative senescence."[42,43] Furthermore the expression of CD28 is also downregulated in long term cultures of $CD4^+$ T-cell clones undergoing in vitro replicative senescence, supporting the conclusion that these $CD28^{null}$ T-cells represent senescent T-cells.[15,44]

The accumulation of $CD8^+CD28^{null}$ cells in the elderly is associated with T-cell repertoire shrinkage, as a result of the reduction of naïve $CD8^+$ T-cell-output and the oligoclonal expansion of $CD8^+$ T-cells. The possible significance of persistent viral infections such as cytomegalovirus infection in the accumulation of $CD8^+CD28^{null}$ T-cells has been recently underscored and will be discussed later.[7,8,45-47]

The phenotypic analysis of T-cell subsets that express NKRs (CD16, CD56, CD57, CD161, CD94, NKG2A) in healthy elderly individuals demonstrates that the majority of these cells are included in the $CD8^+CD28^{null}$ subset. Furthermore, most $CD8^+CD28^{null}$ T-cells express other NK markers such as CD224 or KLRG1, supporting the role of this cell subpopulation as cytolytic effector cells. Other NK receptors include killer immunoglobulin receptors (KIR) that constitute a family encoded in multiple gene loci with multiple allelic variants and are specific for MHC class I molecules. The expression of KIR on T-cells is more restricted than the expression of other NKR and they are only found on memory T-cells, primarily senescent or presenescent $CD28^{null}$ cells.

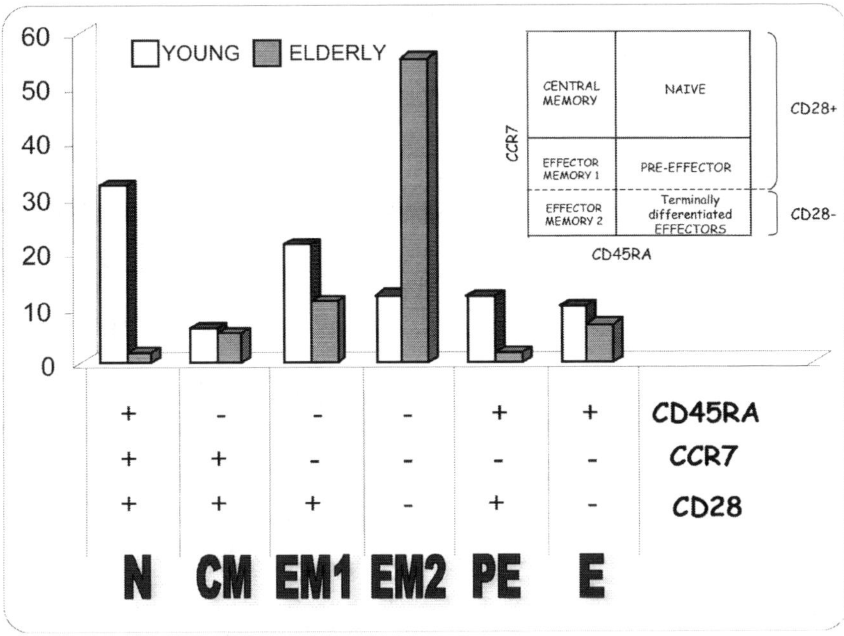

Figure 2. Remodelling of CD8 T-cell subsets in young and elderly individuals. The expression of CCR7, CD45RA and CD28 was analyzed by multiparametric flow cytometry within the CD8$^+$ T-cells from young (white bars) and old healthy donors (black bars). Six different subsets (inserted figure) can be identified based on the expression of CD28, CCR7 and CD45RA: naïve (N) that are CD28$^+$CCR7$^+$CD45RA$^+$, central memory (CM), CD28$^+$CCR7$^+$ CD45RA-, effector-memory 1 (EM1), CD28$^+$CCR7$^-$ CD45RA$^-$, effector-memory 2 (EM2), CD28$^-$CCR7$^-$ CD45RA$^-$, pre-effector cells (PE), CD28$^-$CCR7$^-$CD45RA$^+$ and terminally differentiated effectors (E) also called effector-memory CD45RA$^+$ cells, CD28$^-$CCR7$^-$ CD45RA$^+$. The dramatic decrease in the percentage of naïve ($p < 0.001$) and pre-effector cells ($p < 0.01$) observed in the elderly is associated with a significant expansion ($p < 0.001$) of the effector-memory 2 T-cell subset.

Although the expression of NKRs on T-cells is mainly restricted to the CD8 subset, CD4 T-cells can also express significant levels of CD161 (NKRP1), both in young and elderly individuals. Low levels of other NKR as CD94/NKG2 dimers or KIRs can also be expressed on CD4 T-cells in some clinical conditions such as rheumatoid arthritis.[15,26,29,44,48]

The expression of CCR7, a chemokine receptor that controls homing to secondary lymphoid organs, can be used to divide both CD4 and CD8 human memory T-cells into two functionally distinct subsets: (a) those expressing the CCR7 receptors, termed central memory cells, that also express lymph-node homing receptors and lack immediate effector function and (b) another subset that does not express CCR7, expresses receptors for migration to inflamed tissues, displays immediate effector function, termed effector memory cells. Both CCR7-positive and CCR7-negative T-cells differentiate in a step-wise manner from naïve T-cells and persist for long periods after immunization. The division into two functionally distinct subsets favours memory specialization in the immune response.[49] In the CD8$^+$ subpopulation an additional subset exists that corresponds to differentiated effector cells. They are CCR7$^-$CD45RA$^+$ and contain high levels of perforin, are equivalent to the cytotoxic effector cells induced by antigen stimulation and have evolved through extensive rounds of division.[50,51] This subset likely corresponds to the subpopulation of T-cells described by Hoflich et al[52] as "recently activated effector T-cells" that are $CD45RA^{+}CD11^{bright}$ $CD28^{null}CD57^{+}$, produce IFN-γ and TNF-α, contain high levels of perforin and express CD95. It also includes the effector $CD8^{+}$ T-cell pool defined by the co-expression of CD8 and CD56.[21] Moreover, the expression of CD28 can be used to further discriminate additional differentiation stages of CD8 T-cells. Thus, according to the schematic model shown in Figure 2, six differentiation subsets based on the expression of CD45RA, CCR7 and CD28 can be defined. Within the $CD28^{+}$ cells, the $CCR7^{+}CD45RA^{+}$ cells are considered naïve (N), the $CCR7^{+}CD45RA^{-}$ cells are central memory (CM), the $CCR7^{-}CD45RA^{-}$ cells are effector memory 1 (EM1) and the $CCR7^{-}CD45RA^{+}$ cells are pre-effector (pE) cells. In contrast, within the $CD28^{null}$ cells, those that are $CCR7^{-}CD45RA^{-}$ are considered effector memory 2 (EM2) and the $CCR7^{-}CD45RA^{+}$ are considered the terminally differentiated effector (E) subset. A similar model has been proposed by using CD27 instead of CD28.[50,53]

Our results indicate that both in young and elderly individuals NKRs are mainly expressed on the EM2 and the terminally-differentiated effector subsets of CD8 T-cells. Moreover, we have demonstrated that in the elderly, there is a dramatic increase of these subsets (in particular the EM2 subset) indicating that the increased expression of NKRs found in elderly individuals is a consequence of the expansion of these T-cell populations. These results suggest that the $CD8^{+}CD28^{null}$ T-cells with an EM2 phenotype found in old individuals represent a subset of senescent cells that have undergone a process of replicative senescence.

Taken together the results summarized in this section support the conclusion that the expression of NK receptors on T-lymphocytes is the consequence of the accumulation of $CD8^{+}CD28^{null}$ effector-memory 2 cells, accumulated after multiple rounds of division in response to persistent chronic activation (Fig. 3). The reasons why these cells accumulate in vivo might be related to their resistance to apoptosis.[41] Thus it has been shown that long-term activation of $CD8^{+}$ cells in vitro leads these cells to senescence. T-lymphocytes that reach replicative senescence show loss of CD28 expression, shortened telomeres and undetectable levels of telomerase. These cells are also resistant to apoptosis and have diminished caspase 3 activity in response to apoptotic stimuli,[41] suggesting that the progressive accumulation of T-cells showing many of the markers of replicative senescence during aging reflect the diminished capacity of such cells to undergo normal programmed cell death.

CMV-Specific CD8 T-Cells Are Expanded in the Elderly Expression of NK Associated Receptors

The decrease of naïve CD8 cells in the elderly, likely as a consequence of thymus involution, accompanied by the expansion of effector and effector-memory cells is well established. The possibility that this abnormal subset distribution is the consequence of chronic antigen stimulation by latent viruses such as CMV has been previously proposed by ourselves and others.[8,54,55] The relevance of CMV in this process is underlined by the demonstration of oligoclonal expansion of CMV-specific CD8 T-cells (Fig. 4). However in CMV-seronegative elderly individuals, the response against other viruses such as EBV can induce clonal expansions similar to those found in CMV infection.[46]

However, it is unclear to what extent the accumulation of CMV-specific CD8 T-cells is a major factor contributing to the phenotypic changes found in CD8 T-cells described above. The phenotypic analysis of CMV-specific CD8 cells has demonstrated that the proportion of cells coexpressing CD27 and CD28 is strongly decreased in the elderly when compared with young individuals. Furthermore, the analysis of the differentiation stages defined by the combined use of CCR7 and CD45RA also showed that in elderly donors CMV-specific CD8 T-cells exhibit a phenotype associated with effector-memory ($CCR7^{-}$ CD45RA low) or effector ($CCR7^{-}$ $CD45RA^{+}$) T-cells, whereas in young individuals a significant proportion of CMV-specific CD8 cells are included in the naïve subpopulation ($CCR7^{+}CD45RA^{+}$).[7,8] These results indicate that the majority of CMV-specific CD8 cells in elderly individuals have effector-memory 2 and terminally differentiated effector phenotypes (Fig. 4). These cells also have an increased expression of NK-associated

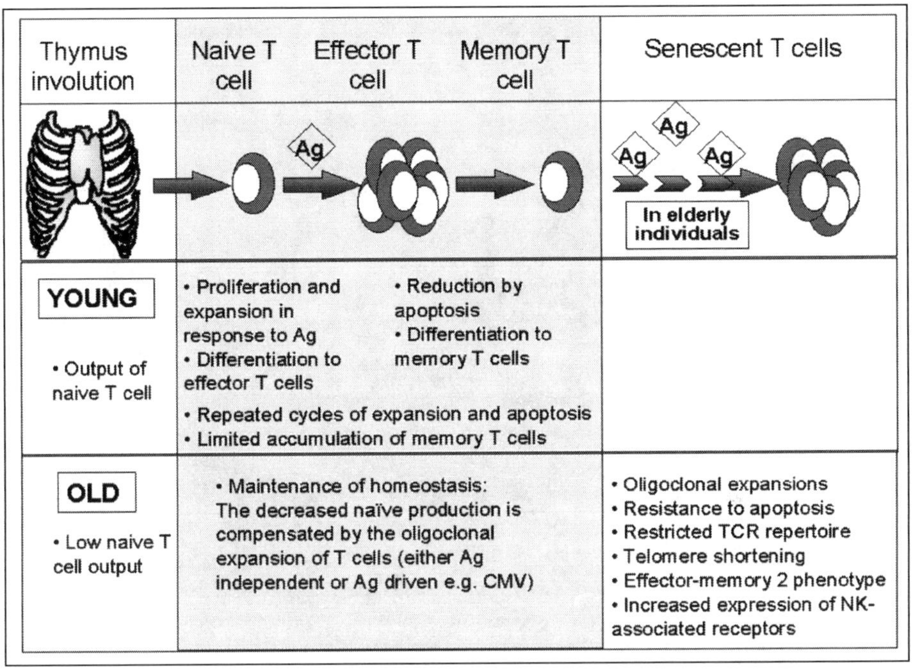

Figure 3. Schematic representation of the CD8 T-cell compartment in young and elderly individuals. Most age-associated alterations observed in CD8 T-cells can be explained by the combined effect of thymus involution leading to a decreased naïve T-cell pool and the lifetime exposure to Ags that trigger naïve cell differentiation into memory cells. In young individuals the repeated cycles of expansions and apoptosis in response to antigens leads to a limited increase of memory CD8 T-cells. However, in the elderly, the depletion of naïve CD8 T-cells is compensated by the accumulation of homeostatic Ag-independent expansions of naïve and memory T-cell subpopulations and expansions of T-cells driven by common persistent pathogens.

receptors.[8] Furthermore whereas essentially all cells in the elderly are KLRG-1-positive, significantly fewer are in the young.[8,47,56] The expression of KLRG-1 on CD28^{null} cells can be considered a marker of end-stage differentiation and apoptosis resistance whereas, on the contrary, CD28^{+} cells are still capable of proliferation despite the expression of KLRG-1.[57] The functional capacity of CMV-specific T-cells is also affected by ageing. Although CMV-specific T-cells release IFN-γ in response to mitogenic stimulation, they do not respond to antigenic stimulation.[47,56,58] It is of interest to note that this characteristic phenotype associated with ageing seems to be restricted to CMV-specific T-cells. Thus, it has been reported that EBV-specific CD8 T-cells from the same individual maintain expression of CD28 and have a lower expression of CD45RA than observed in CMV-specific CD8 T-cells.[55]

In other clinical situations involving chronic viral infection such as HIV infection, an accumulation of CD8^{+}CD28^{null} cells with poor proliferative potential and shortened telomeres has also been observed, indicating that in these patients CD8 T-cells have also undergone a process of replicative senescence, probably as a consequence of chronic antigenic stimulation.[59] As for CMV-specific CD8^{+} T-cells from elderly individuals, the great majority of HIV-specific CD8^{+} T-cells are included in the CD28^{null} subset, whereas CD8^{+} T-cells from the same individuals that recognize other virus, such as EBV or influenza, are essentially CD28^{+}.[60] This finding supports the hypothesis that HIV-specific CD8 T-cells have undergone a process of replicative senescence in

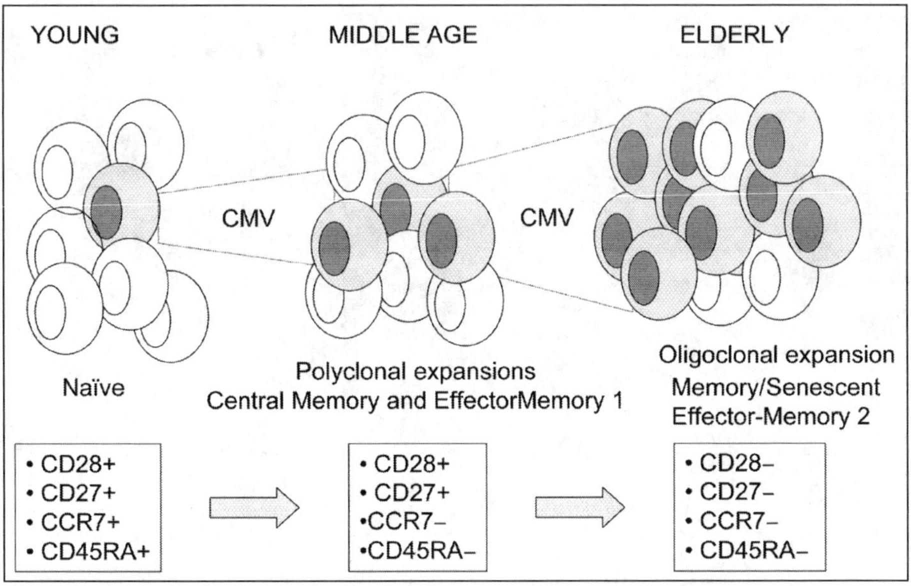

Figure 4. CMV is a major force leading to the oligoclonal accumulation of senescent T-cells and the repertoire shrinkage observed in the aged. In young and middle-aged individuals, CMV infection triggers repeated cycles of expansion and apoptosis leading to a limited accumulation of memory T-cells. In the elderly, large dysfunctional oligoclonal populations of CMV-specific T-cells are found, probably as the consequence of the lifetime chronic stimulation of CD8 T-cells by this persistent virus.

HIV-infected individuals. CD8 T-cells from HIV-infected individuals also show a lower expression of CD56 that correlates with the decrease in the number of CD4 T-lymphocytes.[28] HIV-specific CD8$^+$ T-cells, however, maintain the expression of CD27, express low levels of perforin and mediate decreased specific lysis ex vivo, suggesting an impaired maturation of HIV-specific CTLs in these patients.[61]

Altogether, the findings that the CMV-specific CD8 T-cell phenotype in elderly individuals is similar to the predominant phenotype of CD8 T-cells as a whole and that the accumulation of CD28^{null} T-cells expressing CD57 and CD56 is preferentially observed in CMV-seropositive elderly,[62,63] suggest that latent infection with CMV can be considered a major factor contributing to the differentiation of CD8 T-cells to poorly functional senescent cells with an effector-memory 2 phenotype.

Concluding Remarks and Future Prospects

In conclusion, immunosenescence is a complex series of alterations that are dependent not only on chronological ageing itself, but also on exogenous factors such as persistent antigenic stress leading to chronic activation of the immune system. Ageing is associated with immunological changes in the T-cells primarily due to thymus involution resulting in a decreased production of naïve cells. Moreover, recent evidence also support the idea that many alterations observed in the CD8 T-cell compartment can be explained by the chronic activation of the immune system by latent viruses such as CMV. Whereas at some stages the oligoclonal expansion of virus-specific cells observed in elderly individuals could contribute to lymphocyte homeostasis by maintaining absolute T-cell numbers, at a later stage T-cell alterations will be deleterious for an adequate response to infection. Thus, age-associated changes to CD8 T-cells such as TCR repertoire shrinkage, differentiation to EM2 phenotype, increased NKR expression or altered cytokine production may or may not be detrimental. These changes are observed in the majority of elderly individuals, but only a combination of parameters characterised by the dramatic expansion of $CD8^+CD28^{null}$ cells resulting in an inverted CD4/CD8 ratio that define the "immune risk phenotype" is associated with an increased risk of death.[64,65] Many age-associated alterations of T-cells are similar to those observed in certain circumstances in which other sources of chronic antigenic stimulation are present. This may be happening in cancer and autoimmunity, as well as in other chronic infections where T-cells with an immunosenescent phenotype are accumulated likely as a consequence of long term activation by tumour antigens, autoantigens or viral antigens. We suggest that elucidation of the causes underlying CD8 alterations is necessary to develop future strategies to improve protective immunity in the elderly. Advances in T-cell immunosenescence research will also be of interest to better understand chronic immune-mediated diseases and to provide the basis to design novel alternative therapies.

Acknowledgements

This work was supported by grants QLRT- 2001-00668 (Outcome and Impact of Specific Treatment in European Research on Melanoma, OISTER) and QLK6-CT2002-02283 (T-cells in Ageing, T-CIA) from the 5th Framework Program of the European Union, FIS03/1383 and FIS06/1630 grants (to R.S.), and REIPI (RD06/0008) network (to J.T.) from Ministry of Health (Spain), SAF2003-05184 (to R.T.) from Ministry of Science and Technology (Spain) and by Junta de Andalucia (RS) and Junta de Extremadura (RT). MLP is supported by a grant of the Mexican government, (PROMEP, UdeG490).

References

1. Pawelec G, Barnett Y, Forsey R et al. T-cells and aging, 2002 update. Front Biosci 2002; 7: d1056-d1183.
2. Weng NP. Aging of the immune system: how much can the adaptive immune system adapt? Immunity 2006; 24:495-499.
3. Pawelec G. Working together for robust immune responses in the elderly. Nat Immunol 2000; 1:91.
4. Pawelec G, Effros RB, Globerson A. A multidisciplinary approach to immunity and ageing: ImAgin Eering Mech Ageing Dev 2000; 20(121):1-4.
5. Delarosa O, Pawelec G, Peralbo E et al. Immunological biomarkers of ageing in man: changes in both innate and adaptive immunity are associated with health and longevity. Biogerontology 2006; 7:471-481.
6. Solana R, Pawelec G, Tarazona R. Aging and innate immunity. Immunity 2006; 24:491-494.
7. Koch S, Solana R, Rosa OD et al. Human cytomegalovirus infection and T-cell immunosenescence. Mech Ageing Dev 2006; 127:538-543.
8. Pawelec G, Akbar A, Caruso C et al. Human immunosenescence: is it infectious? Immunol Rev 2005; 205:257-268.
9. Tarazona R, Solana R, Ouyang Q et al. Basic biology and clinical impact of immunosenescence. Exp Gerontol 2002; 37:183-189.
10. Franceschi C, Monti D, Barbieri D et al. Successful immunosenescence and the remodelling of immune responses with ageing. Nephrol Dial Transplant 1996; 11 Suppl 9:18-25.
11. Franceschi C, Cossarizza A. Introduction: the reshaping of the immune system with age. Int Rev Immunol 1995; 12:1-4.
12. Pawelec G, Solana R, Remarque E et al. Impact of aging on innate immunity. J Leukoc Biol 1998; 64:703-712.
13. Solana R, Alonso MC, Pena J. Natural killer cells in healthy aging. Exp Gerontol 1999; 34:435-443.
14. Solana R, Mariani E. NK and NK/T-cells in human senescence. Vaccine 2000; 18:1613-1620.
15. Effros RB, Pawelec G. Replicative senescence of T-cells: does the Hayflick Limit lead to immune exhaustion? Immunol Today 1997; 18:450-454.
16. Pawelec G, Solana R. Immunoageing—the cause or effect of morbidity. Trends Immunol 2001; 22:348-349.
17. Haynes L, Eaton SM, Burns EM et al. Newly generated CD4 T-cells in aged animals do not exhibit age-related defects in response to antigen. J Exp Med 2005; 201:845-851.
18. Swain S, Clise-Dwyer K, Haynes L. Homeostasis and the age-associated defect of CD4 T-cells. Semin Immunol 2005; 17:370-377.
19. Fulop T, Larbi A, Wikby A et al. Dysregulation of T-cell function in the elderly: scientific basis and clinical implications. Drugs Aging 2005; 22:589-603.

20. Sadighi Akha AA, Miller RA. Signal transduction in the aging immune system. Curr Opin Immunol 2005; 17:486-491.
21. Pittet MJ, Speiser DE, Valmori D et al. Cutting edge: cytolytic effector function in human circulating CD8+ T-cells closely correlates with CD56 surface expression. J Immunol 2000; 164:1148-1152.
22. Borrego F, Robertson MJ, Ritz J et al. CD69 is a stimulatory receptor for natural killer cell and its cytotoxic effect is blocked by CD94 inhibitory receptor. Immunology 1999; 97:159-165.
23. McNerlan SE, Rea IM, Alexander HD et al. Changes in natural killer cells, the CD57CD8 subset and related cytokines in healthy aging. J Clin Immunol 1998; 18:31-38.
24. Miyaji C, Watanabe H, Minagawa M et al. Numerical and functional characteristics of lymphocyte subsets in centenarians. J Clin Immunol 1997; 17:420-429.
25. Rea IM, McNerlan SE, Alexander HD. CD69, CD25 and HLA-DR activation antigen expression on CD3+ lymphocytes and relationship to serum TNF-alpha, IFN-gamma and sIL-2R levels in aging. Exp Gerontol 1999; 34:79-93.
26. Abedin S, Michel JJ, Lemster B et al. Diversity of NKR expression in aging T-cells and in T-cells of the aged: the new frontier into the exploration of protective immunity in the elderly. Exp Gerontol 2005; 40:537-548.
27. Galiani MD, Aguado E, Tarazona R et al. Expression of killer inhibitory receptors on cytotoxic cells from HIV-1-infected individuals. Clin Exp Immunol 1999; 115:472-476.
28. Tarazona R, Delarosa O, Casado JG et al. NK-associated receptors on CD8 T-cells from treatment-naïve HIV-infected individuals: defective expression of CD56. AIDS 2002; 16:197-200.
29. Vallejo AN, Weyand CM, Goronzy JJ. T-cell senescence: a culprit of immune abnormalities in chronic inflammation and persistent infection. Trends Mol Med 2004; 10:119-124.
30. Casado JG, Soto R, Delarosa O et al. CD8 T-cells expressing NK associated receptors are increased in melanoma patients and display an effector phenotype. Cancer Immunol Immunother 2005; 54:1162-1171.
31. Tarazona R, Casado JG, Soto R et al. Expression of NK-associated receptors on cytotoxic T-cells from melanoma patients: a two-edged sword? Cancer Immunol Immunother 2004; 53:911-924.
32. Michel JJ, Turesson C, Lemster B et al. CD56-expressing T-cells that have features of senescence are expanded in rheumatoid arthritis. Arthritis Rheum 2007; 56:43-57.
33. Lanier LL. NKG2D in innate and adaptive immunity. Adv Exp Med Biol 2005; 560:51-56.
34. Lanier LL. NK cell recognition. Annu Rev Immunol 2005; 23:225-274.
35. Voehringer D, Koschella M, Pircher H. Lack of proliferative capacity of human effector and memory T-cells expressing killer cell lectinlike receptor G1 (KLRG1). Blood 2002; 100:3698-3702.
36. Vallejo AN. CD28 extinction in human T-cells: altered functions and the program of T-cell senescence. Immunol Rev 2005; 205:158-169.
37. Vallejo AN, Weyand CM, Goronzy JJ. Functional disruption of the CD28 gene transcriptional initiator in senescent T-cells. J Biol Chem 2001; 276:2565-2570.
38. Vallejo AN, Brandes JC, Weyand CM et al. Modulation of CD28 expression: distinct regulatory pathways during activation and replicative senescence. J Immunol 1999; 162:6572-6579.
39. Tarazona R, Delarosa O, Alonso C et al. Increased expression of NK cell markers on T-lymphocytes in aging and chronic activation of the immune system reflects the accumulation of effector/senescent T-cells. Mech Ageing Dev 2000; 121:77-88.
40. Scheuring UJ, Sabzevari H, Theofilopoulos AN. Proliferative arrest and cell cycle regulation in CD8(+)CD28(-) versus CD8(+)CD28(+) T-cells. Hum Immunol 2002; 63:1000-1009.
41. Spaulding C, Guo W, Effros RB. Resistance to apoptosis in human CD8+ T-cells that reach replicative senescence after multiple rounds of antigen-specific proliferation. Exp Gerontol 1999; 34:633-644.
42. Batliwalla F, Monteiro J, Serrano D et al. Oligoclonality of CD8+ T-cells in health and disease: aging, infection, or immune regulation? Hum Immunol 1996; 48:68-76.
43. Monteiro J, Batliwalla F, Ostrer H et al. Shortened telomeres in clonally expanded CD28-CD8+ T-cells imply a replicative history that is distinct from their CD28+CD8+ counterparts. J Immunol 1996; 156:3587-3590.
44. Pawelec G, Mariani E, Bradley B et al. Longevity in vitro of human CD4+ T-helper cell clones derived from young donors and elderly donors, or from progenitor cells: age-associated differences in cell surface molecule expression and cytokine secretion. Biogerontology 2000; 1:247-254.
45. Hadrup SR, Strindhall J, Kollgaard T et al. Longitudinal studies of clonally expanded CD8 T-cells reveal a repertoire shrinkage predicting mortality and an increased number of dysfunctional cytomegalovirus-specific T-cells in the very elderly. J Immunol 2006; 176:2645-2653.
46. Wikby A, Ferguson F, Forsey R et al. An immune risk phenotype, cognitive impairment and survival in very late life: impact of allostatic load in Swedish octogenarian and nonagenarian humans. J Gerontol A Biol Sci Med Sci 2005; 60:556-565.

47. Ouyang Q, Wagner WM, Zheng W et al. Dysfunctional CMV-specific CD8(+) T-cells accumulate in the elderly. Exp Gerontol 2004; 39:607-613.
48. van BJ, Thompson A, van der SA et al. Phenotypic and functional characterization of CD4 T-cells expressing killer Ig-like receptors. J Immunol 2004; 173:6719-6726.
49. Sallusto F, Lenig D, Forster R et al. Two subsets of memory T-lymphocytes with distinct homing potentials and effector functions. Nature 1999; 401:708-712.
50. Hamann D, Baars PA, Rep MH et al. Phenotypic and functional separation of memory and effector human $CD8^+$ T-cells. J Exp Med 1997; 186:1407-1418.
51. Hamann D, Kostense S, Wolthers KC et al. Evidence that human $CD8^+CD45RA^+CD27^-$cells are induced by antigen and evolve through extensive rounds of division. Int Immunol. 1999;11:1027-1033.
52. Hoflich C, Docke WD, Busch A et al. CD45RA(bright)/CD11a(bright) $CD8^+$ T-cells: effector T-cells. Int Immunol 1998; 10:1837-1845.
53. Tomiyama H, Matsuda T, Takiguchi M. Differentiation of human CD8(+) T-cells from a memory to memory/effector phenotype. J Immunol 2002; 168:5538-5550.
54. Pawelec G. Hypothesis: loss of telomerase inducibility and subsequent replicative senescence in cultured human T-cells is a result of altered costimulation. Mech Ageing Dev 2000; 20(121):181-185.
55. Vescovini R, Telera A, Fagnoni FF et al. Different contribution of EBV and CMV infections in very long-term carriers to age-related alterations of $CD8^+$ T-cells. Exp Gerontol 2004; 39:1233-1243.
56. Ouyang Q, Wagner WM, Voehringer D et al. Age-associated accumulation of CMV-specific $CD8^+$ T-cells expressing the inhibitory killer cell lectin-like receptor G1 (KLRG1). Exp Gerontol 2003; 38:911-920.
57. Ibegbu CC, Xu YX, Harris W et al. Expression of killer cell lectin-like receptor G1 on antigen-specific human $CD8^+$ T-lymphocytes during active, latent and resolved infection and its relation with CD57. J Immunol 2005; 174:6088-6094.
58. Ouyang Q, Wagner WM, Wikby A et al. Large numbers of dysfunctional $CD8^+$ T-lymphocytes bearing receptors for a single dominant CMV epitope in the very old. J Clin Immunol 2003; 23:247-257.
59. Effros RB, Allsopp R, Chiu CP et al. Shortened telomeres in the expanded $CD28^-CD8^+$ cell subset in HIV disease implicate replicative senescence in HIV pathogenesis. AIDS 1996; 10:F17-F22.
60. Dalod M, Sinet M, Deschemin JC et al. Altered ex vivo balance between $CD28^+$ and. Eur J Immunol 1999; 29:38-44.
61. Appay V, Nixon DF, Donahoe SM et al. HIV-specific CD8(+) T-cells produce antiviral cytokines but are impaired in cytolytic function. J Exp Med 2000; 192:63-75.
62. Wikby A, Johansson B, Olsson J et al. Expansions of peripheral blood CD8 T-lymphocyte subpopulations and an association with cytomegalovirus seropositivity in the elderly: the Swedish NONA immune study. Exp Gerontol 2002; 37:445-453.
63. Olsson J, Wikby A, Johansson B et al. Age-related change in peripheral blood T-lymphocyte subpopulations and cytomegalovirus infection in the very old: the Swedish longitudinal OCTO immune study. Mech Ageing Dev 2000; 20(121):187-201.
64. Ferguson FG, Wikby A, Maxson P et al. Immune parameters in a longitudinal study of a very old population of Swedish people: a comparison between survivors and nonsurvivors. J Gerontol A Biol Sci Med Sci 1995; 50:B378-B382.
65. Wikby A, Maxson P, Olsson J et al. Changes in CD8 and CD4 lymphocyte subsets, T-cell proliferation responses and nonsurvival in the very old: the Swedish longitudinal OCTO-immune study. Mech Ageing Dev 1998; 102:187-198.

Chapter 4

Telomeres, Telomerase and CD28 in Human CD8 T-Cells: Effects on Immunity during Aging and HIV Infection

Steven R. Fauce and Rita B. Effros*

Abstract

The immune system undergoes major alterations during aging, many of which have been implicated in the increased morbidity and mortality associated with infection, as well as the high incidence of cancer in the elderly. Although mouse models have provided important insights into immunosenescence, there are certain facets of human immunological history that cannot be modeled in experimental animals. Here, we focus on the process of replicative senescence in human CD8 T-cells, which seems to be driven by the extensive and long-term cell proliferation required to control certain latent viral infections. Replicative senescence has been well-characterized in cell culture fand is now recognized as an underlying mechanism for shaping the memory T-cell pool in humans. This chapter will focus on the complex relationship between telomeres, telomerase and the T-cell costimulatory receptor, CD28, in modulating the process of CD8 T-cell replicative senescence and the impact of this process on aging and HIV disease.

Introduction

The innate barrier to unlimited proliferation, known as replicative (or cellular) senescence, restricts the behavior of most normal human somatic cells, both in cell culture and in vivo. Replicative senescence is characterized by the initiation of irreversible cell cycle arrest and drastic changes in function, due to alterations in gene expression. Despite these fundamental changes, senescent cells remain viable and metabolically active. Although it was once assumed that T-cells did not undergo replicative senescence, based on reports of unlimited proliferative potential in cell culture,[1,2] it is now known that replicative senescence does, in fact, constitute the final stage of differentiation in normal human T-cells.[3] However, in order to accommodate the extensive amount of proliferation that is required for clonal expansion of CD8 T-cells during primary and even secondary immune responses, T-cells have a greater capacity to divide compared to most other somatic human cells, due to their ability to upregulate the telomere-extending enzyme, telomerase.

The process of T-cell replicative senescence has been studied extensively in cell culture[4-6] and is characterized by such distinct features as an inability to proliferate, resistance to apoptosis,[7] shortened telomeres (5-7 kb)[8] and loss of gene and protein expression of the CD28 costimulatory molecule.[9] The growth arrest cannot be reversed by exposure to antigen or to increasing doses of

*Corresponding Author: Rita B. Effros—Department of Pathology and Laboratory Medicine, David Geffen School of Medicine at UCLA, 10833 Le Conte Ave, Los Angeles, CA 90095-1732, U.S.A. Email: reffros@mednet.ucla.edu

Immunosenescence, edited by Graham Pawelec. ©2007 Landes Bioscience and Springer Science+Business Media.

IL-2.[10,11] Memory CD8 T-cells with very similar features have also been identified and characterized in vivo.[12] Because the process of CD8 T-cell replicative senescence seems to be driven by certain persistent viral infections, it is unique to humans and is not observed in laboratory mice. This chapter will review some of the key features of human CD8 T-cell replicative senescence and highlight the interactions between telomeres, telomerase and CD28 that seem to be central to this process.

Cause and Effect of Cellular Senescence in CD8 T-Cells

It is well documented that there is a dramatic decline in immune function, as well as an increase in disease severity, that occurs with age in humans. One of the most significant defects of the human immune system relates to CD8 T-cell function. Studies have shown that there is a strong correlation between age and a decrease in antigen- or mitogen-induced CD8 T-cell proliferation.[13-15] The widely accepted explanation for the diminished proliferative potential is based on the fact that over a lifetime, humans are exposed to an increasing number of pathogens, many of which are encountered more than once. Both chronic viral infection or repeated encounters with variant forms of the same virus requires extensive cell turnover by the relevant antigen-specific CD8 T-cells, which can eventually lead to diminished proliferative potential and replicative senescence.[16] Indeed, clinical evidence suggests that the increase in severity of viral infections may be related to the increase in the number of senescent CD8 T-cells in elderly persons. This is based, in part, on the fact that there is a significant correlation between high proportions of senescent CD8 T-cells and poor response to vaccines.[17,18]

The idea that the high proportion of senescent CD8 T-cells present in many elderly people is due to antigen-induced CD8 T-cell turnover is supported by the frequent clonal nature of the CD8 T-cell populations.[12] These oligoclonally-expanded cells, which are poorly proliferative and lack CD28 expression, often comprise a large proportion of the overall CD8 T-cell repertoire in the elderly. The clonal expansions, combined with the diminished output of naïve T-cells by the progressively involuting thymus, result in a significant reduction in the overall spectrum of antigenic specificities within the CD8 T-cell pool.[16] With this diminished range of CD8 T-cell antigen specificities, the elderly are unable to successfully combat certain viral infections and will therefore show greater disease-related morbidity compared to their younger counterparts. In addition, it has been suggested that the mere presence of a large population of senescent CD8 T-cells has a detrimental effect on the immune system in general. Several studies have described functional characteristics of senescent CD8 T-cells suggesting that they may exert suppressive influences on the activity of other immune cells.[19,20] It is also believed that the physical presence of these senescent CD8 T-cells will influence the homeostatic mechanisms that regulate the peripheral naive T-cell pool.[21,22] Thus, not only will the peripheral T-cell pool have a less varied T-cell antigen-receptor (TCR) repertoire, but there will also be fewer naive CD8 T-cells to respond to new antigenic challenges.

Characteristics of Senescent CD8 T-Cells: Telomeres and CD28

Two of the prominent features of senescent CD8 T-cells involve telomeres and the CD28 cell surface costimulatory receptor. CD8 T-cells that lack CD28 expression, when tested immediately ex vivo, have much shorter telomeres as compared to the rest of the CD8 T-cell population from that same individual.[23,24] Telomeres are the repetitive hexameric sequences (usually 10-12 kilobases), located at the ends of eukaryotic chromosomes, which function in chromosome stabilization, protection from exonucleolytic degradation and prevention of end-to-end fusion.[8,25,26]

The first report that showed a connection between telomere length and cellular senescence involved a mutation in yeast that resulted in both accelerated telomere shortening and premature senescence.[27] It had been known that telomeres progressively shorten in length during each round of cell division[28,29] due to the "end-replication problem" that was formulated by Olovnikov[30] and Watson.[31] After multiple rounds of division, the telomeres eventually reach a critically short length, which can no longer ensure chromosomal integrity, thereby triggering cell cycle arrest and the induction of replicative senescence.[32,33]

It is thought the critically short telomere length is perceived by the cell as DNA damage, causing an upregulation in the expression of the p16 and/or p21 cell cycle inhibitors, eventually leading to cell cycle arrest.[34-36] Under cell culture conditions, telomeres of T-cells shorten by approximately 100 base pairs per population doubling and reach a size of 5-7 kilobases at senescence.[8,37] Similar telomere shortening is observed in T-cells in vivo, where memory T-cells have shorter telomeres and reduced proliferative potential compared to naive T-cells from the same person, consistent with the distinct proliferative history of these two populations.[38] Interestingly, in experiments on blood samples from individuals from different age groups, telomere loss was much greater in CD8 T-cells versus CD4 T-cells,[39] which may relate to the extensive expansion involved in generating a large population of cells required for direct cell contact-induced cytotoxicity. These and other studies lend support to the generally accepted idea that the telomere shortening that occurs with each cell division is the major factor contributing to the finite replicative life span of all eukaryotic cells.[40]

CD8 T-cell telomere shortening has been documented in vivo during the natural process of aging as well as during chronic viral infection.[19,41] Telomere length has an effect not only on proliferation of T-cells, but also appears to be correlated to immune cell function. A study by Cawthon et al evaluated telomere length in blood samples from a group of 60 year olds, then determined the relationship between this measure and subsequent mortality.[42] Interestingly, the mortality rate from infectious disease was increased 8-fold for individuals in the bottom 25% of cell telomere lengths as compared to individuals from the upper 75%.[42] Constant T-cell turnover induced by chronic viral infection or by multiple exposures to the same pathogen can lead to telomere length attrition and, eventually, to replicative senescence. Significant telomere shortening has been observed in Epstein-Barr virus (EBV)-specific CD8 T-cells isolated from individuals who have been chronically infected with the virus for many years.[43] In such cases of chronic infection, this telomere shortening will lead to replicative senescence of many of the EBV-specific CD8 T-cells, which has been suggested to play a role in the emergence of EBV-induced lymphoma.[44]

Along with the critically short telomere length, the loss of the CD28 molecule from the cell surface of CD8 T-cells is a signature biomarker of replicative senescence. CD8 T-cells that are followed in culture show a gradual decrease in the proportion of cells expressing CD28 as they progress through multiple rounds of antigen-driven proliferation, with less than 10% of the population maintaining CD28 expression after 7 rounds of stimulation.[45] CD28 is a costimulatory molecule found on T-cells that provides a second activation signal following engagement of the T-cell antigen receptor.[46] It is involved in a broad range of functions, including glucose metabolism, apoptosis, mRNA stabilization, IL-2 gene transcription and cell adhesion.[47-50] Although the expression of most T-cell markers reflecting lineage, activation, memory status and adhesion does not change as a CD8 T-cell reaches senescence, this is not the case for CD28.[9]

The complete and irreversible loss of CD28 expression appears to be restricted to end-stage, senescent T-cells and is not related to the normal up- and down-modulation in the level of CD28 expression during activation events and proliferation.[51] Costimulation is critical for full T-cell activation, which is dependent on CD28 binding to the B7.1 and B7.2 molecules on the surface of an activated antigen presenting cell (APC).[46,52-54] If this secondary signal is blocked (by antibodies to the CD28 ligands, for example), T-cells become anergic; they are unable to proliferate or to kill infected target cells. Therefore, the loss of CD28 expression is thought to be a crucial step in the induction of replicative senescence in CD8 T-cells.

The first suggestion of a connection between T-cell replicative senescence and aging came from cross-sectional studies that evaluated the proportion of T-cells lacking CD28 in donors of different ages. Neonates usually have less than 1% CD28-negative T-cells, whereas adults have approximately 15% and elderly donors have 30-50%.[41,55] The majority of the CD28-negative T-cells found in peripheral blood samples are within the CD8, rather than the CD4, T-cell subset.[12] Importantly, it has clearly been demonstrated that T-cells lacking CD28 are derived from a population of cells that were previously CD28-positive and are not part of a separate lineage.[12,56] Similar to cultures of senescent CD8 T-cells, $CD8^+CD28^-$ T-cells, tested immediately ex vivo, are unable to proliferate when stimulated either with activating antibodies and IL-2 or by mitogens that bypass cell surface receptors.[24,55] In addition, these cells, have increased expression of the bcl2 protein and are resistant to superantigen-induced apoptosis,[57] also reminiscent of CD8 T-cells that have undergone replicative senescence in culture.[7]

There is a strong association between high proportions of $CD8^+CD28^-$ T-cells present in the peripheral blood and chronic viral infection,[19,58] suggesting that extensive antigen-driven T-cell turnover is the cause of CD28 loss in vivo. This notion is consistent with the observations that, within elderly humans, the proportion of CD28-negative cells within the oligoclonal expansions is much greater than in rest of the CD8 T-cell population.[59] There is also a strong correlation between the percentage of $CD8^+CD28^-$ T-cells and poor response to influenza vaccine.[17,18] These data suggest that, in addition to chronic infection, multiple exposures to a particular pathogen may also be capable of driving the responding T-cells to undergo extensive proliferation, subsequently leading to replicative senescence. Alternatively, it is possible that the suppressor cell function that has been ascribed to $CD8^+CD28^-$ T-cells may be the underlying cause for their correlation with the poor response to influenza vaccination.[20]

Telomerase: Connections between CD28 and Telomeres

Although telomeres do shorten during each round of replication in all dividing eukaryotic cells, T-cells possess the ability to combat this problem. T-cells are one of the few types of normal somatic cell that have an active enzyme, known as telomerase, that is capable of adding telomere sequences to the ends of chromosomes after each round of replication.[60-62] This enzyme is a ribonucleoprotein consisting of a catalytic subunit, known as hTERT in humans and an RNA template component.[61] In multiple cell types, it has been shown that hTERT gene transfection and the resultant constitutive telomerase activity stabilizes telomere length and increases proliferative potential, thereby either preventing, or at least delaying, replicative senescence.[32,33,63,64]

Many tumor cells express active telomerase, which is thought to prevent telomere shortening and allow for unlimited cell division and rapid tumor growth.[65,66] In CD8 T-cells, telomerase is expressed in a tightly controlled manner during development[52] and also following activation by mitogen, activating antibodies,[67] or antigen.[68-71] For example, high telomerase activity and actual lengthening of telomeres in antigen-specific CD8 T-cells directed toward EBV has been observed in patients during acute infectious mononucleosis.[72] This transient telomerase expression after antigen activation is believed to increase the replicative potential of CD8 T-cells, allowing for the extensive clonal expansion necessary to mount an effective immune response. Endogenous telomerase expression is thought to also be crucial to the long in vivo life span of T-cells.[73] Interestingly, the capacity of CD8 T-cells to upregulate telomerase activity does not appear to be affected by donor age, since activity in naive T-cells from elderly donors is similar to that of younger donors when activated in vitro.[74]

Even though CD8 T-cells are capable of expressing active telomerase, the ability to upregulate this enzyme is lost after repeated encounters with antigen in vitro. Peripheral blood T-cells show a rapid decline in telomerase activation after approximately 10 population doublings and activity is completely undetectable at senescence[75] It is possible that this loss of telomerase inducibility may be a protective mechanism to prevent excessive proliferation and resultant possible mutation and/or transformation. In culture, it has been documented that the loss of telomerase activity was associated with the loss of CD28 from the surface of CD8 T-cells that were progressing toward senescence.[45] These results may explain observations on cells tested immediately ex vivo, which show that CD8 T-cells that lack CD28 expression have shorter telomeres than $CD8^+CD28^+$ T-cell cells from the same donor.[23] In the cell culture setting, the more rapid loss of CD28 on the surface of CD8 T-cells versus CD4 T-cells was accompanied by a similarly divergent pattern of active telomerase expression, further suggesting a strong association between CD28 and telomerase activity.

Maintenance of telomerase activity, by hTERT transduction of naive or antigen-specific memory CD8 T-cell clones, extends the proliferative life span without any discernable alterations in growth characteristic, phenotype or function.[76,77] This same result was reported for polyclonal

CD8 T-cells directed against specific pathogens.[78] Although hTERT transduction did slow the loss of CD28 from the membranes of CD8 T-cells in culture, it did not completely prevent it. It has therefore been hypothesized that telomerase induction in T-cells is dependent upon CD28 signal transduction,[69] a notion consistent with the demonstration that telomerase upregulation in T-cells requires both TCR and CD28 engagement.[45,69] Thus, it is possible that the permanent loss of CD28 expression may be one of the first steps leading toward the telomere-based induction of replicative senescence.

Replicative Senescence of CD8 T-Cells in HIV Disease

It is well documented that CD8 T-cells play a crucial role in combating HIV disease,[79] both during acute infection[80,81] and chronic infection.[82] CD8 T-cells lyse HIV-infected cells by release of perforin and granzyme proteases; they also secrete antiviral cytokines such as IFN-γ and TNF-α and produce soluble factors that can suppress HIV viral replication.[79,83-85] After the acute HIV infection phase, the immune system (specifically, the CD8 T-cell subset) is usually able to control the virus for 8 to 10 years before the disease progresses to full-blown AIDS, at which point the viral load increases exponentially.[81,86,87] It has been shown that CD8 T-cell defects are largely responsible for the eventual failure of the immune system to suppress HIV infection[88] and that loss of CD8 T-cell activity coincides with progression to AIDS.[89,90] Although a person with AIDS will have oligoclonal expansions of HIV-specific CD8 T-cells, many of these cells are nonfunctional and have characteristics suggestive of replicative senescence, such as inability to proliferate, short telomeres and absence of CD28.[91-94] This same phenomenon has been observed in other chronic viral infections such as CMV and EBV. For instance, there is a similar association between CMV seropositivity and the presence of expanded populations of senescent CMV antigen-specific CD8 T-cells in the elderly.[19,95]

One hypothesis for the eventual progression to AIDS is based on the theory that after years of constant exposure to virally infected cells, the HIV-specific CD8 T-cell population has undergone so many rounds of division that it enters the state of replicative senescence.[96] At this point, the cells become nonfunctional and are no longer able to control the virus. There is also data suggesting that the rapid turnover by the HIV-specific CD8 T-cells can lead to proliferation of bystander memory CD8 T-cells that are not involved in the specific antiviral response and that some of these cells also reach replicative senescence.[97] These and other studies have led to the conclusion that chronic HIV disease basically involves a premature and very rapid aging of the immune system.[98] This theory is supported by the observation that persons who have been infected with HIV for many years often have large populations of nonproliferative $CD8^+CD28^-$ HIV-specific T-cells that have shortened telomeres as compared to the CD8 T-cell population as a whole.[24,99-101] In fact, the telomere lengths of many of these HIV-specific CD8 T-cells are similar to those of centenarian lymphocytes. Consistent with the accelerated immunological aging aspect of HIV disease, it has been shown that the causes of death in many AIDS patients are similar to those seen in the elderly, such as certain viral infections and cancers.[102-104]

The importance of telomere shortening as a marker of HIV disease progression was demonstrated in a study that showed a significant correlation between maintenance of telomere length and long term survival.[105] It had previously been reported that telomere lengths of total PBMC[106] and CD8 T-cells[107] from HIV-infected individuals shortened more rapidly than those of seronegative controls. Telomere loss was most significant for disease progressors as compared to asymptomatic individuals, but both groups had much greater telomere loss than the healthy controls. In a separate study, telomere length of CD8, but not CD4, T-cells was shorter in HIV-infected versus uninfected identical twins, ruling out any possibility that the telomere length differences in other studies might have been due to variability in outbred human populations.[99] Similar to studies on long-term T-cell cultures, the shortened telomeres could be accounted for by the CD8 T-cells that no longer expressed CD28.[24]

Based on the telomerase dynamics observed during cell culture experiments, it seems likely that the chronic activation of HIV-specific CD8 T-cells in persons who have been infected for several years leads to eventual loss of the ability to upregulate telomerase. The scenario seen in cell culture suggests that as CD8 T-cells from HIV infected persons proliferate and begin to lose CD28 expression, they also lose the ability to induce telomerase activity, leading to rapid telomere shortening. The importance of maintaining continuous telomerase activity was demonstrated experimentally in studies showing that transduction of HIV-specific CD8 T-cells (isolated from infected individuals) with hTERT extended the cells' proliferative capacity and stabilized telomere length.[34,78] Gene transduction of CD8 T-cells with hTERT also reduced the expression of senescence-associated cell-cycle inhibitors, retarded the loss of CD28 expression, enhanced the inhibition of HIV viral replication, increased production of IFN-γ and TNF-α and increased antigen-specific lytic activity.[34] Most strikingly, the telomerized cells continued to proliferate in vitro for about 2.5 years (when the experiment was stopped), completing more than 60 population doublings, compared to control cultures which underwent replicative senescence in less than 30 population doublings.

Concluding Remarks

It is now generally accepted that replicative senescence of T-cells during aging and/or chronic infection is a major contributing factor to immunological failure. Cultures of senescent CD8 T-cells, resulting from extensive in vitro proliferation driven by multiple rounds of antigenic stimulation, have numerous features that distinguish them from other T-cells. The most easily identifiable characteristic is the inability to proliferate when stimulated by antigen, antibodies, mitogens, or cytokines. However, other features of senescent cells are equally important to their function in vivo. These include resistance to apoptosis, shortened telomeres and complete loss of expression of the CD28 costimulatory receptor.

The connection between telomere length and CD28 relates to the mechanism involved in activation of the telomerase enzyme. Several studies have documented the critical role of CD28-mediated signaling, in concert with T-cell activation, in the optimal upregulation of telomerase. CD8 T-cells that are constantly being activated to proliferate lose CD28 expression, at which point telomerase can no longer be induced when antigen is encountered. This phenomenon may be responsible for causing telomeres to shorten with each round of replication. When telomeres shorten to a critical length, the senescence cascade is initiated.

The underlying cause of the induction of senescence in CD8 T-cells in the elderly appears to be repeated encounters with certain antigens over a period of many years. However, it is probably not aging per se that is responsible, since a similar effect on virus-specific CD8 T-cells is observed in younger individuals harboring chronic viral infections, such as HIV, CMV, EBV and Hepatitis C. In HIV disease, the effect on CD8 T-cell replicative senescence may be exacerbated by the high mutation rate of the HIV-1 virus, together with the deleterious effect of the virus on other components of the immune system. The HIV-specific CD8 T-cell immune response is able to control the virus for many years, but throughout this time, the virus is never completely eliminated and remains latent within CD4 T-cells, macrophages and dendritic cells. The continuous presence of the virus results in chronic activation/proliferation of virus-specific CD8 T-cells, leading to replicative senescence and the eventual exhaustion of the protective response, which ultimately results in opportunistic infections and/or cancer.

The similarity, in terms of replicative senescence, between chronic HIV infection and aging suggests that with the growing proportion of older persons who are infected with HIV, a synergistic effect of aging and chronic infection may be operating with respect to the generation of senescent CD8 T-cells. This notion is consistent with the more rapid disease progression seen in older persons infected with HIV. Strategies to prevent or retard replicative senescence, by genetic or pharmacologic methods, would, therefore, seem to be practical and important immunotherapeutic approaches to enhance immune function during aging and HIV disease.

Acknowledgements

The research described in this chapter has been supported by the following NIH grants: AG 023720 and AI 060362 (RBE) and AI 52031 (SRF).

References

1. Ruscetti FW, Morgan DA, Gallo RC. Functional and morphologic characterization of human T-cells continuously grown in vitro. Journal of Immunology 1977; 119:131-8.
2. Paul WE, Sredni B, Schwartz RH. Long-term growth and cloning of nontransformed lymphocytes. Nature 1981; 294:697-9.
3. Perillo NL, Walford RL, Newman MA et al. Human T-lymphocytes possess a limited in vitro life span. Experimental Gerontology 1989; 24:177-87.
4. Chiu CP, Harley CB. Replicative senescence and cell immortality: the role of telomeres and telomerase. Proceedings Of The Society For Experimental Biology And Medicine. Society For Experimental Biology And Medicine 1997; 214:99-106.
5. Adibzadeh M, Pohla H, Rehbein A et al. Long-term culture of monoclonal human T-lymphocytes: models for immunosenescence? Mechanisms of Ageing and Development 1995; 83:171-83.
6. Grubeck-Loebenstein B, Lechner H, Trieb K. Long-term in vitro growth of human T-cell clones: can postmitotic 'senescent' cell populations be defined? International Archives Of Allergy And Immunology 1994; 104:232-9.
7. Spaulding C, Guo W, Effros RB. Resistance to apoptosis in human CD8+ T-cells that reach replicative senescence after multiple rounds of antigen-specific proliferation. Experimental Gerontology 1999; 34:633-44.
8. Bryant JE, Hutchings KG, Moyzis RK et al. Measurement of telomeric DNA content in human tissues. Biotechniques 1997; 23:476-8, 480, 482, passim.
9. Effros RB, Boucher N, Porter V et al. Decline in CD28$^+$ T-cells in centenarians and in long-term T-cell cultures: A possible cause for both in vivo and in vitro immunosenescence. Experimental Gerontology 1994; 29:601-9.
10. Effros RB, Pawelec G. Replicative senescence of T-cells: does the Hayflick Limit lead to immune exhaustion? Immunology Today 1997; 18:450-4.
11. Perillo NL, Naeim F, Walford RL et al. The in vitro senescence of human T-lymphocytes: Failure to divide is not associated with a loss of cytolytic activity or memory T-cell phenotype. Mechanisms of Ageing and Development 1993; 67:173-85.
12. Posnett DN, Sinha R, Kabak S et al. Clonal populations of T-cells in normal elderly humans: the T-cell equivalent to "benign monoclonal gammapathy". Journal of Experimental Medicine 1994; 179:609-18.
13. Thoman ML, Weigle WO. The cellular and subcellular bases of immunosenescence. Advances In Immunology 1989; 46:221-61.
14. Murasko DM, Weiner P, Kaye D. Decline in mitogen induced proliferation of lymphocytes with increasing age. Clincal and Experimental Immunology 1987; 70:440-8.
15. Grossmann A, Ledbetter JA, Rabinovitch PS. Reduced proliferation in T-lymphocytes in aged humans is predominantly in the CD8$^+$ subset and is unrelated to defects in transmembrane signaling which are predominantly in the CD4$^+$ subset. Experimental Cell Research 1989; 180:367-82.
16. Franceschi C, Bonafe M, Valensin S. Human immunosenescence: the prevailing of innate immunity, the failing of clonotypic immunity and the filling of immunological space. Vaccine 2000; 18:1717-20.
17. Goronzy JJ, Fulbright JW, Crowson CS et al. Value of immunological markers in predicting responsiveness to influenza vaccination in elderly individuals. Journal Of Virology 2001; 75:12182-7.
18. Saurwein-Teissl M, Lung TL, Marx F et al. Lack of antibody production following immunization in old age: association with CD8($^+$)CD28(-) T-cell clonal expansions and an imbalance in the production of Th1 and Th2 cytokines. Journal Of Immunology 2002; 168:5893-9.
19. Looney RJ, Falsey A, Campbell D et al. Role of Cytomegalovirus in the T-cell Changes Seen in Elderly Individuals. Clinical Immunology 1999; 90:213-9.
20. Suciu-Foca N, Manavalan JS, Scotto L et al. Molecular characterization of allospecific T suppressor and tolerogenic dendritic cells: review. International Immunopharmacology 2005; 5:7-11.
21. Rocha B, Dautigny N, Pereira P. Peripheral T-lymphocytes: expansion potential and homeostatic regulation of pool sizes and CD4/CD8 ratios in vivo. European Journal of Immunology 1989; 19:905-11.
22. Freitas AA, Agenes F, Coutinho GC. Cellular competition modulates survival and selection of CD8$^+$ T-cells. European Journal of Immunology 1996; 26:2640-9.
23. Monteiro J, Batliwalla F, Ostrer H et al. Shortened telomeres in clonally expanded CD28-CD8$^+$ T-cells imply a replicative history that is distinct from their CD28$^+$CD8$^+$ counterparts. Journal Of Immunology 1996; 156:3587-90.

24. Effros RB, Allsopp R, Chiu CP et al. Shortened telomeres in the expanded CD28-CD8+ cell subset in HIV disease implicate replicative senescence in HIV pathogenesis. AIDS 1996; 10:F17-F22.
25. Blackburn EH. Structure and function of telomeres. Nature 1991; 350:569-73.
26. Allsopp RC. Models of initiation of replicative senescence by loss of telomeric DNA. Experimental Gerontology 1996; 31:235-43.
27. Lundblad V, Szostak JW. A mutant with a defect in telomere elongation leads to senescence in yeast. Cell 1989; 57:633-43.
28. Allsopp RC, Vaziri H, Patterson C et al. Telomere length predicts replicative capacity of human fibroblasts. Proceedings of the National Academy of Sciences 1992; 89:10114-8.
29. Harley CB, Futcher AB, Greider CW. Telomeres shorten during ageing of human fibroblasts. Nature 1990; 345:458-60.
30. Olovnikov AM. [Principle of marginotomy in template synthesis of polynucleotides]. Dokl Akad Nauk SSSR 1971; 201:1496-9.
31. Watson JD. Origin of concatemeric T7 DNA. National New Biology 1972; 239:197-201.
32. Bodnar AG, Ouellette M, Frolkis M et al. Extension of life-span by introduction of telomerase into normal human cells. Science 1998; 279:349-52.
33. Vaziri H, Benchimol S. Reconstitution of telomerase activity in normal human cells leads to elongation of telomeres and extended replicative life span. Current Biology 1998; 8:279-82.
34. Dagarag M, Evazyan T, Rao N et al. Genetic manipulation of telomerase in HIV-specific CD8+ T-cells: enhanced antiviral functions accompany the increased proliferative potential and telomere length stabilization. Journal of Immunology 2004; 173:6303-11.
35. Pawelec G, Wagner W, Adibzadeh M et al. T-cell immunosenescence in vitro and in vivo. Experimental Gerontology 1999; 34:419-29.
36. Erickson S, Sangfelt O, Heyman M et al. Involvement of the Ink4 proteins p16 and p15 in T-lymphocyte senescence. Oncogene 1998; 17:595-602.
37. Vaziri H, Schachter F, Uchida I et al. Loss of telomeric DNA during aging of normal and trisomy 21 human lymphocytes. American Journal Of Human Genetics 1993; 52:661-7.
38. Weng NP, Levine BL, June CH et al. Human naive and memory T-lymphocytes differ in telomeric length and replicative potential. Proceedings Of The National Academy Of Sciences 1995; 92:11091-4.
39. Son NH, Murray S, Yanovski J et al. Lineage-specific telomere shortening and unaltered capacity for telomerase expression in human T and B-lymphocytes with age. Journal Of Immunology 2000; 165:1191-6.
40. Campisi J. Cellular senescence as a tumor-suppressor mechanism. Trends in Cell Biology 2001; 11: S27-S31.
41. Boucher N, Dufeu-Duchesne T, Vicaut E et al. CD28 Expression in T-cell Aging and Human Longevity. Experimental Gerontology 1998; 33:267-82.
42. Cawthon RM, Smith KR, O'Brien E et al. Association between telomere length in blood and mortality in people aged 60 years or older. Lancet 2003; 361:393-5.
43. Plunkett FJ, Soares MV, Annels N et al. The flow cytometric analysis of telomere length in antigen-specific CD8+ T-cells during acute Epstein-Barr virus infection. Blood 2001; 97:700-7.
44. Rangan SR, Armatis P. Enhanced frequency of spontaneous B cell lines from Epstein-Barr virus (EBV) seropositive donors 80 years and older. Experimental Gerontology 1991; 26:541-7.
45. Valenzuela HF, Effros RB. Divergent telomerase and CD28 expression patterns in human CD4 and CD8 T-cells following repeated encounters with the same antigenic stimulus. Clinical Immunology 2002; 105:117-25.
46. Jenkins MK, Taylor PS, Norton SD et al. CD28 delivers a costimulatory signal involved in antigen-specific IL-2 production by human T-cells. Journal Of Immunology 1991; 147:2461-6.
47. Sozou PD, Kirkwood TBL. A Stochastic Model of Cell Replicative Senescence Based on Telomere Shortening, Oxidative Stress and Somatic Mutations in Nuclear and Mitochondrial DNA. Journal of Theoretical Biology 2001; 213:573-86.
48. Shimizu Y, van Seventer GA, Ennis E et al. Crosslinking of the T-cell-specific accessory molecules CD7 and CD28 modulates T-cell adhesion. Journal of Experimental Medicine 1992; 175:577-82.
49. Frauwirth KA, Riley JL, Harris MH et al. The CD28 signaling pathway regulates glucose metabolism. Immunity. 2002; 16:769-77.
50. Verweij CL, Geerts M, Aarden LA. Activation of interleukin-2 gene transcription via the T-cell surface molecule CD28 is mediated through an NF-kB-like response element. The Journal Of Biological Chemistry 1991; 266:14179-82.
51. Vallejo AN, Brandes JC, Weyand CM et al. Modulation of CD28 expression: distinct regulatory pathways during activation and replicative senescence. Journal Of Immunology 1999; 162:6572-9.
52. June CH, Bluestone JA, Nadler LM, Thompson CB. The B7 and CD28 receptor families. Immunology Today 1994; 15:321-31.

53. Azuma M, Cayabyab M, Phillips JH et al. Requirements for CD28-dependent T-cell-mediated cytotoxicity. Journal Of Immunology 1993; 150:2091-101.
54. Tan P, Anasetti C, Hansen JA et al. Induction of alloantigen-specific hyporesponsiveness in human T-lymphocytes by blocking interaction of CD28 with its natural ligand B7/BB1. The Journal Of Experimental Medicine 1993; 177:165-73.
55. Azuma M, Phillips JH, Lanier LL. CD28- T-lymphocytes. Antigenic and functional properties. Journal Of Immunology 1993; 150:1147-59.
56. Effros RB. Replicative senescence in the immune system: impact of the Hayflick limit on T-cell function in the elderly. American Journal of Human Genetics 1998; 62:1003-7.
57. Posnett DN, Edinger JW, Manavalan JS et al. Differentiation of human CD8 T-cells: implications for in vivo persistence of $CD8^+$ CD28- cytotoxic effector clones. International Immunology 1999; 11:229-41.
58. Pawelec G, Akbar A, Caruso C et al. Is immunosenescence infectious? Trends in Immunology 2004; 25:406-10.
59. Schwab R, Szabo P, Manavalan JS et al. Expanded $CD4^+$ and $CD8^+$ T-cell clones in elderly humans. Journal of Immunology 1997; 158:4493-9.
60. Chan SR, Blackburn EH. Telomeres and telomerase. Philosophical Transactions of the Royal London Society B: Biological Sciences 2004; 359:109-21.
61. Blackburn EH. Telomerases. Annual Review of Biochemistry 1992; 61:113-29.
62. Greider CW. Telomeres, telomerase and senescence. Bioessays 1990; 12:363-9.
63. Yang J, Chang E, Cherry AM et al. Human endothelial cell life extension by telomerase expression. The Journal Of Biological Chemistry 1999; 274:26141-8.
64. Luiten RM, Pene J, Yssel H et al. Ectopic hTERT expression extends the life span of human $CD4^+$ helper and regulatory T-cell clones and confers resistance to oxidative stress-induced apoptosis. Blood 2003; 101:4512-9.
65. Counter CM, Avilion AA, LeFeuvre CE et al. Telomere shortening associated with chromosome instability is arrested in immortal cells which express telomerase activity. The EMBO Journal 1992; 11:1921-9.
66. Kim NW, Piatyszek MA, Prowse KR et al. Specific association of human telomerase activity with immortal cells and cancer. Science 1994; 266:2011-5.
67. Igarashi H, Sakaguchi N. Telomerase Activity Is Induced by the Stimulation to Antigen Receptor in Human Peripheral Lymphocytes. Biochemical and Biophysical Research Communications 1996; 219:649-55.
68. Broccoli D, Young JW, de Lange T. Telomerase activity in normal and malignant hematopoietic cells. Proceedings Of The National Academy Of Sciences 1995; 92:9082-6.
69. Weng NP, Levine BL, June CH et al. Regulated expression of telomerase activity in human T-lymphocyte development and activation. Journal of Experimental Medicine 1996; 183:2471-9.
70. Hiyama K, Hirai Y, Kyoizumi S et al. Activation of telomerase in human lymphocytes and hematopoietic progenitor cells. Journal of Immunology 1995; 155:3711-5.
71. Weng N, Levine BL, June CH et al. Regulation of telomerase RNA template expression in human T-lymphocyte development and activation. Journal Of Immunology 1997; 158:3215-20.
72. Maini MK, Soares MV, Zilch CF et al. Virus-induced $CD8^+$ T-cell clonal expansion is associated with telomerase up-regulation and telomere length preservation: a mechanism for rescue from replicative senescence. Journal Of Immunology 1999; 162:4521-6.
73. Roth A, Yssel H, Pene J et al. Telomerase levels control the lifespan of human T-lymphocytes. Blood 2003; 102:849-57.
74. Weng NP, Palmer LD, Levine BL et al. Tales of tails: regulation of telomere length and telomerase activity during lymphocyte development, differentiation, activation and aging. Immunology Review 1997; 160:43-54.
75. Bodnar AG, Kim NW, Effros RB et al. Mechanism of telomerase induction during T-cell activation. Experimental Cell Research 1996; 228:58-64.
76. Rufer N, Migliaccio M, Antonchuk J et al. Transfer of the human telomerase reverse transcriptase (TERT) gene into T-lymphocytes results in extension of replicative potential. Blood 2001; 98:597-603.
77. Hooijberg E, Ruizendaal JJ, Snijders PJ et al. Immortalization of human $CD8^+$ T-cell clones by ectopic expression of telomerase reverse transcriptase. Journal of Immunology 2000; 165:4239-45.
78. Dagarag M, Ng H, Lubong R et al. Differential impairment of lytic and cytokine functions in senescent human immunodeficiency virus type 1-specific cytotoxic T-lymphocytes. Journal of Virology 2003; 77:3077-83.
79. Yang OO, Walker BD. $CD8^+$ cells in human immunodeficiency virus type I pathogenesis: cytolytic and noncytolytic inhibition of viral replication. Advanced Immunology 1997; 66:273-311.
80. Borrow P, Lewicki H, Hahn BH et al. Virus-specific $CD8^+$ cytotoxic T-lymphocyte activity associated with control of viremia in primary human immunodeficiency virus type 1 infection. Journal of Virology 1994; 68:6103-10.

81. Koup RA, Safrit JT, Cao Y et al. Temporal association of cellular immune responses with the initial control of viremia in primary human immunodeficiency virus type 1 syndrome. Journal of Virology 1994; 68:4650-5.
82. Ogg GS, Jin X, Bonhoeffer S et al. Quantitation of HIV-1-specific cytotoxic T-lymphocytes and plasma load of viral RNA. Science 1998; 279:2103-6.
83. Bailer RT, Holloway A, Sun J et al. IL-13 and IFN-gamma secretion by activated T-cells in HIV-1 infection associated with viral suppression and a lack of disease progression. Journal of Immunology 1999; 162:7534-42.
84. Garzino-Demo A, Moss RB, Margolick JB et al. Spontaneous and antigen-induced production of HIV-inhibitory beta-chemokines are associated with AIDS-free status. Proceedings Of The National Academy Of Sciences 1999; 96:11986-91.
85. Buseyne F, Fevrier M, Garcia S et al. Dual function of a human immunodeficiency virus (HIV)-specific cytotoxic T-lymphocyte clone: inhibition of HIV replication by noncytolytic mechanisms and lysis of HIV-infected $CD4^+$ cells. Virology 1996; 225:248-53.
86. Daar ES, Moudgil T, Meyer RD et al. Transient high levels of viremia in patients with primary human immunodeficiency virus type 1 infection. The New England Journal of Medicine 1991; 324:961-4.
87. Clark SJ, Saag MS, Decker WD et al. High titers of cytopathic virus in plasma of patients with symptomatic primary HIV-1 infection. The New England Journal of Medicine 1991; 324:954-60.
88. Lieberman J, Shankar P, Manjunath N et al. Dressed to kill? A review of why antiviral CD8 T-lymphocytes fail to prevent progressive immunodeficiency in HIV-1 infection. Blood 2001; 98:1667-77.
89. Cao Y, Qin L, Zhang L et al. Virologic and immunologic characterization of long-term survivors of human immunodeficiency virus type 1 infection. The New England Journal Of Medicine 1995; 332:201-8.
90. Harrer T, Harrer E, Kalams SA et al. Cytotoxic T-lymphocytes in asymptomatic long-term nonprogressing HIV-1 infection. Breadth and specificity of the response and relation to in vivo viral quasispecies in a person with prolonged infection and low viral load. Journal Of Immunology 1996; 156:2616-23.
91. Shankar P, Russo M, Harnisch B et al. Impaired function of circulating HIV-specific $CD8(^+)$ T-cells in chronic human immunodeficiency virus infection. Blood 2000; 96:3094-101.
92. Borthwick NJ, Bofill M, Gombert WM et al. Lymphocyte activation in HIV-1 infection. II. Functional defects of CD28- T-cells. AIDS 1994; 8:431-41.
93. Lewis DE, Tang DS, Adu-Oppong A et al. Anergy and apoptosis in $CD8^+$ T-cells from HIV-infected persons. Journal Of Immunology 1994; 153:412-20.
94. Brinchmann JE, Dobloug JH, Heger BH et al. Expression of costimulatory molecule CD28 on T-cells in human immunodeficiency virus type 1 infection: functional and clinical correlations. The Journal Of Infectious Diseases 1994; 169:730-8.
95. Khan N, Shariff N, Cobbold M et al. Cytomegalovirus seropositivity drives the CD8 T-cell repertoire toward greater clonality in healthy elderly individuals. Journal Of Immunology 2002; 169:1984-92.
96. van Baarle D, Tsegaye A, Miedema F et al. Significance of senescence for virus-specific memory T-cell responses: rapid ageing during chronic stimulation of the immune system. Immunology Letters 2005; 97:19-29.
97. Tripp RA, Hou S, McMickle A et al. Recruitment and proliferation of $CD8^+$ T-cells in respiratory virus infections. Journal of Immunology 1995; 154:6013-21.
98. Appay V, Rowland-Jones SL. Premature ageing of the immune system: the cause of AIDS? Trends in Immunology 2002; 23:580-5.
99. Palmer LD, Weng N, Levine BL et al. Telomere length, telomerase activity and replicative potential in HIV infection: analysis of $CD4^+$ and $CD8^+$ T-cells from HIV-discordant monozygotic twins. Journal of Experimental Medicine 1997; 185:1381-6.
100. Wolthers KC, Miedema F. Telomeres and HIV-1 infection: in search of exhaustion. Trends in Microbiology 1998; 6:144-7.
101. Bestilny LJ, Gill MJ, Mody CH et al. Accelerated replicative senescence of the peripheral immune system induced by HIV infection. AIDS 2000; 14:771-80.
102. Lane HC, Laughon BE, Falloon J et al. NIH conference. Recent advances in the management of AIDS-related opportunistic infections. Annals of Internal Medicine 1994; 120:945-55.
103. Kalayjian RC, Cohen ML, Bonomo RA et al. Cytomegalovirus ventriculoencephalitis in AIDS. A syndrome with distinct clinical and pathologic features. Medicine 1993; 72:67-77.
104. Chitale AR. Cancer and AIDS. Indian Journal of Pathology and Microbiology 2005; 48:151-60.
105. Tucker V, Jenkins J, Gilmour J et al. T-cell telomere length maintained in HIV-infected long-term survivors. HIV Medicine 2000; 1:116-22.
106. Wolthers KC, Bea G, Wisman A et al. T-cell telomere length in HIV-1 infection: no evidence for increased $CD4^+$ T-cell turnover. Science 1996; 274:1543-7.
107. Pommier JP, Gauthier L, Livartowski J et al. Immunosenescence in HIV Pathogenesis. Virology 1997; 231:148-54.

Chapter 5

A Matter of Life and Death of T-Lymphocytes in Immunosenescence

Sudhir Gupta*

Abstract

Aging is associated with progressive decline in T-cell functions. A number of mechanisms have been proposed to explain immunosenescence. In this chapter I will discuss a role of apoptosis of T-lymphocytes in immunosenescence. Molecular signaling of different pathways of apoptosis and their alterations in human aging will be reviewed.

Introduction

Cell death occurs by necrosis, autophagy and apoptosis. Apoptosis or programmed cell death is a physiological form of cell death, which plays an important role in cellular homeostasis, selection of T-cell repertoire in the thymus, deletion of self-reactive T-and B-lymphocytes, regulation of immunological memory and in the killing of target cells by cytotoxic T-lymphocytes and natural killer cells.[1-4] There are two major signaling pathways of apoptosis (Fig. 1):the extrinsic or death receptor pathway[5-10] and the intrinsic or mitochondrial pathway.[6,11-14] In the death receptor signaling pathway, signal is provided by an interaction between the ligand and death receptor, recruitment of adapter proteins and activation of proximal and executioner caspases. In the mitochondrial signaling pathway, a number of molecules are released from the mitochondria intermembrane space into the cytoplasm where they interact with adapter proteins and activate a distinct initiator caspase, which then activates common executioner caspases, resulting in apoptosis. In this chapter, I will briefly review different specific pathways of apoptosis and their alterations in human aging.

Death Receptor Pathway of Apoptosis

Death receptors belong to a large family of tumor necrosis factor receptors (TNFR), including CD95, TNFR, TRAIL and others.[10] Among them, CD95 (Fas)-mediated and TNFR-mediated apoptosis have been extensively studied, especially in relation to aging. There are basic differences in in vitro-induced apoptosis during activation-induced cell death (AICD), CD95-CD95L interaction and TNF-TNFR signaling. In AICD, pre-activated T-cells are re-activated with the same stimulus, whereas in the CD95-CD95L system, pre-activated T-cells are stimulated with anti-CD95 monoclonal antibodies or soluble CD95L and in the TNF-TNFR system, pre-activated T-cells are stimulated with TNF-α.

Activation-Induced Cell Death (AICD)

AICD plays an essential role in both central and peripheral clonal deletion events involved in tolerance and homeostasis.[15] In AICD, activation occurs through proper engagement of T-cell receptors (TCRs) by specific antigens bound to the MHC molecule and is influenced by antigen

*Sudhir Gupta—Medical Sciences I, C-240, University of California, Irvine, CA 92697, U.S.A. Email: sgupta@uci.edu

Immunosenescence, edited by Graham Pawelec. ©2007 Landes Bioscience and Springer Science+Business Media.

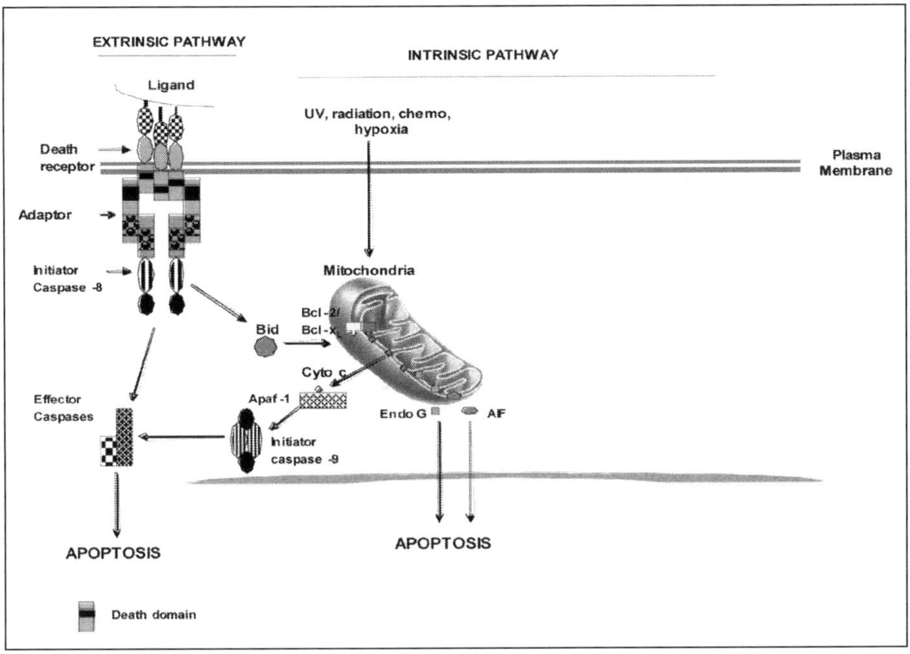

Figure 1. Two different pathways of apoptosis. Extrinsic pathway (death receptor pathway) is mediated by interaction between death receptor and death receptor ligand. Intrinsic pathway is mediated via the mitochondrial pathway. There is a cross-talk between various pathways of apoptosis.

concentration and costimulatory signals.[16] AICD appears to be primarily mediated by Fas (CD95)-Fas ligand (CD95L) interaction,[17-19] and several investigators have demonstrated that CD95-CD95L interaction was necessary for AICD in mature T-cells in vitro[20,21] and in vivo for peripheral T-cell deletion.[22,23]

CD95-Mediated Apoptosis

CD95 is a type I transmembrane receptor that is constitutively expressed on lymphocytes; however, CD95 ligand (CD95L), a type II transmembrane protein, displays more restricted expression and is lacking from resting lymphocytes. CD95L is induced upon activation of lymphocytes and can be cleaved from the cell surface by metalloproteases. Therefore, CD95L may be found in the soluble form in vivo and can trigger apoptosis.[24] The steps of the CD95-mediated apoptosis signaling pathway are shown in Figure 2. Upon ligation with CD95L or anti-CD95 monoclonal antibodies, CD95 undergoes trimerization. Since the cytoplasmic domain does not have intrinsic enzymatic activity, it recruits and interacts with an adapter protein, the fas-associated death domain (FADD), via homologous death domain (DD) by protein-protein interaction. FADD also contains a death-effector domain (DED) and by protein-protein interaction recruits and binds to procaspase-8 (Flice) to form a death-inducing signaling complex (DISC). Procaspase-8 is activated by homodimerization and active caspase-8 is released from the DISC into the cytoplasm where it cleaves downstream executioner/effector pro-caspases to generate active executioner caspases (caspase 3, caspase-6 and caspase-7). These activated effector caspases cleave a number of substrates, including transcription factors, enzymes (involved in DNA repair, cell cycle progression and DNA cleavage) and structural proteins[8] responsible for characteristic morphological and biochemical characteristics of apoptosis.

Apoptosis mediated by CD95-CD95L interaction is regulated by other DED-containing molecules, the FLIP (Flice-inhibitory protein). This protein contains two DEDs. Cellular FLIP

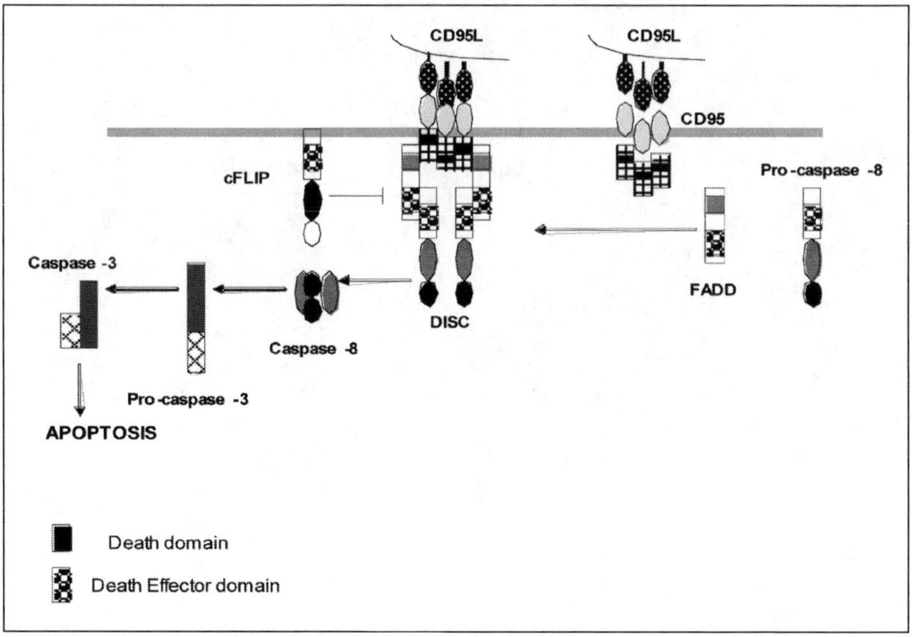

Figure 2. CD95-CD95L pathway of apoptosis. Interaction of CD95 with CD95L leads to oligomerization of CD95 death domain, recruitment of adapter protein (FADD) and pro-caspase-8 and formation of death-inducing signaling complex (DISC), which becomes a platform for activation of effector caspases (caspase-3, -6, and -7) and induction of apoptosis.

(cFLIP) is present in two alternatively spliced isoforms, the long ($FLIP_L$) and short ($FLIP_S$) forms.[25] FLIP inhibits apoptosis by two distinct mechanisms; [a] DED of FLIP binds to CD95-FADD complexes and inhibits the recruitment and activation of procaspases-8[26,27] and [b] FLIP promotes the activation of NF-κB and Erk signaling pathways by recruiting adapter proteins (including RIP, TRAF-1, 2, 3, Raf) and binding to IKKγ.[28,29]

TNFR-Mediated Apoptosis

TNF-α exerts its biological activity by binding to type I and type II receptors (TNFR-I and TNFR-II) and activating several signaling pathways.[5-8,30-33] TNFRs are type I transmembrane proteins with one to five cysteine-rich repeats in their extracellular domains and a DD in the cytoplasmic tail of TNFR-I (but TNFR-II lacks a DD). Both cell survival and cell death signals mediated by TNF-α require distinct sets of adapters and other downstream signaling molecules.

Steps of TNF-α-induced signaling are shown in Figure 3. Upon ligation with TNF-α, TNFR-I undergoes trimerization of its receptor DD, which in turn recruits an adapter protein, TNFR-associated death domain (TRADD). TRADD then may recruit FADD. Procaspase-8 is recruited to FADD and then undergoes dimerization to convert it into active caspase-8. The remaining downstream signaling steps are similar to those described above for CD95-mediated apoptosis. Alternatively, TRADD may recruit distinct sets of adapter proteins, TRAF-2 (TNFR-associated factor-2) and receptor interactive protein (RIP). TRAF-2 and RIP stimulate pathways leading to activation of MAP kinase and NFκB respectively. MAPK may inhibit[34] or promote[35] apoptosis. TRADD along with RIP and TRAF2 form a signaling complex that activates NF-κB resulting in the induction of anti-apoptotic genes and suppression of apoptosis.[36-40] Both RIP and TRAF-2 are required for NF-κB activation. TRAF-2 recruits IKK, whereas RIP activates IKK.[41] IKK stimulates NF-κB by catalyzing phosphorylation of IκB.[42,43] IκB is phosphorylated at two specific serine residues. This phosphorylation is a signal for ubiquitination and degradation of IκB by the 26S proteosome.[42] Free NF-κB dimers are released and translocated to the nucleus, where they activate transcription of target genes. The anti-apoptotic genes that are upregulated by NF-κB activation include *cIAP1, cIAP2, XIAP, Gadd45β, Bcl-x$_L$, A20, TRAF-1, TRF-2 and FLIP.*[39,40,44] An inhibition of NF-κB is associated with upregulation of Bax, suggesting that Bax is negatively regulated by NF-κB.[45]

The outcome of TNF signaling (death versus survival) is determined by the balance between NF-κB and JNK activation. JNK activation enhances TNF-induced apoptosis.[46] Deng et al[47] demonstrated that TNF-α-induces apoptosis via sustained activation of JNK, which cleaves Bid in a caspases-8-independent manner to yield a unique 21kDa Bid cleaved product (jBid), which is different from caspase-8-dependent cleaved Bid (tBid) of 15kDa (Fig. 3). jBid translocates to the mitochondria and preferentially releases Smac/Diablo from the mitochondria, which may disrupt TRAF-2-cIAP1 complex formation and its inhibition of caspases-8 activation. More recently, it has been demonstrated that TNF-induced activation of JNK accelerates turnover of c-FLIP. JNK-mediates phosphorylation and activation of E3 ubiquitin ligase Itch, which specifically ubiquitinates c-FLIP and induces its proteasomal degradation.[48]

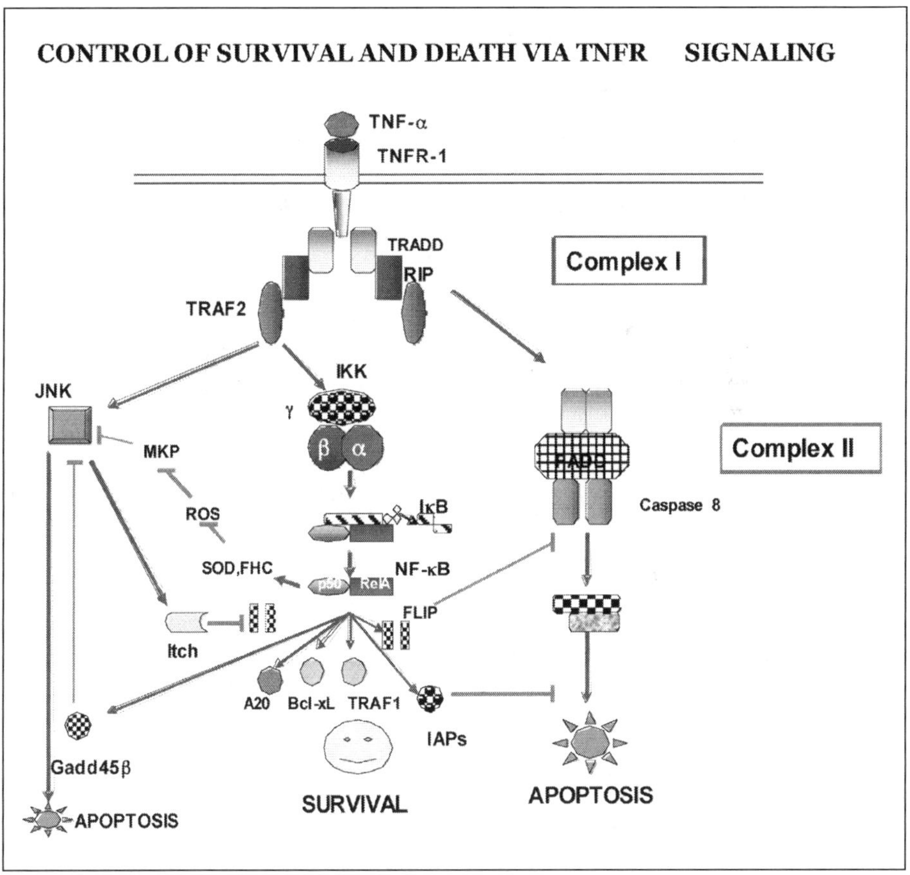

Figure 3. TNF-TNFR signaling pathway. Upon ligation with TNF-α, TNFR-I mediates both survival signal (via NF-κB activation) and death signal (via FADD), whereas TNFR-II mediated predominantly a survival signal by recruitment of different set of adapter proteins (TRAF-2 and RIP) and activation of NF-κB. See the text for abbreviations.

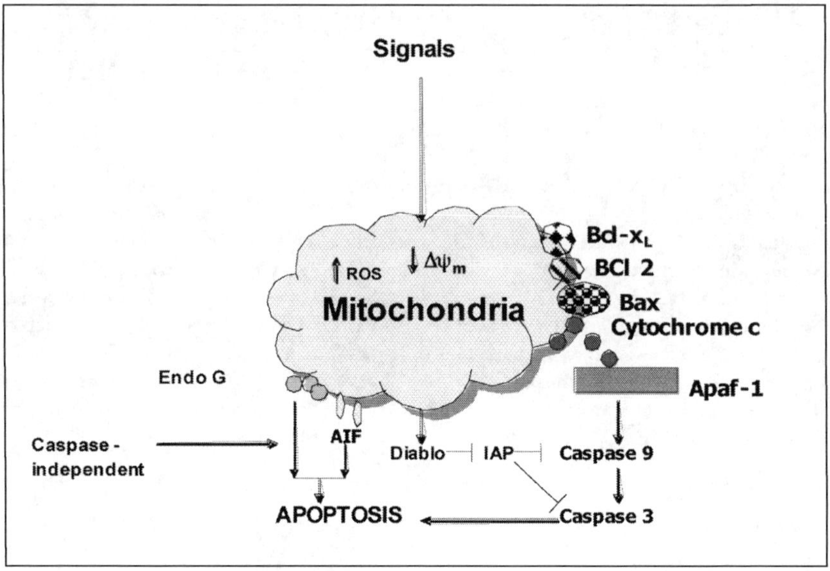

Figure 4. Mitochondrial pathway of apoptosis. Both caspases-dependent dependent and caspases-independent pathways are shown. Caspase-dependent pathway is mediate by the release of cytochrome c from the mitochondria, which binds to Apaf-1 and recruits pro-caspase-8 to form an Apoptosome. Active caspases-9 activates effector caspases resulting in apoptosis. AIF and Endo G, once release from the mitochondria, migrate to the nucleus and cause DNA fragmentation and apoptosis in a caspases-independent manner.

Mitochondrial Pathway of Apoptosis

A number of stimuli, including chemotherapeutic agents, UV radiation, stress molecules (reactive oxygen and reactive nitrogen species) and growth factor withdrawal, appear to mediate apoptosis via the mitochondrial pathway.[11-14,76-81] Since increased oxidative stress is a universally accepted change during aging, this pathway of apoptosis is especially relevant to aging. Mitochondria contain two well-defined compartments: the matrix, surrounded by the inner membrane (IM) and the intermembrane space, which is surrounded by the outer membrane (OM). The IM contains various molecules, including ATP synthase, electron transport chain and adenine nucleotide translocator (ANT). Under physiological conditions these molecules allow the respiratory chain to create an electrochemical gradient (membrane potential). The OM contains a voltage-dependent anion channel (VDAC). Bcl-2 is located on the IM and appears to play an important role in the maintenance of the mitochondrial membrane potential ($\Delta\Psi$m). The intermembrane space contains holocytochrome C, certain pro-caspases, adenylate kinase 2, Endo G, Daiblo/Smac and apoptosis-inducing factor (AIF). The permeabilization of the OM therefore results in the release of these molecules into the cytoplasm. IM permeabilization leads to changes in $\Delta\Psi$m. Once released from the mitochondria, cytochrome c binds to an adapter molecule Apaf-1 (Apoptotic protease-activating factor) in the presence of ATP/dATP and recruits pro-caspase 9 to form apoptosome (Fig. 4). Procaspase-9 is dimerized and activated without undergoing cleavage and active caspase-9 activates executioner caspases to orchestrate apoptosis.

There is evidence to suggest that certain molecules present in the mitochondrial intermembrane space can promote apoptosis in a caspase-independent manner. AIF is a caspase-independent death effector, which upon induction of apoptosis translocates from the intermembrane space of the mitochondria to the nucleus. Once in the nucleus, AIF causes chromatin condensation and large scale DNA fragmentation to fragments of ~50 kbp.[52,53] EndoG nuclease is also capable of mediating

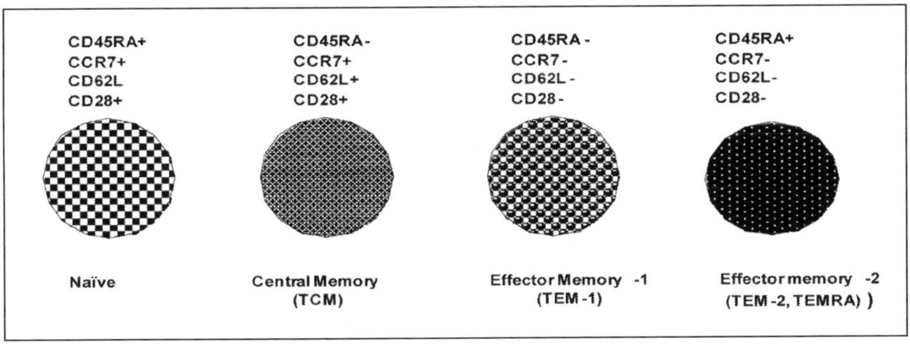

Figure 5. Naïve and different subsets of memory T-cells as defined by cell surface antigens.

caspase-independent apoptosis. Endo G, upon its release from the mitochondrial intermembrane space, appears to directly mediate nuclear DNA fragmentation.[54]

Mitochondrial membrane permeabilization (MMP) is controlled by a variety of members of the Bcl-2 family.[11-14,55] The Bcl-2 family members are divided into three groups:anti-apoptotic (Bcl-2, Bcl-xL, Mcl-1, Bcl-w and A1), pro-apoptotic "BH3 only" (Bid, Bim, Bik, Bmf, Bad, Hrk, BNIP3) and pro-apoptotic "BH-123" (Bax, Bak and Bok) proteins.

Several of the pro-apoptotic members of the Bcl-2 family, including Bax, Bak, Bad, Bid and Bim, initiate MMP by forming what appears to be a channel. In order to exert their effects, the members of Bcl-2 pro-apoptotic family must dock onto the mitochondrial OM. During apoptosis, Bax, which is present in the cytoplasm in a monomer form, is translocated to the mitochondrial membrane to form a dimer or high order oligomers. Bak can also loosely associate with OM. During apoptosis, Bim, a calcium-dependent proapoptotic molecule which is present in microtubules also translocates to OM. Bad, in its phosphorylated form (inactive form), is sequestered in the cytoplasm by interacting with 14-3-3 protein. During apoptosis, Bad is dephosphorylated and then translocates to the OM of mitochondria where it interacts and heterodimerizes with Bcl-x_L to block the anti-apoptotic function of Bcl-x_L. Bcl-2 and Bcl-x_L inhibit cytochrome C release.

Apoptosis in Subsets of T-Lymphocytes

Naïve T-cells (T_N) following exposure to an antigen undergo clonal expansion followed by clearance of antigen eg. the virus. This phase is followed by a phase of contraction during which virus-specific T-cells undergo apoptosis and then a small number of antigen-specific T-cells are retained as memory T-cells.[56] The memory T-cells display differential expression of adhesion molecules (CD62L) and chemokine receptors (CCR-7), which allow them to home into lymph nodes (central memory), nonlymphoid tissue and mucosal sites -(effector memory) and also to respond to microbes at peripheral tissue sites.[57,58] These subpopulations of naïve, central and effector memory T-cells are identified by a number of cell surface proteins.[56,59-61] Effector memory $CD8^+$ T-cells are further subdivided into two subsets. One subset of effector memory (T_{EM}) is $CCR7^-CD45RA^-$, where as a second set of effector memory $CD8^+$ T-cells re-express CD45RA (T_{EMRA}) Figure 5 shows phenotypic characteristics of naïve and various memory $CD8^+$ T-cells in humans. Although generally it is considered that T_{EMRA} subset is lacking from $CD4^+$ T-cells, in analyzing data of Salusco et al[62] there is clear evidence of a small population of T_{EMRA} $CD4^+$ T-cells as well, which appears to be increased in aging (unpublished observation).

There is limited information regarding relative sensitivity of central and effector memory $CD8^+$ T-cells to apoptosis in humans. We have examined relative sensitivity/resistance of various memory subsets of $CD4^+$ and $CD8^+$ T-cells to death-receptor and mitochondrial signaling pathway of apoptosis.

Death-Receptor-Induced Apoptosis in Naïve and Memory $CD4^+$ and $CD8^+$ T-Cells

Recently, we have reported relative sensitivity of naïve and various memory $CD8^+$ T-cell subsets to TNF-α-induced apoptosis.[63,64] Our data show that T_N and T_{CM} $CD8^+$ T-cells were sensitive, whereas T_{EM} and T_{EMRA} $CD8^+$ T-cells were resistant to TNF-α-induced apoptosis. The apoptosis profile correlated with the activation of caspase-8 and caspase-3. Since no correlation was observed between the relative sensitivity of four $CD8^+$ T-cell subsets to TNF-α-induced apoptosis and the expression of TNFR-I or TNFR-II,[63] we examined downstream signaling molecules, including phosphorylation of IκB and NF-κB activity following activation with TNF-α. IκB phosphorylation and NF-κB activity were higher in T_{EM} and T_{EMRA} $CD8^+$ T-cells, as compared to T_N and T_{CM} $CD8^+$ T-cells. In addition, expression of Bcl-2 was higher and Bax expression (measured by Western blotting) was lower in T_{EM} and T_{EMRA} $CD8^+$ T-cells as compared to T_N and T_{CM} $CD8^+$ T-cells. This is consistent with positive regulation of Bcl-2 and negative regulation of Bax by NF-κB. In summary, anti-apoptotic molecules are upregulated and pro-apoptotic molecules are downregulated in T_{EM} and T_{EMRA} $CD8^+$ T-cells, which are relatively resistant to apoptosis. These data suggest that signaling molecules downstream of TNFRs may be responsible for differential sensitivity among subsets of $CD8^+$ T-cells to TNF-α-induced apoptosis. Recently, we have also observed that similar to $CD8^+$ T-cells, T_N and T_{CM} $CD4^+$ T-cells ($T_{CM} > T_N$) are sensitive to TNF-α-induced apoptosis, whereas T_{EM} and T_{EMRA} $CD4^+$ T-cells are resistant to TNF-α-induced apoptosis.[65] However, the relative differences among subsets of $CD4^+$ T-cells are less striking than those observed with the subsets of $CD8^+$ T-cells.

More recently, we have also examined relative sensitivity of naïve and memory subsets of $CD4^+$ and $CD8^+$ T-cells to apoptosis via CD95-mediated pathway. Similar to TNFR-mediated apoptosis, T_N and T_{CM} subsets of $CD8^+$ T-cells and $CD4^+$ T-cells were sensitive, whereas T_{EM} and T_{EMRA} subsets were relatively resistant to apoptosis. The apoptosis profile corresponds to activation of caspase-8 and caspase-3; however, no correlation was observed between the sensitivity/resistance to apoptosis and the expression of CD95 (manuscript in preparation).

Mitochondrial Pathway of Apoptosis in Subsets of $CD8^+$ and $CD4^+$ T-Cells

We also investigated whether the relative sensitivity/resistance among naïve and various memory subsets was applicable to the mitochondrial pathway of apoptosis as well. To investigate this, we examined the effect of H_2O_2 on apoptosis in naïve and various subsets of $CD8^+$ and $CD4^+$ T-cells. We observed that both T_N and T_{CM} $CD4^+$ and $CD8^+$ T-cells ($T_{CM} > T_N$) display sensitivity to H_2O_2-induced apoptosis, whereas T_{EM} and T_{EMRA} $CD8^+$ and $CD4^+$ T-cells are resistant. Apoptosis of T_N and T_{CM} subsets of $CD4^+$ and $CD8^+$ T-cells is associated with the release of cytochrome C and AIF and activation of both caspase-9 and caspase-3. In addition, H_2O_2 decreased intracellular glutathione (GSH) in T_N and T_{CM} $CD4^+$ and $CD8^+$ T-cells and exogenous GSH inhibited H_2O_2-induced apoptosis of T_N and T_{CM} $CD4^+$ and $CD8^+$ T-cells. These data demonstrate that H_2O_2 induces apoptosis predominantly in human T_N and T_{CM} $CD4^+$ and $CD8^+$ T-cells, which is associated with release of cytochrome c and AIF and activation of caspase-9 and caspase-3. Intracellular GSH, at least in part, appears to play a role in H_2O_2-induced apoptosis of T_N and T_{CM} $CD4^+$ and $CD8^+$ T-cells (manuscript in preparation).

Apoptosis of Subsets of $CD4^+$ and $CD8^+$ T-Cells in Human Aging

Unlike mice, human aging is associated with progressive T-cell lymphopenia, which is shared by both $CD4^+$ and $CD8^+$ T-cells, albeit more pronounced $CD8^+$ T-cell lymphopenia.[66,67]

In aging, there is a significant reduction in naïve $CD8^+$ T-cells[67] and $CD8^+$ $CD28^+$ T-cells, which contain both naïve and central memory $CD8^+$ T-cells.[68] In addition, there is an accumulation of $CD8^+CD28^-$ T-cells, which are oligoclonal and show characteristics of cellular senescence (i.e., short telomere length indicative of long replicative history) and increased IFN-γ production.[69-71] These $CD8^+$ $CD28^-$ T-cells are comprised of two subpopulations of effector memory $CD8^+$ T-cells, namely T_{EM} and T_{EMRA} $CD8^+$ T-cells. Our study shows a marked decrease in the absolute numbers, of T_N and T_{CM} and a significant increase in T_{EMRA} $CD8^+$ T-cells. Fagnoni et al[67] also observed an increase in primed $CD8^+CD28^-CD45RA^+$ (equivalent to TEMRA) in aged humans.

Although thymic output of naïve T-cells in aging is decreased,[72] our study shows that increased apoptosis of naïve and central memory T-cells may also contribute to their peripheral lymphopenia.

Activation-Induced Cell Death and CD95-Mediated Apoptosis in Naïve and Memory Subsets of $CD4^+$ and $CD8^+$ T-Cells in Aging

Apoptosis in T-lymphocytes and their subsets in human aging has been studied primarily via death receptor signaling, which will be briefly reviewed.

There is a general agreement that apoptosis of T-cells is increased during human aging.[73-80] Several investigators have reported increased AICD in human aging.[77-80] A role for increased AICD of naïve T ($CD45RO^-$) T-cells has been suggested to contribute to age-associated T-cell deficiency.[81] Brezinska et al[82] concluded that AICD (as measured by DNA content and caspase-3 activation) in $CD8^+CD28^+$ (containing T_N and T_{CM}) and $CD8^+CD28^-$ (containing T_{EM} and T_{EMRA}) was comparable. However, these investigators presented data from a single middle-aged individual. We have reported that in aged humans, both $CD45RA^+$ (naïve) and $CD45RO^+$ (memory) $CD4^+$ and $CD8^+$ T-cells were more sensitive to anti-CD95-induced apoptosis as compared to young subjects.[73] Furthermore, $CD45RO^+$ cells displayed greater sensitivity to anti-CD95-induced apoptosis as compared to $CD45RA^+$ $CD4^+$ and $CD8^+$ T-cells in both young and aged subjects. Miyawaki et al[83] also reported that healthy adult memory T-cells are more susceptible to anti-CD95-induced apoptosis as compared to naïve T-cells. We reported decreased expression of Bcl-2 in both $CD4^+$ and $CD8^+$ T-cells from aged humans as compared to young subjects; however, we did not examine Bcl-2 expression in naïve and memory subsets.[73] Shinohara et al[84] demonstrated decreased Bcl-2 expression in memory subsets of $CD4^+$ and $CD8^+$ T-cells in healthy adults. This is consistent with our observation of increased sensitivity of memory T-cell subsets to death-receptor-mediated apoptosis as compared to naïve T-cell subsets. There appears to be a better correlation between AICD and the expression of CD95L expression rather than with CD95 expression.

We and others have reported increased CD95-mediated apoptosis in $CD4^+$ and $CD8^+$ T-cells in aged healthy subjects,[72,73,85] which is associated with increased and early activation of both caspase-8 and caspase-3.[85] Furthermore, both $CD4^+$ and $CD8^+$ T-lymphocytes from aged humans display increased expression of CD95 and CD95L and of an adapter molecule the FADD. In addition, CD95-mediated apoptosis in both $CD45RA^+$ (naïve) and CD45 RA^- (memory) $CD4^+$ and $CD8^+$ from aged subjects was increased as compared to young subjects (manuscript in preparation). Since these subsets as defined by CD45 expression are heterogeneous, we have investigated CD95-mediated apoptosis in T_N, T_{CM}, T_{EM} and T_{EMRA} $CD4^+$ and $CD8^+$ T-cells in aged humans. Our data show that both T_N and T_{CM} $CD4^+$ and $CD8^+$ T-cells from aged subjects are more sensitive to anti-CD95-induced apoptosis, which is associated with greater caspase-8 and caspase-3 activation as compared to young subjects. In contrast, T_{EM} and T_{EMRA} subsets were comparably resistant to apoptosis (manuscript in preparation).

TNFR-Mediated Apoptosis in Naïve and Memory Subsets of $CD4^+$ and $CD8^+$ T-Cells in Aging

Since TNF-α production is increased in aging,[86] we investigated susceptibility of CD4 and CD8 cells from aged subjects to TNF-α-induced apoptosis.[66,74-87] Both subsets exhibited increased sensitivity to TNF-α-induced apoptosis, which was associated with increased activation of both caspase-8 and caspase-3. In contrast to our observations, Salvoni et al[88] using freshly isolated T-cell subsets and using TNF-α and cyclohexamide to induce apoptosis, observed resistance of aged $CD4^+$ T-cells to TNF-α-induced-apoptosis; however, they demonstrated increased susceptibility of aged $CD8^+$ T-cells to apoptosis as detected by Annexin V staining. The externalization of phosphatidyl serine (which binds to Annexin V) is mediated by the scramblase enzyme, which is sensitive to calcium.[89] Therefore, significant changes in intracellular calcium may result in a cell being positive for Annexin V without undergoing apoptosis; calcium signaling is different among $CD4^+$ and $CD8^+$ T-cells and among young and aged T-cells (Our unpublished observations). In our study, we used a model of in vivo activation in which no cyclohexamide was added. The sensitivity of T-cells to TNF-α-induced apoptosis appears to be age-dependent as cord blood lymphocytes are least sensitive,[90] whereas aged T-cells are most sensitive to TNF-α-induced apoptosis.[74] We observed an increased expression of TRADD and FADD in lymphocytes from aged subjects, both at the level of mRNA and protein.[73,74]

Since FADD is a common conduit for both CD95- and TNFR-mediated apoptosis and apoptosis of $CD4^+$ and $CD8^+$ T-cells in aging via CD95 and TRFRs is increased, we examine the role of increased FADD expression on increased apoptosis in aging. T-cells from aged humans transfected with dominant negative FADD resulted in decreased TNF-α-induced apoptosis to a level comparable to young T-cells, whereas wild type FADD resulted in increased apoptosis in both young and aged T-cells, albeit to a greater extent in young T-cells, to a level comparable to aged T-cells, thus establishing a role of increased FADD in increased apoptosis in aged T-cells.[91]

NF-κB plays an important role in cell survival. Several investigators have reported a decrease in TNF-α-induced DNA-binding activity of NF-κB in lymphocytes from aged humans as determined by supershift gel mobility assay and recently developed ELISA assay[84,92,93] and we have noted translocation of the p65 subunit of NF-κB to the nucleus by confocal microscopy (unpublished observations). Furthermore, we demonstrated that the expression of upstream molecules IKKβ and phosphorylation of IκBα in T-cells from aged humans were decreased and overexpression of IKKβ resulted in an increased phosphorylation of Iκ-B and decreased TNF-α-induced apoptosis of T-cells to a level comparable to that of young subjects. This was associated with an upregulation of Bcl-2 and cIAP2.[87] A decreased activation of NF-κB due to decreased proteasome-mediated degradation of IκB has also been suggested.[93] These observations provide evidence for a mechanism by which decreased NF-κB plays an important role in increased sensitivity of aging T-cells to TNF-α-induced apoptosis.

We have also examined TNF-α-induced apoptosis in both naïve and memory subsets of $CD4^+$ and $CD8^+$ T-cells, using TUNEL assays and flow cytometry and have observed that both $CD45RA^+$ naïve and $CD45RA^-$ memory $CD4^+$ and $CD8^+$ T-cells from aged individuals were more sensitive to TNF-α-induced apoptosis.[74]

As discussed above, naïve T-cells, when defined only by the presence of CD45RA, also contain T_{EMRA} $CD8^+$ T-cells (and a very small population of T_{EMRA} $CD4^+$ T-cells) and $CD45RA^-$ ($CD45RO^+$) contain both T_{CM} and T_{EM} $CD8^+$ T-cells. Therefore, we have examined the relative sensitivity of T_N, T_{CM}, T_{EM}, T_{EMRA} $CD8^+$ and $CD4^+$ T-cell subsets to TNF-α-induced apoptosis. In aged humans, we observed that T_N and T_{CM} $CD8^+$ T-cells displayed increased TNF-α-induced apoptosis as compared to young subjects, which is associated with increased caspase-8 and caspase-3 activation. In contrast, T_{EM}, T_{EMRA} $CD8^+$ T-cells are comparably resistant to TNF-α-induced apoptosis and display minimal caspase activation in both young and aged subjects.[94] Therefore, it appears that during aging, the decrease in T_N $CD8^+$ T-cells is due to both decreased thymic output as well as increased apoptosis. We have also observed increased apoptosis in T_N and T_{CM} ($T_{CM} > T_N$) $CD4^+$ T-cells in aged humans as compared to young subjects; however, no significant difference was observed in the apoptosis of T_{EM} and T_{EMRA} $CD4^+$ T-cells between aged and young humans; both were resistant to apoptosis.[65]

Why are central memory cells more sensitive to apoptosis as compared to naïve T-cells and why are effector memory cells relatively resistant to apoptosis and accumulate during aging? Are $CD8^+CD28^-$ T-cells in aging represented by $CD8^+CD28^-$ T-cells generated in vitro by repeated activation (replicative senescence), or exhausted cells? Can $CD8^+CD28^-$ T-cells from aged and additionally from young subjects, be rescued and made to proliferate again and how? Since T_{CM}

cells have a high replicative capacity (more than naïve T-cells), increased apoptosis may be critical to make space for new T_{CM} $CD4^+$ and $CD8^+$ T-cells and to maintain homeostasis of T_{CM} cells. In contrast, low replicative capacity of T_{EM} and T_{EMRA} cells does not allow for the creation of an "immunological nitch", or they are exhausted and therefore T_{EM} and T_{EMRA} cells are resistant to apoptosis. A large number of studies have been reported on $CD8^+CD28^-$ T-cells generated after repeated stimuli (as a model of aging) and indicate that they possess features of replicative senescence (low proliferative potentials and resistance to apoptosis). However, Brzezinska et al[82] have reported that aged $CD8^+CD28^-$ proliferate more than adult counterparts. We have observed that both T_{EM} and T_{EMRA} $CD8^+$ T-cells from young and aged subjects can proliferate well in the presence of exogenous IL-2 and IL-15 (unpublished observation). We have also observed increased expression of the IL-15 gene in $CD8^+$ T-cells from aged humans (by gene array). These observations suggest that $CD8^+CD28^-$ T-cells generated by repeated activation in vitro are not a true model for $CD8^+CD28^-$ T-cells in aged humans. Furthermore, increased accumulation of $CD8^+CD28^-$ T-cells in aged humans may be due to an increased growth rather than due to differences in apoptosis between aged and young humans. Finally it would be interesting to study PD1-PD1-L interactions in effector memory $CD8^+$ T-cells and to use this interaction to target and rescue these oligoclonal cells from exhaustion, which may provide a unique opportunity for aged individuals to respond better to vaccinations.

Acknowledgement

The work cited is in part supported by a grant from UPHS AG-18313.

References

1. Hengartner MO. The biochemistry of apoptosis. Nature 2002; 407:770-776.
2. Gupta S. Suicidal journey in the Fas(t)track. Recent Res Dev Immunol 2000; 2:11-19.
3. Gupta S. Molecular steps of cell suicide:An insight into immune senescence. J Clin Immunol 2000; 20:229-239.
4. Krammer PH. CD95's deadly mission in the immune system. Nature 2000; 407:789-795.
5. Ashkanazi A, Dixit VM. Death receptors:signaling and modulation. Science 1998; 281:1305-1308.
6. Gupta S. Molecular steps of death receptor and mitocondrial pathways of apoptosis. Life Sci 2000; 69:2957-2964.
7. Gupta S. Molecular steps of TNF receptor-mediated apoptosis. Curr Mol Med 2001; 1:299-306.
8. Gupta S. Decision between life and death during TNF-induced signaling. J Clin Immunol 2002; 22: 270-278.
9. Scaffidi C, Fulda S, Srinivasan A et al. Two CD95 (Apo-1/Fas) signaling pathways. EMBO J 1998; 17:1675-1687.
10. Locksley RM, Kileen N, Lenardo MJ. The TNF and TNF receptor superfamilies:interating mammalian biology. Cell 2001; 104:487-501.
11. Green DR, Evan GI. A matter of Life and Death. Cancer Cell 2002; 1:19-30.
12. Kroemer G, Reed JC. Mitochondrial control of cell death. Nature Med 2000; 6:513-519.
13. Martinou J-C, Green DR. Breaking the mitochondrial barrier. Nature Rev Mol Cell Biol 2001; 2:63-67.
14. Zamzami N, Kroemer G. The mitochondrion in apoptosis:how pandora's box opens. Nature Rev Mol Cell Biol 2001; 2:67-71.
15. Green DR, Droin N, Pinkoski M. Activation-induced cell death in T-cells. Immunol Rev 2003; 193:70-81.
16. Di Somma MM, Somma F, Montani MSG et al. TCR engagement regulates differential responsiveness of human memory T-cells to Fas (CD9)-mediated apoptosis. J Immunol 1999; 162:3851-3858.
17. Dhein J, Walczak H, Baumler C et al. Autocrine T-cell suicide mediated by APO-1/(fas-CD95). Nature 1995; 373:438-441.
18. Brunner T et al. Cell autonomous Fas (CD95)/Fas ligand interaction mediates activation-induced apoptosis in T-cell hybridomas. Nature 1995; 373:441-444.
19. Ju ST et al. Fas (CD95/FasL interactions are required for programmed cell death after T-cell activation. Nature 1995; 373:444-448.
20. Anderson MR et al. Fas ligand mediates activation-induced cell death in human T-lymphocytes. J Exp Med 1995; 181:71-77.
21. Mixter PF, Russell JQ, Budd RC. Delayed kinetics of T-lymphocyte anergy and deletion in lpr mice. J Autoimmunity 1994; 7:697-710.

22. Mogil, RJ et al. Fas (CD95) participates in peripheral T-cell deletion and associated apoptosis in vivo. Int Immunol 1995; 7:1451-1458.
23. Renno T, Hahne M, Tschopp J et al. Peripheral T-cells undergoing superantigen-induced apoptosis in vivo express B220 and upregulate Fas and Fas ligand. J Exp Med 1996; 183:431-437.
24. Tanaka M, Suda T, Takahashi T et al. Expression of the functional fas ligand in activated lymphocytes. EMBO J 1995; 14:223-239.
25. Irmler M, Thome M, Hahne M et al. Inhibition of death receptor signals by cellular FLIP. Nature 1997; 388:190-195.
26. Thome, M, Schneider P, Hofmann C et al. Viral Flice-inhibitory proteins (FLIPs) prevent apoptosis induced by death receptors. Nature 1997; 386:517-521.
27. Yeh WC, Itie A, Elia AJ et al. Requirement of casper (c-FLIP) in regulation of death receptor-induced apoptosis and embryonic development. Immunity 2000; 12:633-642.
28. Kataoka T, Budd RC, Holler N et al. The caspase-8 inhibitor FLIP promotes activation of NF-κB and Erk signaling pathways. Curr Biol 2000; 10:640-648.
29. Golks A, Brenner D, Krammer PH et al. The c-FLIP-NH2 terminus (p22-FLIP) induces NF-κB activation. J Exp Med 2006; 203:1295-1305.
30. Screaton G, Xu X-N. T-cell life and death signaling via TNF-receptor family members. Curr Opin Immunol 2000; 12:316-3222.
31. Thomas B, Grell M, Pfizenmaier K et al. Identification of a 60-kDa tumor necrosis factor (TNF) receptor as the major signal transducing component in TNF responses. J Exp Med 1990; 172:1019-1023.
32. Darnay BG, Aggarwal BB. Early events in TNF signaling:a story of associations and dissociations. J Leukocyte Biol 1997; 61:559-66.
33. Wallach D, Boldin M, Varfolomeev E et al. Cell death induction by receptors of the TNF family:towards a molecular understanding. FEBS Lett 1997; 410:96-106.
34. Natoli G, Costanzo A, Ianni A et al. Activation of SAPK/JNK by TNF receptor 1 through a noncytotoxic TRAF-2-dependent pathway. Science 1997; 275:200-203.
35. Ichijo N, Nishida E, Irie K et al. Induction of apoptosis by ASK1, a mammalian MAPKKK that activates SAPK/JNK and p38 signaling pathways. Science 1997; 275:90-94.
36. Beg AA, Baltimore D. An essential role for NF-κB in preventing TNF-α-induced cell death. Science 1996; 274:782-784.
37. Ghosh S, May MJ, Kopp EB. NF-κB and rel proteins:evolutionarily conserved mediators of immune responses. Annu Rev Immunol 1998; 16:225-260.
38. Baldwin AS. The NF-κB and IκB proteins:new discoveries and insights. Annu Rev Immunol 1996; 14:649-681.
39. Karin M, Lin A. NF-κB at the crossroads of life and death. Nature Immunol 2002; 3:221-227.
40. Ghosh S, Karin M. Missing pieces in the NF-kB puzzle. Cell 2002; 109:S81-S96.
41. Devin A, Cook A, Lin Y et al. The distinct role of TRAF2 and RIP in IKK activation by TNFR1: TRAF2 recruits IKK to TNFR-1 while RIP mediates IKK activation. Immunity 2000; 12:419-429.
42. Zandi E, Chen YI, Karin M. Direct phosphorylation of IκB by IKKα and IKKβ:Discrimination between free and NF-κB-bound substrate. Science 1998; 281:1360-1363.
43. Brown K, Gerstberger S, Carlson L et al. Control of I κB-α proteolysis by site-specific, signal induced phosphorylation. Science 1995; 281:1360-1363.
44. Pahl HL. Activators and target genes of Rel/NF-kB transcription factors. Oncogene 1999; 18: 6855-6866.
45. Bentires-Alj M, Dejardin E, Viatour P et al. Inhibition of the NF-κB transcription factor increases Bax expression in cancer cell lines. Oncogene 2001; 20:2805-2813.
46. Varfolomeev EE, Ashkenazi A. Tumor necrosis factor: An apoptosis JunKie. Cell 2004; 116:491-497.
47. Deng Y, Ren X, Yang L et al. A JNK-dependent pathway is required for TNF-α-induced apoptosis. Cell 2003; 115:61-70.
48. Chang L, Kamata H, Solinas G et al. The E3 ubiquitin ligase Itch couples JNK activation to TNF-α-induced cell death by inducing c-$FLIP_L$ turnover. Cell 2006; 124:601-613.
49. Gupta S. Molecular signaling in death receptor and mitochondrial pathways of apoptosis. Internat J Oncol 2003; 22:15-20.
50. Hegde R, Srinivasula SM, Zhang Z et al. Identification of Omi/HtrA2 as a mitochondrial apoptotic serine protease that disrupts inhibitor of apoptosis protein-caspase interaction. J Biol Chem 2002; 277:432-438.
51. Suzuk i Y, Imai Y, Nakayama H et al. A serine protease, HtrA2, is released from the mitochondria and interacts with XIAP, inducing cell death. Mol Cell 2001; 8:613-621.
52. Lorenzo HK, Susin SA Penninger J et al. Apoptosis inducing factor (AIF):a physiologically old, caspases-independent effector of cell death. Cell Death Diff 1999; 6:516-524.

53. Loeffler M, Daugas E, Susin SA et al. Dominant cell death induced by extramitochondrially targeted apoptosis-inducing factor. FASEB J 2001; 15:758-767.
54. Li LY, Luo X, Wang X. Endonuclease G is an apoptotic DNAase when released from mitochondria. Nature 2001; 412:95-99.
55. Reed JC. Double identity for protein of Bcl-2 family. Nature 1997; 387:773-778.
56. Kaech SM, Ahmed R. Memory CD8$^+$ T-cell differentiation:initial antigen encounter triggers a developmental program in naïve cells. Nature Immunol 2001; 2:415-422.
57. Moser B, Loetscher P. Lymphocyte traffic control by chemokines. Nature Immunol 2001; 2:123-128.
58. Schluns KS, Lefrancois L. Cytokine control of memory T-cell development and survival. Nat Rev Immunol 2003; 3:269-279.
59. Sallusto F, Lenig D, Forster R et al. Two subsets of memory T-lymphocytes with distinct homing potentials and effector functions. Nature 1999; 401:708-712.
60. Monteiro J, Baltiwala F, Ostere H et al. Shortened telomere in clonally expanded CD28-CD8$^+$ T-cells imply a replicative history that is distinct from there CD28$^+$CD8$^+$ counterparts. J Immunol 1996; 162:6572-6579.
61. Weninger W, Crowley MA, Manjunath N et al. Migratory properties of naïve, effector and memory CD8($^+$) T-cells. J Exp Med 2001; 194:953-966.
62. Sallusto F, Geginat J, Lanzavecchia A. Central memory and effector memory T-cell subsets:Function, generation and maintenance. Ann Rev Immunol 2004; 22:745-763.
63. Gupta S, Su H, Bi R et al. Differential sensitivity of naïve and memory subsets of human CD8$^+$ T-cells to TNF-α-induced apoptosis. J Clin Immunol 2006; 26:193-203.
64. Gupta S, Gollapudi S. Molecular Mechanisms of TNF-α-induced apoptosis in naïve and memory T-cell subsets. Autoimmunity Rev 2006; 5:264-268.
65. Gupta S, Bi R, Gollapudi S. Central memory and effector memory subsets of human CD4$^+$ and CD8$^+$ T-cells display differential sensitivity to TNF-α-induced apoptosis. NY Acad Sci 2005; 1050: 108-114.
66. Gupta S. Tumor necrosis factor-α-induced apoptosis in T-cells from aged humans:a role of TNFR-I and downstream signaling molecules. Exp Gerontol 2002; 37:293-299.
67. Fagnoni FF, Vescovini R, Paserri G et al. Shortage of circulating naïve CD8$^+$ T-cells provides new insights on immunodeficiency in aging. Blood 2002; 95:2860-2868.
68. Aggarwal S, Gupta S. Increased apoptosis of T-cell subsets in aging humans:Altered expression of Fas (CD95), Fas ligand, Bcl-2 and Bax. J Immunol 1998; 160:1627-1637.
69. Aggarwal S, Gollapudi S, Gupta S. Increased TNF-α-induced apoptosis in lymphocytes from aged humans:changes in TNF-α receptor expression and activation of caspases. J Immunol 1999; 162: 2154-2161.
70. Gupta S, Chiplunkar S, Kim C et al. Effect of age on molecular signaling of TNF-α-induced apoptosis in human lymphocytes. Mech Ageing Dev 2003; 124:503-509.
71. Gupta S. A road to ruins:An insight into immunosenescence. Adv Cell Aging Gerontol 2003; 13: 169-185.
72. Phelouzat MA, Arbogast A, Laforge T et al. Excessive apoptosis of mature T-lymphocytes is a characteristic feature of human immune senescence. Mech Ageing Dev 1996; 88:25-38.
73. Phelouzat MA, Laforge T, Abrogast A et al. Suspetibility to apoptosis of T-lymphocytes from elderly humans is associated with increased in vivo expression of functional fas receptors Mech Ageing Dev 1997; 96:35-46.
74. Lechner H, Amort M, Steger MM et al. Regulation of CD95 (Apo-1) expression and the induction of apoptosis of human T-cells:changes in old age. Int Arch Allergy Immunol 1996; 110:238-243.
75. Potestio M, Caruso C, Gervasi F et al. Apoptosis and aging. Mech Ageing Dev 1998; 102:221-237.
76. Aggarwal S, Gupta S. Increased activity of caspase-3 and caspase-8 during Fas-mediated apoptosis in lymphocytes from aging humans. Clin Exp Immunol 1999; 117:285-290.
77. Fagiola U, Cossarizza A, Scala E et al. Increased cytokine production in mononuclear cells of healthy elderly people. Eur J Immunol 1993; 23:2375-2378.
78. Gupta S, Bi R, Kim C et al. Role of NF-κB signaling pathway in increased tumor necrosis factor-α-induced apoptosis of lymphocytes in aged humans. Cell Death Diff 2005; 12:177-183.
79. Savioli S, Capri M, Scarcella E et al. Age-dependent changes in the susceptibility to apoptosis of peripheral blood CD4$^+$ and CD8$^+$ T-lymphocytes with virgin or memory phenotype. Mech Ageing Dev 2003; 124:409-418.
80. Orrenius S, Zhivotovsky B, Nicotera P. Regulation of cell death:The calcium-apoptosis link. Nature Rev Mol Cell Biol 2003; 4:552-564.
81. Aggarwal S, Gollapudi S, Yel L et al. TNF-α-induced apoptosis in neonatal lymphocytes:TNFRp55 expression and downstream pathways of apoptosis. Genes Immunity 2000; 1:271-279.
82. Gupta S, Kim C, Yel L et al. A role of Fas-associated death domain (FADD) in increased apoptosis in aged humans. J Clin Immunol 2004; 24:24-29.

83. Trebilcock GU, Ponnappan U. Evidence for lowered induction of nuclear factor kappa B in activated human T-lymphocytes during aging. Gerontology 146; 42:137-146.
84. Ponnappan U, Zhong M, Trebilcock GU. Decreased proteosome-mediated degradation in T-cells from the elderly: A role in immune senescence. Cell Immunol 1999;192:167-174.
85. Gupta S, Bi R, Su K et al. Characterization of naïve, memory and effector $CD8^+$ T-cells: Effect of age. Exp Gerontol 2004; 39:545-550.

CHAPTER 6

T-Cell Signalling, a Complex Process for T-Cell Activation Compromised with Aging:
When Membrane Rafts Can Simplify Everything

Tamas Fulop,* Graham Pawelec, Carl Fortin, Anis Larbi

Abstract

Aging is associated with altered immune responsiveness, termed "immunosenescence". It is now well accepted that both arms of the immune system, innate as well as adaptive, undergo immunosenescence. However, the adaptive immune response and especially T-cells are the most affected by aging. Aging is associated with both changes in lymphocytes subpopulations and, importantly, functional changes within these subsets. Indeed, T-cells present functional modifications resulting in a decreased clonal expansion and interleukin-2 production as well as a shift in Th1/Th2 response with aging. Identifying alterations in the activation process involving the TCR, CD28 and IL-2 receptor signalling cascades are crucial to understanding immunoseneescence. The putative reasons for this altered activation of T-cells with aging will be reviewed here, based on our own recent work and international collaborations.

Introduction

It is accepted that aging is associated with altered immunity, termed immunosenescence.[1,2] Both arms of the immune system, innate as well as adaptive, undergo age-associated changes.[3,4] These are not always associated with a loss of function, but with a general deregulation of the immune response with aging. The adaptive arm and especially T-cells are the most affected by aging,[5] but the basic causes of dysfunction are not clear. The most invoked cause is the involution of the thymus[6] which is associated with an altered T-cell subpopulation distribution leading to a general decrease of naïve cells and an increase of memory cells most specifically among the $CD8^+$ T-cells.[7] More recently, longitudinal studies involving very old subjects have associated immunosenescence with the accumulation of anergic T-cells, mainly $CD8^+$ T-cells specific for antigens from cytomegalovirus (CMV).[8] These can represent >20% of the whole peripheral blood CD8 repertoire. This accumulation could certainly contribute to, but not completely explain, all the changes that are the hallmarks of immunosenescence. Aging is associated with changes in lymphocytes subpopulations, but the functional changes within these subsets may be more important to depict. Indeed, T-cells from the elderly are clearly not all fully functional. Thus there is likely to be an alteration in the activation processes of T-cells with aging. These are briefly reviewed here.

*Corresponding Author: Tamas Fulop—Research Center on Aging, Immunology Program, Geriatric Division, Faculty of Medicine, University of Sherbrooke, 1036 rue Belvedere sud, Sherbrooke J1H 4C4, Quebec, Canada. Email: tamas.fulop@usherbrooke.ca

Immunosenescence, edited by Graham Pawelec. ©2007 Landes Bioscience and Springer Science+Business Media.

T-Cell Functional Changes with Aging

T-cells are the backbone of the cellular immune response. During the encounter with foreign antigen, presented by specialised antigen presenting cells (APC), T-cells become activated by signals transduced via their T-cell receptor (TCR) as well as via their coreceptors.[9] Clonal expansion, assured by their proliferation, must then follow in order to generate sufficient T-cells of the same specificity. This proliferation needs interleukin-2 (IL-2), which was originally named T-cell growth factor. Aging leads to a decline in the ability to mount a rigorous and efficient T-cell response to newly encountered as well as to recall antigens.[5] This decline manifests as a decrease in delayed type hypersensitivity response, diminished ability to respond to vaccination and increased susceptibility to virulent viral and bacterial infections. Aging is also associated with altered T-cell apoptosis susceptibility.[10,11] These functional changes are reflected in T-cell inability to mount an effective proliferative response leading to clonal expansion, along with the decreased IL-2 secretion.[12,13] Both $CD4^+$ and $CD8^+$ T-cell proliferation is decreased with aging, mainly of the naïve T-cells as observed in aged mice as well as in aged humans. Recently, it was shown that even memory $CD8^+$ T-cells lacking CD28 expression which is necessary for T-cell activation, are able to proliferate under certain circumstances. The T-cells are accumulating during aging are mostly $CD8^+CD28^-$[14] and express CD57 as well as inhibitory receptors such as KLRG-1. One over-riding change is the shift of the balance of subsets in the periphery from naïve to memory T-cells during aging. There is a consensus that the number of naïve cells decreases with age while memory cells (including central memory, effector memory) increase (Fig. 1). However, the role of each T-memory subset in immunosenescence is not well-known.

Antigenic Stimulation of T-Cells with Aging

Receptors Required for an Optimal Response and Their Changes with Aging

The most important receptors implicated in the clonal expansion of T-cells are the T-cell receptor (TCR), the coreceptors including CD28 and the IL-2 cytokine receptor (IL-2R). These receptors function via an intracellular signalling cascade assuring the specificity and the fidelity of the response. T-cells need a first signal priming them for a full response to a specific

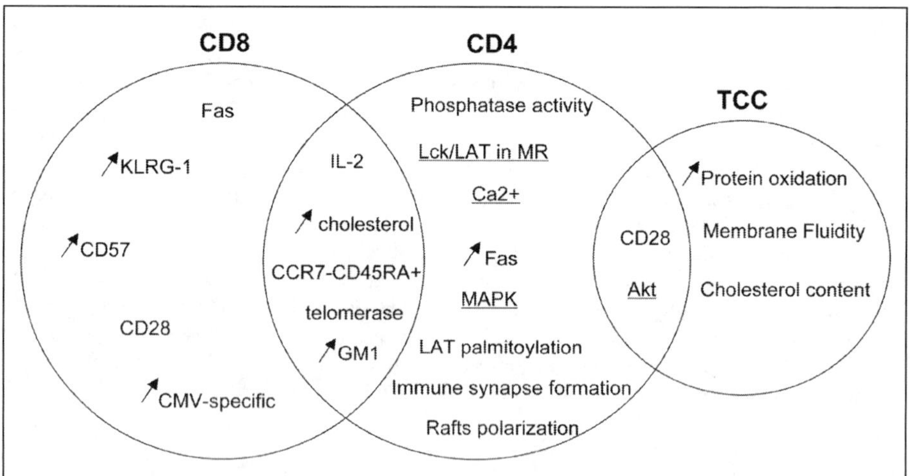

Figure 1. Changes in T-cell properties with aging. The age-related changes in T-cell properties are depicted here for $CD4^+$ and $CD8^+$ T-cells. The in vitro aging of $CD4^+$ T-cell clones (TCC) is also included. Arrows indicate an increase of the corresponding parameter while no arrow indicates a decrease. The changes which are equivalent between cell types are enclosed in the intersections.

antigen presented in the frame of self-MHC molecules on APC (signal 1).[15] This first signal causes the signalling machinery to become assembled in the membrane, in order to be able to proceed to the next stage allowing the sustained activation of the cell. This is assured by various coreceptors, among which CD28 is very important, which deliver the 2nd signal.[16] Signals transmitted by these receptors allow the formation of an immunological synapse (IS) responsible for sustained T-cell activation. The immune response is eventually terminated by T-cell inactivation and regulation of homeostasis via the initiation of activation-induced cell death (AICD). Signal 1 plus signal 2 delivered via two different receptors leads to a common outcome, IL-2 production and clonal expansion. Thus, IL-2 and its receptor (IL-2R) form an autocrine loop representing the third signal completing the requirements for clonal expansion. These receptors must act together for full activation of T-cells assuring a controlled response to a specific antigen.

Several studies have shown that the number of TCR is not changed with aging. However, the CD28 number seems to decrease, mainly in certain T-cell subpopulations, such as the memory $CD8^+$ T-cells.[17] These cells are present in large numbers primarily as a result of chronic stimulation by antigens probably of viral origin such as CMV, Epstein Barr virus and other herpes viruses.[18] It is also thought that they are the result or indeed perhaps to some extent the cause of the low grade chronic inflammation commonly observed in the elderly and termed "Inflam-aging"[19] (Fig. 2). The question which naturally arises is whether this is a normal process related to aging, to age-related diseases or to the progressing frailty syndrome occurring in certain groups of elderly subjects. The $CD4^+$ T-cell subpopulations do not display such a marked decrease in CD28 coreceptor number, which is prevalent in the $CD8^+$ T-cell.[20] While there is a consensus on CD28 expression in aging T-cells, changes in IL-2 receptor expression with aging are still controversial; our own work suggests their number is not changing significantly with healthy aging.

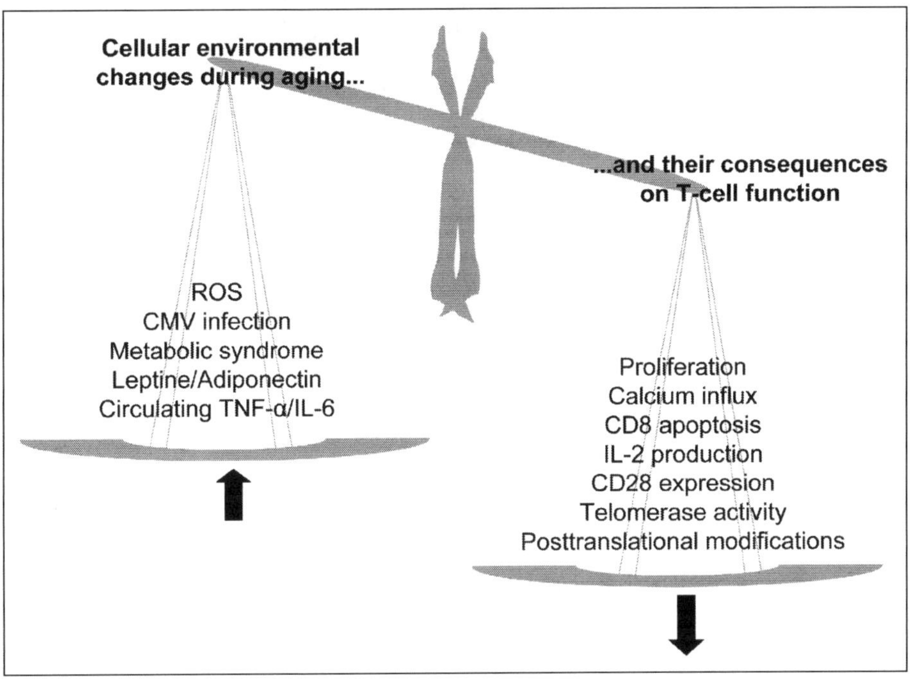

Figure 2. Age-related changes in T-cell environment and their consequences on T-cell function. The equilibrium between pro-inflammatory/anti-inflammatory stimuli is not maintained in aging and results in a diminution in several T-cell functions.

Signal Transduction and Its Fate with Aging

For an adequate T-cell response elicited by a ligand via a receptor the fidelity of signal transduction is even more important than the receptor number. Each individual receptor has a specific signalling mechanism, but much cross-talk exists between them to ensure that the response will take place (Fig. 3). The first step in receptor-mediated signalling is commonly the activation of different tyrosine kinases, leading to the tyrosine phosphorylation of several downstream molecules.[21] In the case of the TCR, one of the first events is the phosphorylation of Lck, via recruitment of ZAP-70 and leading to many downstream, but still early events, including the phosphorylation of the Linker of activated T-cell (LAT). This is a very tightly controlled process that involves phosphatases such as CD45, as well as regulatory molecules such as Csk and Cbp/PAG. CD45 is a receptor-like protein tyrosine phosphatase expressed on all haematopoietic cells; it acts by dephosphorylating the negative regulatory C-terminal residue of Lck. It is now well-documented that other early events related to protein tyrosine phosphorylation following TCR activation are altered, such as the generation of myo-inositol 1,4,5-trisphosphate, intracellular free calcium mobilization and protein kinase C (PKC) translocation to the membrane. It was shown that defects in translocation of PKC following TCR stimulation are present in T-cells of old humans and also mice. This activation finally leads to the activation of transcription factors such as NF-AT and NF-κB, resulting in the production of IL-2 which is consequently also altered with aging.

The CD28 coreceptor, in contrast, is mainly linked to the phosphatidyl-inositol-3-kinase (PI-3K)/Akt/IκB kinase (IKK)/NF-κB and the PI-3K/PDK-1/PKC-θ/IKK/NF-κB pathways.[22]

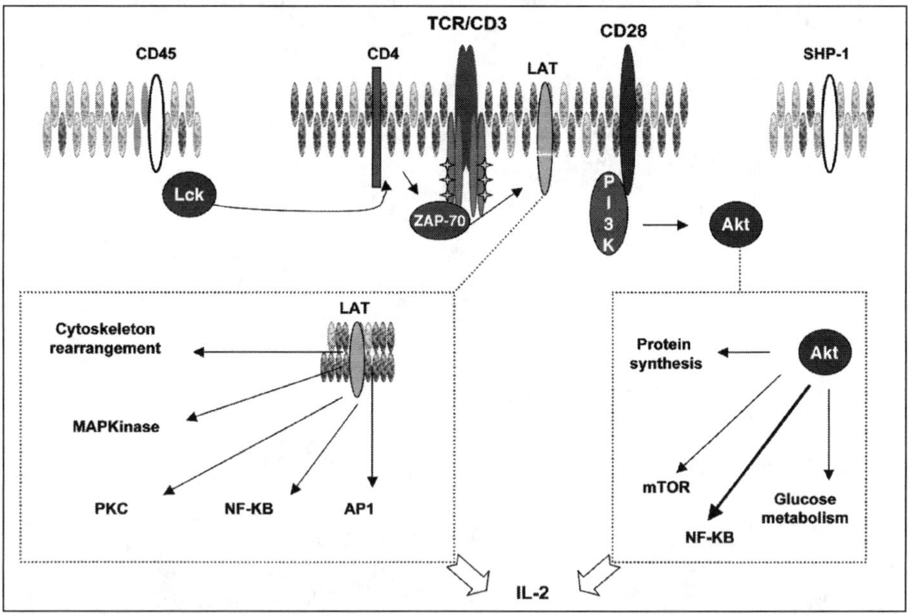

Figure 3. The main T-cell signalling pathways. TCR and CD28 pathways are depicted. TCR ligation induces Lck activation in membrane rafts via its interaction with CD45. The phosphorylation of the immuno-tyrosine-based activation motifs (ITAMs) at the zeta chain of the TCR/CD3 complex induces ZAP-70 recruitment and the subsequent activation of LAT. The adaptor LAT has no intrinsic activity but binds to several molecules which allows the activation of several pathways as described in the figure. The ligation of TCR coreceptor, CD28, induces the activation of Akt. Akt is the keystone for the activation of many metabolic pathways. Both TCR and CD28 signalling pathways have the same outcome, converging via NF-κB activation on IL-2 production needed for clonal expansion.

This is linked to other signalling machinery including the complex Carma1/Bcl10/Malt1. All these pathways converge to the activation of NF-κB, thereby assuring the production of IL-2. One example of signalling crosstalk is the phosphorylation of Lck by CD28 and TCR activation of PI3K. Our own recent work indicates that CD28 signalling leading to the phosphorylation of Akt is decreased mainly in $CD4^+$ T-cells from aged individuals.[20] This further contributes to the decreased NF-κB activation already shown to be due to a decreased inactivation of IκB by the proteasome.

For clonal expansion, the IL-2 produced must then be able to transduce signals through its specific receptor via a signalling pathway which involves the Jak/STAT pathway and also the MAPK Erk1/2 pathway. The IL-2R is composed of three subunits: the α subunit (CD25) is of very high affinity for IL-2, the β subunit is of low affinity, while the common γ chain is required for signalling. Among the Jak/STAT members, Jak3 and STAT3/5 need to be activated to provide for adequate proliferation of T-cells. We were able to demonstrate that aging was accompanied by an alteration in the activation and activity of Jak3 under IL-2 stimulation as well as in that of the transcription factors STAT3/5.[23]

Thus, our group and others have shown that there are alterations at almost all levels of the signalling cascade of TCR, CD28 as well as IL-2R. There is a general consensus that the expression of signalling molecules such as Lck, LAT, CD45 does not change at the cellular level with aging. The alteration is seen at the functional level, which means that the phosphorylation status upon activation is decreased with aging. This will certainly influence IL-2 secretion and consequently clonal expansion. The fundamental question remains why aging even in naïve T-cells results in this alteration of signal transduction.

T-Cell Membrane Composition Changes with Aging: Role of Cholesterol

As mentioned above, signalling molecule activation is decreased while their expression at the cellular level is maintained with aging. However, the localisation and interaction of these molecules with the membrane seems to play an even greater role than the cellular expression levels. It was suggested a long time ago that the T-cell membrane from elderly subjects is more rigid than that of young subjects.[24] We recently presented some evidence that an increase in free cholesterol could explain these physico-chemical changes observed about 20 years ago.[20] There is a two-fold increase in the cholesterol content in T-cells with aging. Cholesterol is an essential component of the membrane as it maintains the ordered structure, as it is now well recognized, through the membrane rafts.[25] It is of note that increasing the cholesterol level in the membranes of T-cells from young subjects with free cholesterol to the level of that in T-cells of elderly subjects led to a decreased proliferation capacity and IL-2 secretion (Fülöp et al unpublished data). Hence, increasing cholesterol in the membrane of T-cells from young subjects rendered them functionally aged. In the meantime, the main marker of membrane rafts, ganglioside M1 (GM-1) is also known to be increased with aging. However, there is still no explanation as to why cholesterol is increased in old T-cell membranes, because the serum cholesterol content remains stable in healthy elderly subjects (at least those compliant with the SENIEUR protocol for selecting healthy donors in immuno-gerontological studies). It could be that[1] the cholesterol uptake is dysregulated[2] intracellular cholesterol production via the HMG-CoA reductase is increased[3] the reverse cholesterol transport assured by HDL could be deficient. Our recent data seem to indicate that the latter may apply, i.e., reverse transport of cholesterol by HDL is indeed altered in T-cells from old subjects (Fülöp et al unpublished data).

Membrane Raft Functional Changes with Aging

As mentioned earlier, the functional role of cholesterol is predicated on the recent demonstration that the T-cell membrane is not homogeneous as was supposed by the Nicholson model.[26,27] The membrane contains microdomains, called membrane rafts, composed mainly of cholesterol, glycosphyngolipids and GPI-anchored proteins and, more importantly in the present context, signalling molecules. TCR ligation induces a redistribution of phosphorylated proteins into membrane rafts, which are highly compact relatively small domains (20 to 200 nM) composed of saturated lipids and signalling molecules.[28,29] The saturation of the lipids as well as the enrichment in cholesterol allows the rafts to move through the membrane as discrete units. Their movement will be directed to various poles of the cell and this phenomenon depends on their component such as GM1, GM3 or flotillin-1.[30] The role of membrane rafts is not limited to signal transduction, but also to lipid transport, virus entry, cell movement, as well as cell-cell communication. The accumulation or clustering of signalling molecules via membrane rafts initiates the formation of a signalling platform, also termed the "signalosome", which increases the efficiency of signalling. Sustained T-cell activation via organised membrane raft signalling ultimately leads to the formation of a mature immune synapse needed to achieve full T-cell activation.[31] Hence, the physico-chemical properties of the membrane will directly modulate the formation of such a signalling platform which ultimately influences cellular activation and functions. In T-cells, some of the signalling machinery is constitutively present in the membrane rafts, including the TCR and Lck, while others are recruited during activation, such as CD28, IL-2R, LAT and PI3K. It is of note that $CD4^+$ and $CD8^+$ T-cells possess different activation requirements. The signalling machinery in the $CD4^+$ T-cells is assembled dependent on membrane rafts, but in $CD8^+$ T-cells a certain level pre-assembly of the signalosome has been demonstrated by ourselves and others. This could perhaps help to explain the different fate of these two T-cell subpopulations with aging, as mentioned earlier.

Thus, with aging we have demonstrated that there is an alteration in the function of the membrane rafts as they are almost unable to coalesce in $CD4^+$ T-cells from the elderly.[20] The alterations are less dramatic for $CD8^+$ T-cells. We have demonstrated an alteration in the recruitment and activation of Lck and LAT into membrane rafts.[32] In this context one of the most important findings is that CD28, as well as the IL-2R, cannot be recruited into the membrane rafts of $CD4^+$ T-cells from elderly subjects. This helps to explain alterations in signalling of these receptors with aging, as well as the lack of coalescence of membrane rafts. In contrast, in $CD8^+$ T-cells these receptors are already localized to the membrane rafts prior to stimulation. This could be the consequence of chronic stimulation, such as chronic inflammation or by chronic antigenic stimulation that is seen in the case of CMV. Thus, the age-associated alterations in the properties of membrane rafts include an increase in cholesterol content, impaired coalescence and selective differences in the recruitment of key proteins involved in TCR signalling. Furthermore, the movement of molecules through the membrane and hence their localisation, is dependent on posttranslational modifications including acylation, farnesylation and palmitoylation. Recently, it was demonstrated that LAT phosphorylation was not optimal in antigen-primed anergic $CD4^+$ T-cells after TCR ligation. It is of interest that LAT association with membrane rafts was defective in these $CD4^+$ T-cells and this was partly explained by its impaired palmitoylation.[33] It can be supposed that the posttranslational lipidation of the signalling molecules targeting them to membrane rafts is altered with aging. Alterations in these posttranslational modifications will clearly modulate the intensity and duration of activation. It must also be borne in mind that additional events of the signalling cascade, not yet well-investigated in the context of aging, can also influence T-cell activation, among them the phosphatases which are well-known negative regulators.

Phosphatase Activity Changes in T-Cells

In addition to the molecules signalling positively via their phosphorylation there are regulatory molecules which negatively influence the signalling cascade.[34] Among these, phosphatases negatively regulate several steps of the process, targeting molecules such as Lck and PI3K, modulated by CD45, PTEN, SHIP and the SHP-1/2. These phosphatases are activated in a similar manner to kinases in order to assure that the system is not escaping control. They are also regulated by their mobility in and out of the membrane rafts. Their association to membrane rafts will increase the possibilities to inhibit some activation molecules. The best example is CD45. Its association with membrane rafts has a positive effect on Lck activation but when it is displaced (as it is in the quiescent status) Lck is inactivated. We have recently shown similar phenomena for the SHP-1 molecule in neutrophils.[35]

There are very few data concerning phosphatase activity in relation to T-cell receptor activation. Some data suggest that their expression is not altered with aging. Other data seem to suggest that CD45 activity is altered with aging. Certainly no data exist in relation to their regulation via membrane raft localization. This should be explored in the future.

T-Cell Clones: A Model for T-Cell Aging

One model developed to study the effect of chronic stress and aging is the in vitro culture of T-cell clones (TCC). In vivo, T-cell clones are subjected to chronic stresses and may reach a state of clonal exhaustion after a certain number of cell divisions. This process can be modelled in vitro by long-term cultures of cells which are intermittently exposed to antigen and provided with growth factors. Such T-cell cultures can be studied longitudinally for changes occurring during their finite lifespan.[36] A variety of parameters has been examined in this model including cytokine production, loss of CD28 expression and decreased telomere lengths, enhanced susceptibility to apoptosis, increased DNA damage, different gene expression profiles and proteomic patterns. We studied the signal transduction in TCC derived from various individuals of different ages.[37] We also compared these clones during their lifespan (same clone that underwent different rounds of cell division). Alterations in the signalling could be demonstrated in TCC derived from aged individuals compared to young individuals. However, TCC derived from centenarians display a higher responsiveness than TCC derived from less elderly donors. These alterations are similar to those observed in freshly isolated T-cells obtained from elderly individuals included in the healthy population defined by the SENIEUR protocol. In TCC, CD28 expression is very low. However as described before, the location of receptor is more important than its expression levels. We hypothesized that although the total amount of CD28 was very low, there remained enough localised specifically to the membrane rafts to allow proper signalling and T-cell function. We tested this hypothesis and were able to see significant differences in the phosphorylation of Akt, which is the downstream molecule involved in CD28 signalling. Again, TCC display a similar pattern when compared to peripheral T-cells from young, elderly and centenarians.

Oxidative Stress Increases and Signalling Changes with Aging

As discussed above, impaired signal transduction can be at least partly explained by the quantitative changes in membrane composition leading to altered membrane raft function. One additional explanation for the functional changes of membrane rafts invokes qualitative changes of lipids and proteins. One of the most important experimentally sustained theories of aging is the "free radical theory" put forward by Harman in 1956.[38] Briefly, the quantity of free radicals produced because of the leakage of the mitochondria is increased in aging, while the antioxidant defence is decreased. In addition to serving as a source of ROS, mitochondria are themselves targeted by ROS, leading to further interference with their function. Free radicals have several roles in signalling. A beneficial role for free radicals exists because they are needed for the nuclear translocation of the NF-κB and AP-1.[39] However, unbalanced ROS production either by the T-cells themselves as a result of chronic stimulation by antigens or by the presence at the inflammation sites of myelo-phagocytic cells can have deleterious effects on T-cell signalling.[40,41] It was shown that oxidative stress alone is able to decrease the number of CD28 molecules on T-cells, as well as altering Lck signalling.[42] A similar situation was found for telomere shortening, which results not only from successive cell doublings, but also as a result of damage induced by ROS. Moreover, free radicals attack the macromolecular components of the cell and cell membrane leading to structural and functional alterations of these molecules. As mentioned above, membrane rafts play an important role in T-cell signalling and any change in their composition, function and size might have deleterious effects on T-cell function. Furthermore, a conformational change of proteins could be induced by free radicals. Identical assaults can occur on the lipids e.g., cholesterol, rendering the membrane more rigid. Oxidative stress can also influence posttranslational modification of proteins and thus modulate their localization and interaction. It is hypothesized that because of the different requirement for membrane rafts by CD4 and CD8 cells, as well as their differential susceptibility to apoptosis, $CD4^+$ and $CD8^+$ T-cells are differentially susceptible to free radicals. This idea implies that $CD4^+$ T-cells should be more susceptible than CD8+ T-cells to oxidative stress. Recently, Kim and Nel[43] have shown that memory T-cells from old mice are more resistant to oxidative stress, mitochondrial dysfunction and apoptosis than naïve T-cells because they express more NF-E2-related factor-2 leading to increased glutathione levels via increased Phase II antioxidant enzymes. Because the number of memory T-cells is greatly increased in the CD8+ T-cell compartment it is not surprising that they resist oxidative stress better and survive longer. Thus, age-associated increased oxidative stress could be an important factor contributing directly and indirectly to the altered T-cell activation seen with aging. However, potential therapeutic effects of the various antioxidants is still questionable perhaps because we lack adequate experimental systems to test their role, including time course, concentration and combination.

Role of the Nutrition: Metabolic Syndrome and T-Cells

It is well accepted that both malnutrition and overnutrition (obesity) are detrimental to the immune response.[44] Protein-energy malnutrition has been shown to mimic some of the effects of immunosenescence.[45,46] Moreover, nutritional supplementation of those malnourished older individuals could restore some of the age-related alterations. Another threat recently encountered by the ever growing elderly population is the epidemic of "metabolic syndrome" related to obesity via insulin resistance.[47] These elderly individuals are in a constant inflammatory condition, as shown by macrophage infiltration into the adipose tissue and adipocytes secreting pro-inflammatory cytokines including IL-6 and TNFα or adipokines (leptin, adiponectin) which can modulate the immune response.[48-50] They also secrete free fatty acids modulating insulin receptor signalling. These molecules and lipids induce endoplasmic reticulum stress (ER stress) leading to the production of more cytokines via NF-κB translocation to the nucleus.[44] These effects are well known in muscle cells and in the liver, but no data exist for T-cells. Nevertheless, chronic nutritional stimulation of T-cells by a hostile, pro-inflammatory environment can induce membrane raft changes and lead to the alteration of T-cell activation.

Stulnig et al have shown that the addition of polyunsaturated fatty acids to T-cell cultures in vitro led to modifications of membrane rafts, in particular displacement of LAT.[51,52] This has direct effects on T-cell receptor signalling. Therefore, we have here a real possibility to modulate T-cell and B-cell immune functions via their receptors by nutritional supplementation or by changes in food intake. Using this model, in our own study,[53] healthy young donors were supplemented intravenously for 2 hours with a mixture of lipids (Intralipid 20%) which contains mainly palmitic, oleic and linoleic acids. Blood samples were collected before and after injection and T-cells were isolated for further analysis. This study demonstrated that increases in lipid plasma levels have a direct effect on T-cell functions including signalling following TCR stimulation, IL-2 production and cell proliferation. This is of particular interest when we consider that this lipid supplementation is often given to hospitalized patients. One should reconsider the balance between beneficial and side effects of this supplementation in the case of immuno-depressed patients. It is known that elderly individuals have a very different nutritional intake than young people. Increased lifespan is due to better health services, vaccination and a better quality of life which includes food intake. Nevertheless, this can be improved even more because elderly individuals often have disturbed eating patterns which may not provide optimal nutrition.

Moreover nutrition can also impact on the posttranslational modifications of the signalling proteins which are important for their association to membrane rafts. Most of these modifications involve lipidation such as acetylation, myristoylation, farnesylation, geranyl-geranylation. In this context, Hundt et al[33] have recently shown that LAT palmitoylation was defective in anergized CD4+ T-cells. This can explain its altered association with membrane rafts and with the central supra-molecular activation cluster (c-SMAC) of the immune synapse.

It was demonstrated that in *C. elegans* and in *Drosophila,* nutrition-sensitive IGF-1/Insulin receptor signalling is important for longevity.[54-56] Disruption of this pathway prolongs the life span of these animals. Moreover, the beneficial effect of caloric restriction has been attributed to the modulation of the metabolic effect of the IGF-1/insulin pathway.[57] Thus, nutrition-modulated cellular metabolic pathways could play a role in the maintenance of an adequate immune response with aging.

Role of the Metabolic Pathways in T-Cells

Not only extrinsic nutrition but also intrinsic metabolism plays a fundamental role in T-cell activation. There are very few studies devoted to this aspect of T-cell activation thus far. However, it is known that CD28 is a key receptor controlling underlying metabolic needs for T-cell responses, in addition to its role in immune synapse formation. CD28 acts through the modulation of glucose metabolism via the PI3K/Akt pathway, leading to GSK3 stimulation, as well as via the modulation of protein synthesis by mTOR and Pim1/2 molecules.[58,59] As yet, there are no data demonstrating that CD28 is intervening in the lipid metabolism of T-cells, by controlling e.g., the HMG-CoA reductase but one should consider its putative role in this pathway. Nevertheless, the coordinated modulation of metabolism by CD28 would probably not be sufficient to provide all the metabolic needs of T-cells for clonal expansion. In this connection, 48 to 72 hours after stimulation, T-cells start to express the insulin receptor and which renders them insulin-sensitive and enables them to take up glucose via the GLUT insulin-induced transporter.[60,61] No data are available so far as to whether aging has an effect on this phenomenon. Our own data seem to suggest that the insulin receptor number expressed after 72 hours on T-cells is not changied with aging.

Finally, CD28 also contributes via PI3K and mTOR activation to cyclin-E up-regulation, which is essential for T-cells to transit from G_0 to S phase. The TCR and CD28 together contribute to cell cycle progression by the activation of these two pathways. Thus, an alteration with aging in this specific signalling pathway can also alter T-cell activation already at an early stage. No data are actually available yet concerning this pathway in aging.

Conclusion

One of the most striking aspects of the deleterious age-associated alterations of the immune response collectively designated immunosenescence is altered T-cell activation, the causes of which are not completely elucidated. We can document many changes in molecular events with aging, but we are not yet able to explain all these changes. Recent studies have shed some light on the role of altered TCR, CD28 and IL-2R signal-transduction. The final outcome of protein rafting is the formation of the immunological synapse which is needed for sustained activation resulting in a complete immune response. The ultimate defect in signalling can be explained by the newly discovered alterations in composition, function and size of membrane rafts with aging. These functional and physicochemical properties are influenced by intrinsic as well as extrinsic factors. Understanding the events that lead to changes in the TCR signalling cascade would be of great benefit considering the large number of diseases in which membrane raft dysfunction is thought to play a role.

Acknowledgements

This work was partly supported by a grant-in aid from the Canadian Institute of Health Research (No 63149), ImAginE (EU contract QLK6-CT-1999-02031), ZINCAGE project (EU contract n. FOOD-CT-2003-506850), T-cells and Aging "T-CIA" (QLK6-CT-2002-02283) and the Deutsche Forschungsgemeinschaft (SFB 685-B4).

References

1. Pawelec G, Solana R. Immunosenescence. Trends Immunol 1997; 11:514-516.
2. Miller RA. The aging immune system: primers and prospectus. Science 1996; 273:70-74.
3. Fulop T, Larbi A, Douziech N et al. Signal transduction and functional changes in neutrophils with aging. Aging Cell 2004; 3:217-226.
4. Grubeck-Loebenstein B, Wick G. The aging of the immune system. Adv Immunol 2002; 80:243-284.
5. Fulop T, Larbi A, Wikby A et al. Dysregulation of T-cell function in the elderly: scientific basis and clinical implications. Drugs Aging 2005; 22:589-603.
6. Fagnoni FF, Vescovini R, Passeri G et al. Shortage of circulating naive CD8(+) T-cells provides new insights on immunodeficiency in aging. Blood 2000; 95:2860-2868.

7. Effros RB. Replicative senescence of CD8 T-cells: effect of human aging. Exp Gerontol 2004; 39:517-524.
8. Pawelec G, Akbar A, Caruso C et al. Is immunosenescence infectious? Trends Immunol 2004; 25:406-410.
9. Pawelec G, Hirokawa K, Fulop T. T-cell signalling with aging. Mech Age Dev 2001; 122:1613-1637.
10. Larbi A, Muti E, Giacconi R et al. Role of lipid rafts in activation-induced cell death: the fas pathway in aging. Adv Exp Med Biol 2006; 584:137-55.
11. Gupta S. Molecular and biochemical pathways of apoptosis in lymphocytes and aged humans. Vaccine 2000; 18:1596-1601.
12. Messaoudi I, Warner J, Nikolich-Zugich N et al. Molecular, cellular and antigen requirements for development of age-associated T-cell clonal expansions in vivo. J Immunol 2006; 173:301-308.
13. Douziech N, Serers I, Larbi A et al. Modulation of human lymphocyte proliferative response with aging. Exp Gerontol 2002; 37:369-87.
14. Brzezinska A, Magalska A, Szybinska A et al. Proliferation and apoptosis of human CD8(+)CD28(+) and CD8(+)CD28(-) lymphocytes during aging. Exp Gerontol 2004; 39:539-44.
15. Nel AE, Slaughter N. T-cell activation through the antigen receptor. Part 2: role of signaling cascades in T-cell differentiation, anergy, immune senescence and development of immunotherapy. J Allergy Clin Immunol 2002; 109:901-915.
16. Thomas RM, Gao L, Wells AD. Signals from CD28 induce stable epigenetic modification of the IL-2 promoter. J Immunol 2005; 174:4639-46.
17. Thewisen M, Linsen L, Somers V et al. Premature immunosenescence in rheumatoid arthritis and multiple sclerosis patients. Ann N Y Acad Sci 2005; 1051:255-62.
18. Ouyang Q, Wagner WM, Voehringer D et al. Age-associated accumulation of CMV-specific $CD8^+$ T-cells expressing the killer cell lectine-like receptor G1 (KLRG-1). Exp Gerontol 2003; 38:911-920.
19. De Martinis M, Franceschi C, Monti D et al. Inflamm-ageing and lifelong antigenic load as major determinants of ageing rate and longevity. FEBS Lett 2005; 579:2035-2039.
20. Larbi A, Dupuis G, Khalil A et al. Differential role of lipid rafts in the functions of $CD4^+$ and $CD8^+$ human T-lymphocytes with aging. Cell Signal 2006; 18:1017-30.
21. Kabouridis PS. Lipid rafts in T-cell receptor signalling. Mol Membr Biol 2006; 23:49-57.
22. Matsumoto R, Wang D, Blonska M et al. Phosphorylation of CARMA1 plays a critical role in T-Cell receptor-mediated NF-kappaB activation. Immunity 2005; 23:575-85.
23. Fulop T, Larbi A, Douziech N et al. Cytokine receptor signalling and aging. Mech Ageing Dev 2006; 127:526-37.
24. Rivnay B, Bergman S, Shinitzky M et al. Correlations between membrane viscosity, serum cholesterol, lymphocyte activation and aging in man. Mech Ageing Dev 1980; 12:119-26.
25. Gombos I, Kiss E, Detre C et al. Cholesterol and sphingolipids as lipid organizers of the immune cells' plasma membrane: their impact on the functions of MHC molecules, effector T-lymphocytes and T-cell death. Immunol Lett 2006; 104:59-69.
26. Laude AJ, Prior IA. Plasma membrane microdomains: organization, function and trafficking. Mol Membr Biol 2004; 21:193-205.
27. Simons K, Ikonen E. Functional rafts in cell membranes. Nature 1997; 387:569-72.
28. Kabouridis PS. Lipid rafts in T-cell receptor signalling. Mol Membr Biol 2006; 23:49-57.
29. Manes S, Viola A. Lipid rafts in lymphocyte activation and migration. Mol Membr Biol 2006; 23:59-69.
30. Douglass AD, Vale RD. Single-molecule microscopy reveals plasma membrane microdomains created by protein-protein networks that exclude or trap signaling molecules in T-cells. Cell 2005; 121:937-50.
31. Cemerski S, Shaw A. Immune synapses in T-cell activation. Curr Opin Immunol 2006; 18:298-304.
32. Larbi A, Douziech N, Dupuis G et al. Age-associated alterations in the recruitment of signal transduction proteins to lipid rafts in human T-lymphocytes. J Leuk Biol 2004; 75:373-381.
33. Hundt M, Tabata H, Jeon MS et al. Impaired activation and localization of LAT in anergic T-cells as a consequence of a selective palmitoylation defect. Immunity 2006; 24:513-22.
34. Hermiston ML, Xu Z, Majeti R et al. Reciprocal regulation of lymphocyte activation by tyrosine kinases and phosphatases. J Clin Invest 2002; 109:9-14.
35. Fortin CF, Larbi A, Lesur O et al. Impairment of SHP-1 down-regulation in the lipid rafts of human neutrophils under GM-CSF stimulation contributes to their age-related, altered functions. J Leukoc Biol 2006; 79:1061-72.
36. Pawelec G, Rehbein A, Haehnel K et al. Human T-cell clones in long-term culture as a model of immunosenescence. Immunol Rev 1997; 160:31-42.
37. Pawelec G, Mariaini M, Barnett R et al. Human T-cell clones in long-term culture as models for the impact of chronic antigenic stress In: Conn M, ed. Handbook of Models for human aging. In: Elsevier, 2006: 781-793.

38. Harman D. Aging: a theory based on free radical and radiation chemistry. J Gerontol 1956; 11:298-300.
39. Lee KS, Kim SR, Park SJ et al. Peroxisome proliferator activated receptor-gamma modulates reactive oxygen species generation and activation of nuclear factor-kappaB and hypoxia-inducible factor 1alpha in allergic airway disease of mice. J Allergy Clin Immunol 2006; 118:120-7.
40. Rider DA, Sinclair AJ, Young SP. Oxidative inactivation of CD45 protein tyrosine phosphatase may contribute to T-lymphocyte dysfunction in the elderly. Mech Ageing Dev 2003; 124:191-8.
41. Chakravarti B, Abraham GN. Effect of age and oxidative stress on tyrosine phosphorylation of ZAP70. Mech Ageing Dev 2002; 123:297-311.
42. Ma S, Ochi H, Cui L et al. Hydrogen peroxide induced down-regulation of CD28 expression of Jurkat cells is associated with a change of site alpha-specific nuclear factor binding activity and the activation of caspase-3. Exp Gerontol 2003; 38:1109-18.
43. Kim HJ, Nel AE. The role of phase II antioxidant enzymes in protecting memory T-cells from spontaneous apoptosis in young and old mice. The journal of immunology 2005; 175:(5)2948-2959.
44. Wellen KE, Hotamisligil GS. Inflammation, stress and diabetes. J Clin Invest 2005; 115:1111-1119.
45. Lesourd BM. Nutrition and immunity in the elderly: modification of immune responses with nutritional treatments. Am J Clin Nutr 1997; 66:478S-484S.
46. Lesourd BM. Immune response during disease and recovery in the elderly. Proc Nutr Soc 1999; 58:85-98.
47. Fulop T, Tessier D, Carpentier A. The metabolic syndrome. Pathol Biol 2006 (in press).
48. Hotamisligil GS, Arner P, Caro JF et al. Increased adipose tissue expression of tumor necrosis factor-alpha in human obesity and insulin resistance. J Clin Invest 1995; 95:2409-15.
49. Sonnenberg GE, Krakower GR, Kissebah AH. A novel pathway to the manifestations of metabolic syndrome. Obes Res 2004; 12:180-6.
50. Fantuzzi G. Adipose tissue, adipokines and inflammation. J Allergy Clin Immunol 2005; 115:911-9.
51. Zeyda M, Staffler G, Horejsi V et al. LAT displacement from lipid rafts as a molecular mechanism for the inhibition of T-cell signaling by polyunsaturated fatty acids. J Biol Chem 2002; 277:28418-23.
52. Stulnig TM, Berger M, Sigmund T et al. Polyunsaturated fatty acids inhibit T-cell signal transduction by modification of detergent-insoluble membrane domains. J Cell Biol 1998; 143:637-44.
53. Larbi A, Grenier A, Frisch F et al. Acute in vivo elevation of intravascular triacylglycerol lipolysis impairs peripheral T-cell activation in humans. Am J Clin Nutr 2005; 82:949-56.
54. Rincon M, Rudin E, Barzilai N. The insulin/IGF-1 signaling in mammals and its relevance to human longevity. Exp Gerontol 2005; 40:873-877.
55. Laron Z. Do deficiencies in growth hormone and insulin-like growth factor-1 (IGF-1) shorten of prolong longevity? Mech Ageing Dev 2005; 126:305-307.
56. Franceschi C, Olivieri F, Marchegiani F et al. Genes involved in immune response/inflammation, IGF1/insulin pathway and response to oxidative stress play a major role in the genetics of human longevity: the lesson of centenarians 2005; 126:351-361.
57. Bartke A. Long-lived Klotho mice: new insights into the roles of IGF-1 and insulin in aging. Trends Endocrinol Metab 2006; 17:33-5.
58. Diehn M, Alizadeh AA, Rando OJ et al. Genomic expression programs and the integration of the CD28 costimulatory signal in T-cell activation. Proc Natl Acad Sci USA 2002; 99:11796-801.
59. Bonnevier JL, Yarke CA, Mueller DL. Sustained B7/CD28 interactions and resultant phosphatidylinositol 3-kinase activity maintain G1-->S phase transitions at an optimal rate. Eur J Immunol 2006; 36:1583-97.
60. Stentz FB, Kitabchi AE. Hyperglycemia-induced activation of human T-lymphocytes with de novo emergence of insulin receptors and generation of reactive oxygen species. Biochem Biophys Res Commun 2005; 335:491-5.
61. Stentz FB, Kitabchi AE. De novo emergence of growth factor receptors in activated human $CD4^+$ and $CD8^+$ T-lymphocytes. Metabolism 2004; 53:117-22.

Chapter 7

Immunosenescence, Thymic Involution and Autoimmunity

Wayne A. Mitchell* and Richard Aspinall

Abstract

In recent years life expectancy in Western Societies has dramatically increased with greater numbers of individuals living longer; consequently the prevalence of age-associated diseases such as infections, cancers and autoimmune disease increases. A striking feature of the ageing process is the involution of the thymus. This primary lymphoid organ is instrumental in generating naive T-cell required to successfully defeat against 'foreign' and 'self' antigens. Much effort has been made to find means of reversing the effects of ageing and a variety of factors have been investigated in a quest to maintain a youthful immune system. In this review we examine some features of immunosenescence and the work undertaken, with particular interest to the role of the cytokine interleukin-7, to further our understanding of the relationship between ageing and development of autoimmunity.

Introduction

With advancing age the occurrence of physiological, cellular and functional changes has the effect of altering the health and well-being of an individual. For several years immunologist have actively studied the mechanisms that underpin these changes, within the ageing immune system, in an attempt to improve the physical welfare of the rapidly expanding elderly population. Ageing is characterised by a decline in the ability of the individual to adapt to environmental stress. This continuous and slow process compromises the normal functioning of various organs, apparatuses and systems in both qualitative and quantitative terms and also alters morphological aspects. It means that senescence is not represented by a pre-established moment, but consists of slow and long-lasting preparation of the organism for a morpho-functional involution which in itself is part of the normal biological cycle.[1] Senescence of the immune system, also referred to as 'immunosenescence', describes the dysregulation of the immune function related to the ageing process which in turn contributes to the increased susceptibility to infection, cancer and autoimmune diseases.[2] The aim of this brief review is to examine contribution of immunosenescence in the context of autoimmune disease.

Common Ageing Signature

The concept of ageing and longevity is generally considered in the context of advancing chronological years where different milestones are normally associated with specific timepoints e.g., onset of puberty is usually between 10-13 years, middle age is thought to begin in the early to mid 40's and so on. Succinctly, aging is characterized by changes in appearance, such as a gradual reduction in height and weight loss due to loss of muscle and bone mass, a lower metabolic rate, lower reaction

*Corresponding Author: Wayne A. Mitchell—Imperial College of Science, Technology and Medicine, Faculty of Investigative Sciences, Department of Immunology, Chelsea and Westminster Campus, 369 Fulham Road, London, SW10 9NH, U.K.
Email: w.mitchell@imperial.ac.uk

Immunosenescence, edited by Graham Pawelec.

times, declines in certain memory functions, declines in sexual activity and, in women, menopause, a functional decline in audition, olfaction and vision, declines in kidney, pulmonary and immune functions, declines in exercise performance and multiple endocrine changes.[3] Recently it has been shown that cellular age as defined by genetic profiling may provide a better indication of the ageing process within an individual as opposed to chronological age. This may have implication on the ability of these individual to combat infections and disease and furthermore suggest that the rate that an individuals ages is independent of time. Zahn and colleagues demonstrate that a common gene expression profile exist which correlates not only with chronological but physiological age. By the examination of muscle cells, the differential expression of genes within discrete genetic pathway were found to display age-related changes, these pathways included extracellular matrix, cell growth, complement activation, cytosolic ribosome, chloride transport and mitochondrial electron transport chain.[4] Interestingly, alterations in the mitochondrial electron transports chain were also found to correlate with ageing in mice and flies suggesting a common aging signature.

Immunosenescence

In somatic cells, the process of replicative senescence has been suggested to act as a 'tumour suppressive mechanism' with the aim of preventing cells from acquiring multiple mutations that are needed for malignant transformation. It entails the irreversible arrest of cell proliferation and altered cell function.[5] As a matter of fact, one of the features of cancerous cells is their ability to become 'immortalised' which allows these cells to continually proliferate. Hayflick originally demonstrated the existence of a replicative limit for somatic cell, known as Hayflick limit. As cells approach their replicative limit alteration in their functions was observed, these changes are dependent on number of cell divisions and not time as suggested above.[5] Immunosenescence therefore reflects the senescent changes associated with the cellular components of the immune systems. Cell acquire three characteristics associated with senescence, these being; (1) growth arrest, with cells unable to enter the S-phase of the cell cycle while remaining metabolically active; (2) altered function, where cells resemble terminally differentiated cells and (3) resistance to apoptotic cell death.[5] Summary of cellular features associated with immunosenescence is found in Table 1. An important factor for

Table 1. Cellular features associated with Immunosenescence.

Common Features of Immunosenescence
Thymus Involution with alterations in T-cell subsets
Decline of Naïve T-cell rates
Proliferation and expansion of memory T-cells component
Decrease IL-2 production, IL-2 receptor expression and poor T-cell response to IL-2
Decrease in IL-4 production by Th_2-cells
Loss of costimulatory signal CD28
Increased telomerase activity in $CD28^-$ T-cells
Short telomere length
Diminished B-cell function
Lower levels of TREC containing naïve T-cells
Increased antiapoptotic function and resistance to apoptosis
Increase in immature T-cells
Increased likelihood of mutations
Decrease in ratio of T-helper/T-suppressor cells
Lower primary and secondary responses to immunization in B-cells
Decreased immunoglobulin production

Adapted from Prelog.

consideration is the functional role played by the thymus and its influence on the development of immunosenescence.

Age and Thymic Output

The thymus is required for the development and maturation of thymocytes and for the generation of a diverse T-cell repertoire necessary to protect the host against 'foreign' and 'self' antigen. It is a primary lymphoid organ located in the anterior mediastinum and produces T-cells throughout life although the number of T-cells it produces declines with age. In a young healthy adult (less than 30 years old) there are approximately 2×10^{11} T-cells of which 1-2% can be found within the blood and up to 50% of these cells are contained within the "antigen naïve" population. The decline in T-cell production by the thymus with age is associated with thymic atrophy and is considered to be a forerunner to changes in the immune system where ageing is linked with immune decline and the onset of immune dysfunction. In mice the reduction in T-cell production corresponds to a diminution in thymic size[6] whereas in humans the thymus remains relatively stable in terms of size but its fat content increases as sites of thymopoiesis decrease.[7] A summary of the proposed causes of thymic involution can be found in Table 2.

Whilst the rate of export of T-cells from the thymus drops with age any reduction in T-cell numbers in the body is prevented by the proliferation of constituent members of the peripheral T-cell pool which keeps the total number of cells in the pool within closely defined limits. Because of the restrictions placed on the activation of naive T-cells, this proliferation is most likely to be of memory T-cells accounting for the increase in the memory T-cells which accompanies ageing.[8] This will change the naïve to memory ratio and lead to a rather restricted antigen repertoire. In addition if the replication of the memory T-cells is not of sufficient high fidelity, there will be an accumulation of defects within these cells.[9] Several experimental studies have shown that ageing is

Table 2. Proposed causes of thymic involution. Taken from Mitchell[87]

Proposed Causes of Thymic Involution	References
Inhibition of TCR receptor rearrangements	6
Loss of self-peptides on thymic epithelial MHC	88
Ageing of thymic stroma with loss of trophic cytokines	89-91
Ageing of stem cell population	92,93
Increased expression of certain cytokines as age increases	26,94
• Leukaemia inhibitor factor (LIF)	
• Oncostatin M (OSM)	
• IL-6	
• Stem cell factor	
• Macrophage colony stimulating factor	
Corticosteroid suppression of thymus	95
Action of pituitary ACTH production which drives adrenal corticosteroid production	96
Involution protects against autoimmune disease and cancer	97
Wear and tear model	
Atrophy is due to a loss of the thymic microenviroment function	98,89,90,
The thymus as an energy expensive organ is allowed to involute after a full repertoire has been established in order that energy can be saved and invested for reproduction in germ line cells	91
Production of sex hormones	99-102

Taken from Mitchell WA, Meng I, Nicholson SA, et al. Thymic output, ageing and zinc. Biogerontology. Oct 2006; 7(5-6):461-470.

associated with an increase in the number of cells showing defects in signalling pathways[10] cell cycle progression[11] cytokine production[12] and the expression of key cell surface molecules.[13] Another consequence of this replication is that like other somatic cells T-cells have a limited proliferative lifespan[14] which is a problem when proliferation is at the core of a successful immune response.[15] This combination of narrowed repertoire, increased incidence of defects and reduced replicative ability readily accounts for the increased incidence of infections with common pathogens amongst the elderly but this age associated immunodegeneration has also been proposed to contribute to the development of autoimmune diseases such as rheumatoid arthritis.[16]

T-Cell Development and Thymic Selection and Gender

Intrathymic T-cell development involves an ordered sequence of events involving expansion differentiation and selection of populations of thymocytes. Production of a mature peripheral $\alpha\beta^+$T-cell from the thymus is the final step in the differentiation process which in adults originates with multipotential stem cells in the bone marrow. These seed the thymus and their commitment to the T-cell lineage is accompanied by the rearrangement of the beta chain of the T-cell receptor (TCR) and its expression with a pre-α chain. IL-7, produced by the MHC Class II$^+$ thymic epithelium[17] cells has been suggested as a cofactor in the process of TCRβ chain rearrangement[18] in addition to its role in permitting the survival of the cells undergoing the rearrangement and selection processes.[19,20] A productive pre and TCR coupling at this stage allows the cell to be positively selected and the resulting population expresses both CD4 and CD8 molecules together at the cell surface.[21] During this phase the TCR chain undergoes rearrangement and expression. This includes the excision of the TCR locus from between the flanking rec and J genes located within the TCR locus which has formed the basis for a quantitative assay to analyse thymic output under different pathological conditions in the human[22-26] and recently in the mouse.[27]

In any population of newly produced thymocytes there may be those bearing receptors with an ability to recognise self peptides with a high avidity. Thymocytes bearing such receptors have been shown to be eliminated within the thymus by a selection process termed clonal deletion.[28] Any which slip through this selection process are held in check by mechanisms in the periphery including regulatory T-cells which prevent their activation.[29] These regulatory cells are CD4$^+$CD25$^+$ and represent a unique population crucial for the prevention of autoimmunity. These cells are generated in the thymus, where their production is dependant on MHC Class II$^+$ thymic cortical epithelial cells.[30] Previous work has shown that a small percentage of self reactive T-cells are released from the thymus in early life[31] but this population contains a number of antigen specific regulatory CD4$^+$CD25$^+$ T-cells.[28]

The MHC Class II$^+$ epithelial cells thus have a central role in the process of T-cell development, taking part in the selection process in addition to producing essential cytokines such as IL-7. However their function may be compromised by age. We have previously shown that the greatest difference in thymic output between men and women occurs between the ages of 40 and 60 with females showing a higher thymic output during this period than males.[32] Similarly during the mid-life period in mice, the rate of thymic atrophy is faster in males than in females.[33]

Thymic Rejuvenation

Several physiological and pathological factors are known to interfere with the normal function of the thymus which in turn causes the thymus to experience atrophy, these include; infection, disease, ageing, pregnancy, puberty, physical and emotional stress, environmental conditions, alterations in hormonal and cytokine levels as well as deficiency of nutritional factors such as Zinc.[34] Unlike age-related thymic atrophy many of the factors mentioned are associated with transient or reversible atrophy. This may indicated the extent to which factors within the thymic microenvironment influence the regulation of cellular immunity. Where physiological resources become limited, for example in the case of Zinc deficiency, the immune system may prioritise first line defence function above more luxurious functions i.e., increasing the T-cell repertoire.[35,36] Therefore increasing the likelihood of thymic atrophy unless additional signals are received which prevent this process.

Considering the overwhelming impact of thymic involution on the immune system it is reasonable to hypothesize that if ways can be found to rejuvenated the thymus, thereby increasing its overall function, it may be possible to prevent many of the delirious effects associated with ageing. Potential factors have been reported to prevent or reverse the thymic atrophy, these include; (1) the action of interleukin 7 (IL-7);[20,37-41] (2) administration of dietary supplements such as Zinc [35,42-44] and herbal remedies Ginkgo biloba leaf extract EGb 761[45] and Melatonin;[46] and (3) the activity of steroidal hormones. The ability to rejuvenate thymic output is not only beneficial in the context of ageing but also to individuals requiring reconstitution of their T-cell repertoire due to infections (HIV) or following medical intervention (cancer therapies). The major findings of some of the factors currently being study for thymic regeneration will be discussed prior to looking at the potential implication these may have in the field of autoimmune diseases.

Methods of Thymic Regenerations

IL-7

The cytokine IL-7 has been shown to have a key role in normal T-cell development,[20,47] it is produced in the thymus and bone marrow where normal T-cell precursors develop and studies suggest that the level of IL-7 production may be a critical modulator of T-cell development. Initial studies by Bhatia et al[47] on young mice treated with anti-IL-7 showed that severe thymic atrophy occurred with greater than 99% decrease in thymic cellularity after prolong administration. The similarity between the atrophy seen following treatment with antibodies to IL-7 and that seen in ageing prompted an analysis of IL-7 expression with age in the thymic stromal cells. In the mouse MHC Class II$^+$ epithelial cells have been shown to be the site of IL-7 synthesis within the thymus.[17] Using quantitative PCR one study has shown that IL-7 levels decreased 15-fold by 22 months of age within the thymus, but that keratin-8, a molecule whose expression is associated primarily with cortical epithelial cells only showed a 6-fold decline by 22 months of age.[48] These results echoed an earlier study[38] which showed that the age-associated decline in intrathymic expression of IL-7 was not matched by a similar decline in expression of connexin 43 a molecule associated with gap junction formation in thymic epithelial cells.[49] In situations where IL-7 production is absent or reduced thymic atrophy is induced, resulting in normal levels of DN1 population but a reduction in all other developmental stages. This effect is reversed with the addition of IL-7. Conversely, where IL-7 is expressed at excessive levels a similar bottleneck at the DN1-DN2 developmental stages occurs. Work undertaken by Aspinall et al on aged mice has shown that stimulation by IL-7 can reverse age-related atrophy of the thymus, leading to a restoration of thymic output.[20,39] Phillips[40] and Virts[41] have demonstrated that intrathymic injection of IL-7 secreting S17 cells was also capable of preserving high levels of DN2-DN3 thymocytes in old age compared to age-matched controls with an additional observed increase in the expression of bcl-2 levels.[40,41] These authors also suggest that despite these findings the thymic involution was not diminished with age. One striking features associated with the lack of IL-7 production is the reduced thymopoiesis and export into the periphery. These events may fuel additional complications within the T-cell pool due to the disproportional relationship between the naïve T-cells and memory T-cell fraction. Taken together these results may have implication on the methods used for regenerating the thymic or suggest that additional factors may be required to truly reverse age-related these changes. It may also reflect a deeper level of complexity in the development of thymocytes than simply replacement of a single factor.

Dietary Supplements: Zinc

Several dietary supplements have been suggested as potential boosters for the immune system. Zinc deficiency has been identified in a number of disorders the most notable including sickle cell anaemia and acrodermatitis enteropathica. Individuals suffering from Acrodermatitis enteropathica, an autosomal recessive disease caused by a defect in zinc metabolism, experience thymic atrophy and impaired cell-mediated immunity resulting in increased susceptibility to infection and disease.[50] These symptoms are effectively corrected by supplementation with zinc.

There are several interesting factors associated with Zinc which warrant further investigation to elucidate its contribution to cellular immunity. Firstly, a hallmark of zinc deficiency in animal models is the development of age-independent thymic atrophy.[51] Secondly, individuals with zinc deficiency are known to suffer from increase susceptibility to infection and disease indicative of poor immune function. Third with increasing age there is a decreased ability to absorb Zinc in the gut therefore increasing the likely of individuals become deficient of Zinc.[36] Fourth, studies in aged mice have shown that drinking water supplemented with zinc sulphate can increase thymic mass and possibly thymopoiesis.[36] Fifth, Zn deficiency has been noted as a secondary disorder in disease such as diabetes, AIDS, Down's syndrome and select cancers.[52] Sixth, Zinc supplementation has been shown to increase thymulin secretions in aged mice[42] and human[53] suggesting a beneficial role for thymic function. Collectively these factors provide compelling reasons for investigating the potential impact to be made by Zinc on the immune system of free living old people.

Dietary Supplements: Ginkgo Biloba Leaf Extract EGb 761

Ginko biloba leaves have been used as part of traditional Chinese medicine for several thousand years. EGb761 is the complex chemical mixture extracted from the Ginko biloba leaf and has been shown to have protective and rescue effects on a variety of medical conditions including neurodegenerative disorders,[54] cardiovascular disease[55] and ageing.[56] The functional properties of EGb761 have been attributed to its antioxidant and free radical scavenging activities. Tian and colleagues demonstrated that administration of EGb761 both in vitro and in vivo was capable of protecting thymocytes against the reactive oxygen species. Oral dosage of EGb761 was given for 60 days at 1600 μg/day/mouse to 22 month old C57BL animals. After this time, the mice were sacrifices and the size of their thymus and spleen were assessed. It was found their organs had significantly increased in mass compared to age-matched controls. These mice were also observed to have significant responsiveness to mitogens.[45] Similar results were obtained when investigating the effects of melatonin which is also known to act on reactive oxygen species.[46] This suggests that compounds with anti-oxidant properties may also be important for the rejuvenation of the thymus.

Growth Hormones and Sex Steroids

It has been a long established view that alterations in the ratio of growth hormones to sex steroids are important factors in thymic atrophy. The presence of increasing levels of sex steroids, marking the onset of puberty, has been linked with thymic atrophy.[57,58] When chemical or surgical castration is performed on aged animal, regeneration of the thymus is observed. These effects can be reversed by the administration of synthetic sex steroids.[59-63] Sex steroids act on early thymocyte differentiation, specifically blocking the triple negative stages 1 to 2 (TN1 to TN2 stage).[6,64,65] Progression through the TN development stages is IL-7 dependent and therefore suggests that the castration effects may be mediated by IL-7.[65] A recent report by Min and colleagues,[66] investigated the validity of the hypothesis that low levels of growth hormones (GH) and high sex steroid production accelerate thymic involution. The authors used mice with mutations in the genes encoding for the growth-hormone-releasing factor receptor or gonadotropin-releasing hormone, which leads to a reduction of GH and diminished sex steroid production.[66] The results indicated that changes in the production of GH or sex steroids were not required to initiate or sustain thymic involution. In addition the blocking of sex steroid production did not delay thymic involution. These results are contrary to the finding of other groups which have shown increase in thymic cellularity following castration. It is suggested by Min that these cellular effects are transient and that the thymus still undergoes involution.

These findings highlight the complexity facing those investigating the restoration/ rejuvenation of thymic function. It is unlikely that any single factor will be found capable of restoring thymic function but more conceivable that a combination of the mechanisms describe will all be required to make a functional contribution.

Autoimmunity

The etiology of autoimmune disease is multifactorial with several factors including genetic (IPEX), immunological, hormonal and environmental believed to make important contribution to the onset of these diseases. The exact mechanism or triggers for their development are still not clear.[67] Autoimmune diseases can be classified into two group these being; (1) organ-specific such as thyroid disease, type 1 diabetes and myasthenia gravis or (2) systemic diseases like rheumatoid arthritis and systemic lupus erythematosus. In severe cases, autoimmune disease can be life threatening. In recent years there has been a dramatic increase in the incidence of autoimmune diseases like type 1 diabetes in Western countries although the other disorders such as Rheumatoid Arthritis have remained constant.[68]

A variety of theories have been proposed to explain the autoimmune phenomena which have been extensively reviewed elsewhere.[69-75] The current view of autoimmunity is that the first line of defense against self reactive T-cells is clonal deletion in the thymus. Selection within the developing thymocyte population by thymic epithelial cells ensures that self reactive cells are deleted from the peripheral T-cell repertoire.[76] Any which slip through this selection process are held in check by mechanisms in the periphery including regulatory T-cells which prevent their activation.[29] If such peripheral mechanisms fail then T-cells recognise self and drive a chronic persistent immune response. This implies that the fundamental breakdown which can lead to immune disease lies in the escape of forbidden clones within the recent thymic emigrants .

Alternative theories include 'Hygiene hypothesis' by Rook and Stanford who suggest that a lack of exposure to infectious organisms may result in the suboptimal internal balance of the immune systems, leading to the increased prevalence of immune disorder.[77] Despite several studies there is little evidence to support this idea.[68,78] A final proposal for consideration is that of the 'two hit model of lymphopenia', by Krupica, Fry and Mackall,[79] who suggest that the development of autoimmunity is associated with lymphopenia, a state which renders an immunologically delicate environment that invokes vigorous cycling of self-reactive T-cells and thus provides fertile ground for the possible dysregulation of homeostatic mechanisms and loss of self tolerance. This first hit coupled to a second hit, such as cytokine overproduction, skewing of the Treg/nonTreg ratio or encounter with localised tissue inflammation is enough to overcome the immune system's checks and balances and break self-tolerance.[79]

Autoimmune Responses, Gender and Age

Analysis of the range of autoimmune diseases (see Table 3) shows two striking points; the first is that most sufferers are women and the second is that for most cases the peak time of onset is in the region of 30 to 60 years of age. The gender related skewing of the incidence of autoimmunity has been noted previously and has been put down to a difference in the environment provided

Table 3. Gender differences and age of onset of autoimmune diseases

Autoimmune Disease	Ratio ♀:♂	Age of Onset
Primary Biliary Cirrhosis	9:1	40-50
Chronic Active Hepatitis	6:1	40-60
Pernicious Anemia	3:2	>60
Rheumatoid Arthritis	3:1	
Sjorgens syndrome	9:1	50
Progressive sytemic sclerosis	2:1	30
Autoimmune hypothyroidism	10:1	40-60
Grave's disease	7:1	30
Addison's disease	1.8:1	30
Multiple sclerosis	2:1	30

for the differentiation of T-cells towards a Th1 phenotype in women[80] but this does not account for the presence of forbidden clones in the peripheral T-cell pool. However recently we showed that in humans the output from the thymus is gender dependant with females showing a higher thymic output for longer in their lifespan than males.[32] Similarly the rate of thymic atrophy in the mouse is faster in males than in females[33] If thymic selection changes with age then greater thymic output for longer in females compared with males could lead to more forbidden clones within the recent thymic emigrant pool.

With respect to the current evidence it can be hypothesized that changes in the thymic epithelial cells with age which alter their ability to perform both positive and negative selection at the same level these selection processes occur in the young. Thus at middle age when the difference in thymic output between the genders is the greatest, more self-reactive T-cell leave the thymus in females than in males. When these results are taken in conjunction with the observation from the Norfolk study that the 45-60 age group not only shows a much higher prevalence of rheumatoid arthritis than the 16-44 age group but also the greatest gender difference in disease prevalence, it suggest that it is not the age related proliferation of T-cells in the peripheral T-cell pool which contributes to the development of the disease, but a change in the properties of the cells emigrating from the thymus during this period.

In other words if the thymic selection process changes with age and becomes less discriminatory then this could lead either to more self reactive cells T-cells or fewer $CD4^{+}CD25^{+}$ regulatory T-cells emigrating from the thymus. For females, where production of T-cells by the thymus is greater than males in the mid-life period, this poorly selected population could lead to more functional self-reactive cells and a higher incidence of autoimmune disease which fits with the epidemiological studies. Increases in the numbers of self reactive T-cells means that eventually the conditions exist for them to evade the normal control systems and they become activated and undergo clonal expansion and cause overt disease. The increased incidence of autoimmune disease in females compared with males is therefore a by-product of increased thymic output and poorer selection processes.

The Role of IL-7 and Autoimmunity

To conclude this review we will examine some of the current data regarding the role of IL-7 in the regulation of autoimmunity. We have already established that IL-7 makes a major contribution to the normal process of thymopoiesis and that the reduction of IL-7 appears to be integral in the age-associated atrophy of the thymus. Numerous studies have investigated the regenerative capacity of IL-7 in an attempt to reverse the effects of immunosenscence on the thymus. Likewise recent reports have examined further the contribution made by IL-7 on the pathogenesis of autoimmune diseases. Staton et al[81] demonstrated using the epidermal cell antigen Skn, as a model of autoimmunity, that the IL-7 is critical for modulating lesion development. The group provide evidence that lesion formation is dependent on the mice being in an immunocompromised status i.e., a lymphopenic condition and the skin must have received a mild trauma. Only when mice in this condition were treated with $CD4^{+}$ T-cells via adoptive transfer was a significant decrease in the grade of lesion observed. In addition this was associated with elevation in the IL-7 levels. Further experiments showed that exogenous administration of IL-7 reduced the severity of the lesions whereas administration anti-IL-7 had the reverse effect. Harnaha et al[82] also provide supportive evidence for role of IL-7 by demonstrating that $CD4^{+}CD25^{+}$ T-cells require IL-7 as a survival factor for the activity of immunoregulatory dendritic cells in the suppression of diabetes in mice.

The current evidence from the murine models of autoimmune disease suggests that IL-7 provides a major contribution in the regulation and survival of T-cell required to protect against the development of diseases. In man, the role of IL-7 is still being investigation. Ponchel[83] reported for the first time that serum IL-7 levels were decreased in patients with RA compared to healthy controls and also osteoarthritis patients (IL-7 levels: controls > osteoarthritis > RA). Although the IL-7 levels were decreased there was no difference in its activity when tested within PBMC from these patients. Moreover, they found that the number of cells containing T-cell receptor excision circles, which represents recent thymic output was diminished. This finding was independent of gender and age. They conclude that IL-7 deficiency is likely to be an important contributing factor to the inability of RA patients to reconstitute there T-cell following lymphodepletion therapy.[83,84] In view of this evidence, alternative therapeutic strategies need to be investigated in order to prevent RA patient becoming lymphopenic as this may exacerbate the disease progress.

Administration of IL-7 in Humans

The vast majority of IL-7 studies in the literature have been conducted on murine disease models. Rosenberg et al recently published a study which examined the therapeutic effects of IL-7 administered to humans with metastatic cancer.[85] Patient where subdivided into 4 cohorts and each received a total of 8 subcutaneous injections at 3 day intervals for 21 day at a given dose of IL-7. The dosage given was 3, 10, 30 or 60 µg/kg. Increases were noted in the CD4/CD8 lymphocyte ratio at 10, 30 and 60 µg/kg, which was maintained above baseline values 7 days after the last injection was given at the highest concentration. The immunophenotype indicated an increasing trend towards a higher proportion of naïve relative to memory cell at 60 µg/kg. Analysis of the T regulatory cells as defined by $CD4^{+}CD25^{+}FoxP3$ demonstrated a decrease in expression of these cells both before and after IL-7 administration. Interestingly proportion of these cells did not express the IL-7 receptor (CD127) which may account for the non responsiveness to IL-7 therapy.[85]

Conclusions and Future Perspectives

As the demographic outlook changes to reflect the increase proportion of older individuals within the society, this promises some major challenges to the socio-economic and healthcare planning for the future. Over the past century there has been a dramatic increase in the prevalence of age-related cancers and autoimmune disease. This in turn has lead to a greater desire to find ways of reducing the age-related incidence of these disease hence enabling individuals to live longer healthier lives. The current discussion has highlighted the growing body of evidence in favour of a significant role for IL-7 in restoring and maintaining the thymic function which is vital for ensuring a fully competent immune system. The current human studies of administering IL-7 to metastatic cancers has shown some promising results and it will be of interest to undertake further studies that focus on the effects of administering IL-7 to RA patients and other autoimmune diseases.

Although influential, it is arguably a gross oversimplification to suppose that IL-7 holds the answers to address age-associated thymic changes. Several additional factors have already been described which have shown significant contributions to the makeup of the thymic microenvironment and it will be a major achievement to understand how these different factors work together to provide a healthy outlook for our old age.

Acknowledgements

Work in the authors laboratory is supported by the BBSRC (grant no. 16279) and the EU Zincage project (contract no. FOOD-CT-2003-506850).

References

1. Malaguarnera L, Ferlito L, Imbesi RM et al. Immunosenescence: a review. Archives of Gerontology and Geriatrics 2001; 32(1):1-14.
2. Pawelec G. Immunosenescence: impact in the young as well as the old? Mech Ageing Dev 1999; 108(1):1-7.
3. de Magalhaes JP. What is Ageing? senescence.info: http://www.senescence.info; 1997-2005.
4. Zahn JM, Sonu R, Vogel H et al. Transcriptional profiling of aging in human muscle reveals a common aging signature. PLoS Genet 2006; 2(7):115e.
5. Campisi J. Replicative senescence: an old lives' tale? Cell 1996; 84(4):497-500.
6. Aspinall R. Age-associated thymic atrophy in the mouse is due to a deficiency affecting rearrangement of the TCR during intrathymic T-cell development. J Immunol 1997; 158(7):3037-3045.
7. Kendall MD, Johnson HR, Singh J. The weight of the human thymus gland at necropsy. J Anat 1980; 131(Pt 3):483-497.
8. Kurashima C, Utsuyama M, Kasai M et al. The role of thymus in the aging of Th cell subpopulations and age-associated alteration of cytokine production by these cells. Int Immunol 1995; 7(1):97-104.
9. Kirkwood TL, Kapahi P, Shanley DP. Evolution, stress and longevity. J Anat 2000; 197(Pt 4):587-590.

10. Utsuyama M, Wakikawa A, Tamura T et al. Impairment of signal transduction in T-cells from old mice. Mech Ageing Dev 1997; 93(1-3):131-144.
11. Quadri RA, Arbogast A, Phelouzat MA et al. Age-associated decline in cdk1 activity delays cell cycle progression of human T-lymphocytes. J Immunol 1998; 161(10):5203-5209.
12. Hobbs MV, Weigle WO, Noonan DJ et al. Patterns of cytokine gene expression by $CD4^+$ T-cells from young and old mice. J Immunol 1993; 150(8 Pt 1):3602-3614.
13. Vallejo AN, Brandes JC, Weyand CM et al. Modulation of CD28 expression: distinct regulatory pathways during activation and replicative senescence. J Immunol 1999; 162(11):6572-6579.
14. Weng NP, Levine BL, June CH et al. Human naive and memory T-lymphocytes differ in telomeric length and replicative potential. Proc Natl Acad Sci USA 1995; 92(24):11091-11094.
15. Effros RB, Pawelec G. Replicative senescence of T-cells: does the Hayflick Limit lead to immune exhaustion? Immunol Today 1997; 18(9):450-454.
16. Weyand CM, Fulbright JW, Goronzy JJ. Immunosenescence, autoimmunity and rheumatoid arthritis. Exp Gerontol 2003; 38(8):833-841.
17. Moore NC, Anderson G, Smith CA et al. Analysis of cytokine gene expression in subpopulations of freshly isolated thymocytes and thymic stromal cells using semiquantitative polymerase chain reaction. Eur J Immunol 1993; 23(4):922-927.
18. Tsuda S, Rieke S, Hashimoto Y et al. Il-7 supports D-J but not V-DJ rearrangement of TCR-beta gene in fetal liver progenitor cells. J Immunol 1996; 156(9):3233-3242.
19. von Freeden-Jeffry U, Solvason N, Howard M et al. The earliest T lineage-committed cells depend on IL-7 for Bcl-2 expression and normal cell cycle progression. Immunity 1997; 7(1):147-154.
20. Andrew D, Aspinall R. Il-7 and not stem cell factor reverses both the increase in apoptosis and the decline in thymopoiesis seen in aged mice. J Immunol 2001; 166(3):1524-1530.
21. Fehling HJ, von Boehmer H. Early alpha beta T-cell development in the thymus of normal and genetically altered mice. Curr Opin Immunol 1997; 9(2):263-275.
22. Douek DC, McFarland RD, Keiser PH et al. Changes in thymic function with age and during the treatment of HIV infection. Nature 1998; 396(6712):690-695.
23. Fry TJ, Mackall CL. Current concepts of thymic aging. Springer Semin Immunopathol 2002; 24(1):7-22.
24. Jamieson BD, Douek DC, Killian S et al. Generation of Functional Thymocytes in the Human Adult. Immunity 1999; 10(5):569-575.
25. Markert ML, Boeck A, Hale LP et al. Transplantation of Thymus Tissue in Complete DiGeorge Syndrome. N Engl J Med 1999; 341(16):1180-1189.
26. Sempowski GD, Hale LP, Sundy JS et al. Leukemia inhibitory factor, oncostatin M, IL-6 and stem cell factor mRNA expression in human thymus increases with age and is associated with thymic atrophy. J Immunol 2000; 164(4):2180-2187.
27. Pido-Lopez J, Imami N andrew D et al. Molecular quantitation of thymic output in mice and the effect of IL-7. Eur J Immunol 2002; 32(10):2827-2836.
28. Cozzo C, Larkin J 3rd, Caton AJ. Cutting edge: self-peptides drive the peripheral expansion of $CD4^+CD25^+$ regulatory T-cells. J Immunol 2003; 171(11):5678-5682.
29. Jordan MS, Boesteanu A, Reed AJ et al. Thymic selection of $CD4^+CD25^+$ regulatory T-cells induced by an agonist self-peptide. Nat Immunol 2001; 2(4):301-306.
30. Bensinger SJ, Bandeira A, Jordan MS et al. Major histocompatibility complex class II-positive cortical epithelium mediates the selection of $CD4(^+)25(^+)$ immunoregulatory T-cells. J Exp Med 2001; 194(4):427-438.
31. Riley MP, Cerasoli DM, Jordan MS et al. Graded deletion and virus-induced activation of autoreactive $CD4^+$ T-cells. J Immunol 2000; 165(9):4870-4876.
32. Pido-Lopez J, Imami N, Aspinall R. Both age and gender affect thymic output: more recent thymic migrants in females than males as they age. Clin Exp Immunol 2001; 125(3):409-413.
33. Aspinall R, Andrew D. Gender-related differences in the rates of age associated thymic atrophy. Dev Immunol 2001; 8(2):95-106.
34. Taub DD, Longo DL. Insights into thymic aging and regeneration. Immunol Rev 2005; 205:72-93.
35. Fraker PJ, King LE, Laakko T et al. The dynamic link between the integrity of the immune system and zinc status. J Nutr 2000; 130(5S Suppl):1399S-1406S.
36. Fraker PJ, King LE. Reprogramming of the immune system during zinc deficiency. Annu Rev Nutr 2004; 24:277-298.
37. Imami N, Aspinall R, Gotch F. Role of the thymus in T lymphocyte reconstitution. Transplantation 2000; 69(11):2238-2239.
38. Andrew D, Aspinall R. Age-associated thymic atrophy is linked to a decline in IL-7 production. Exp Gerontol 2002; 37(2-3):455-463.
39. Henson SM, Snelgrove R, Hussell T et al. An IL-7 fusion protein that shows increased thymopoietic ability. J Immunol 2005; 175(6):4112-4118.
40. Phillips JA, Brondstetter TI, English CA et al. IL-7 gene therapy in aging restores early thymopoiesis without reversing involution. J Immunol 2004; 173(8):4867-4874.

41. Virts EL, Phillips JA, Thoman ML. A novel approach to thymic rejuvenation in the aged. Rejuvenation Res. Spring 2006; 9(1):134-142.
42. Mocchegiani E, Fabris N. Age-related thymus involution: zinc reverses in vitro the thymulin secretion defect. Int J Immunopharmacol 1995; 17(9):745-749.
43. Prasad AS. Zinc and immunity. Mol Cell Biochem 1998; 188(1-2):63-69.
44. Bogden JD, Oleske JM, Lavenhar MA et al. Effects of one year of supplementation with zinc and other micronutrients on cellular immunity in the elderly. J Am Coll Nutr 1990; 9(3):214-225.
45. Tian YM, Tian HJ, Zhang GY et al. Effects of Ginkgo biloba extract (EGb 761) on hydroxyl radical-induced thymocyte apoptosis and on age-related thymic atrophy and peripheral immune dysfunctions in mice. Mech Ageing Dev 2003; 124(8-9):977-983.
46. Tian YM, Li PP, Jiang XF et al. Rejuvenation of degenerative thymus by oral melatonin administration and the antagonistic action of melatonin against hydroxyl radical-induced apoptosis of cultured thymocytes in mice. J Pineal Res 2001; 31(3):214-221.
47. Bhatia SK, Tygrett LT, Grabstein KH et al. The effect of in vivo IL-7 deprivation on T-cell maturation. J Exp Med 1995; 181(4):1399-1409.
48. Ortman CL, Dittmar KA, Witte PL et al. Molecular characterization of the mouse involuted thymus: aberrations in expression of transcription regulators in thymocyte and epithelial compartments. Int Immunol 2002; 14(7):813-822.
49. Alves LA, Campos de Carvalho AC, Cirne Lima EO et al. Functional gap junctions in thymic epithelial cells are formed by connexin 43. Eur J Immunol 1995; 25(2):431-437.
50. Oleske JM, Westphal ML, Shore S et al. Zinc therapy of depressed cellular immunity in acrodermatitis enteropathica. Its correction. Am J Dis Child 1979; 133(9):915-918.
51. Prasad AS. Clinical, endocrinological and biochemical effects of zinc deficiency. Clin Endocrinol Metab 1985; 14(3):567-589.
52. Keen CL, Gershwin ME. Zinc deficiency and immune function. Annu Rev Nutr 1990; 10:415-431.
53. Prasad AS, Meftah S, Abdallah J et al. Serum thymulin in human zinc deficiency. J Clin Invest 1988; 82(4):1202-1210.
54. Ramassamy C, Averill D, Beffert U et al. Oxidative damage and protection by antioxidants in the frontal cortex of Alzheimer's disease is related to the apolipoprotein E genotype. Free Radic Biol Med 1999; 27(5-6):544-553.
55. Pietri S, Seguin JR, d'Arbigny P, Drieu K et al. Ginkgo biloba extract (EGb 761) pretreatment limits free radical-induced oxidative stress in patients undergoing coronary bypass surgery. Cardiovasc Drugs Ther 1997; 11(2):121-131.
56. Winter JC. The effects of an extract of Ginkgo biloba, EGb 761, on cognitive behavior and longevity in the rat. Physiol Behav 1998; 63(3):425-433.
57. Hirokawa K, Utsuyama M, Kasai M et al. Understanding the mechanism of the age-change of thymic function to promote T-cell differentiation. Immunol Lett 1994; 40(3):269-277.
58. Utsuyama M, Hirokawa K, Mancini C et al. Differential effects of gonadectomy on thymic stromal cells in promoting T-cell differentiation in mice. Mech Ageing Dev 1995; 81(2-3):107-117.
59. Fitzpatrick FT, Greenstein BD. Effects of various steroids on the thymus, spleen, ventral prostate and seminal vesicles in old orchidectomized rats. J Endocrinol 1987; 113(1):51-55.
60. Fitzpatrick FT, Kendall MD, Wheeler MJ et al. Reappearance of thymus of ageing rats after orchidectomy. J Endocrinol 1985; 106(3):R17-19.
61. Greenstein BD, Fitzpatrick FT, Adcock IM et al. Reappearance of the thymus in old rats after orchidectomy: inhibition of regeneration by testosterone. J Endocrinol 1986; 110(3):417-422.
62. Greenstein BD, Fitzpatrick FT, Kendall MD et al. Regeneration of the thymus in old male rats treated with a stable analogue of LHRH. J Endocrinol 1987; 112(3):345-350.
63. Kendall MD, Fitzpatrick FT, Greenstein BD et al. Reversal of ageing changes in the thymus of rats by chemical or surgical castration. Cell Tissue Res 1990; 261(3):555-564.
64. Thoman ML. The pattern of T lymphocyte differentiation is altered during thymic involution. Mech Ageing Dev 1995; 82(2-3):155-170.
65. Heng TS, Goldberg GL, Gray DH et al. Effects of castration on thymocyte development in two different models of thymic involution. J Immunol 2005; 175(5):2982-2993.
66. Min H, Montecino-Rodriguez E, Dorshkind K. Reassessing the role of growth hormone and sex steroids in thymic involution. Clinical Immunology 2006; 118(1):117-123.
67. Molina V, Shoenfeld Y. Infection, vaccines and other environmental triggers of autoimmunity. Autoimmunity 2005; 38(3):235-245.
68. van Eden W. Immunoregulation of autoimmune diseases. Hum Immunol 2006; 67(6):446-453.
69. Ackerman LS. Sex hormones and the genesis of autoimmunity. Arch Dermatol 2006; 142(3):371-376.
70. Bergstrom CT, Antia R. How do adaptive immune systems control pathogens while avoiding autoimmunity? Trends Ecol Evol 2006; 21(1):22-28.

71. Harel M, Shoenfeld Y. Predicting and preventing autoimmunity, myth or reality? Ann NY Acad Sci 2006; 1069:322-345.
72. Marks DJ, Mitchison NA, Segal AW et al. Can unresolved infection precipitate autoimmune disease? Curr Top Microbiol Immunol 2006; 305:105-125.
73. Selmi C, Lleo A, Zuin M et al. Interferon alpha and its contribution to autoimmunity. Curr Opin Investig Drugs 2006; 7(5):451-456.
74. Shepshelovich D, Shoenfeld Y. Prediction and prevention of autoimmune diseases: additional aspects of the mosaic of autoimmunity. Lupus 2006; 15(3):183-190.
75. Ziegler SF. FOXP3: of mice and men. Annu Rev Immunol 2006; 24:209-226.
76. Caton AJ. Mechanisms, manifestations and failures of self-tolerance. Immunol Res 2003; 27(2-3):161-168.
77. Rook GA, Stanford JL. Give us this day our daily germs. Immunol Today 1998; 19(3):113-116.
78. Ristori G, Buttinelli C, Pozzilli C et al. Microbe exposure, innate immunity and autoimmunity. Immunology Today 1999; 20(1):54-54.
79. Krupica T Jr, Fry TJ, Mackall CL. Autoimmunity during lymphopenia: a two-hit model. Clin Immunol 2006; 120(2):121-128.
80. Whitacre CC, Reingold SC, O'Looney PA. A gender gap in autoimmunity. Science 1999; 283(5406):1277-1278.
81. Staton PJ, Carpenter AB, Jackman SH. IL-7 is a critical factor in modulating lesion development in Skn-directed autoimmunity. J Immunol 2006; 176(7):3978-3986.
82. Harnaha J, Machen J, Wright M et al. Interleukin-7 Is a Survival Factor for $CD4^+$ $CD25^+$ T-Cells and Is Expressed by Diabetes-Suppressive Dendritic Cells. Diabetes 2006; 55(1):158-170.
83. Ponchel F, Verburg RJ, Bingham SJ et al. Interleukin-7 deficiency in rheumatoid arthritis: consequences for therapy-induced lymphopenia. Arthritis Res Ther 2005; 7(1):R80-92.
84. Leonard WJ. Interleukin-7 deficiency in rheumatoid arthritis. Arthritis Res Ther 2005; 7(1):42-43.
85. Rosenberg SA, Sportes C, Ahmadzadeh M et al. IL-7 administration to humans leads to expansion of $CD8^+$ and $CD4^+$ cells but a relative decrease of $CD4^+$ T-regulatory cells. J Immunother 2006; 29(3):313-319.
86. Prelog M. Aging of the immune system: a risk factor for autoimmunity? Autoimmun Rev 2006; 5(2):136-139.
87. Mitchell WA, Meng I, Nicholson SA et al. Thymic output, ageing and zinc. Biogerontology 2006; 7(5-6):461-470.
88. Hartwig M, Steinmann G. On a causal mechanism of chronic thymic involution in man. Mech Ageing Dev 1994; 75(2):151-156.
89. Hirokawa K, Sato K, Makinodan T. Influence of age of thymic grafts on the differentiation of T-cells in nude mice. Clin Immunol Immunopathol 1982; 24(2):251-262.
90. Leiner H, Greinert U, Scheiwe W et al. Repopulation of lymph nodes and spleen in thymus chimeras after lethal irradiation and bone marrow transplantation: dependence on the age of the thymus. Immunobiology 1984; 167(4):345-358.
91. Utsuyama M, Kasai M, Kurashima C et al. Age influence on the thymic capacity to promote differentiation of T-cells: induction of different composition of T-cell subsets by aging thymus. Mech Ageing Dev 1991; 58(2-3):267-277.
92. Tyan ML. Age-related decrease in mouse T-cell progenitors. J Immunol 1977; 118(3):846-851.
93. Kadish JL, Basch RS. Hematopoietic thymocyte precursors. I. Assay and kinetics of the appearance of progeny. J Exp Med 1976; 143(5):1082-1099.
94. Haynes BF, Hale LP, Weinhold KJ et al. Analysis of the adult thymus in reconstitution of T-lymphocytes in HIV-1 infection. J Clin Invest 1999; 103(4):453-460.
95. Wyllie AH. Glucocorticoid-induced thymocyte apoptosis is associated with endogenous endonuclease activation. Nature 1980; 284(5756):555-556.
96. Akita S, Malkin J, Melmed S. Disrupted murine leukemia inhibitory factor (LIF) gene attenuates adrenocorticotropic hormone (ACTH) secretion. Endocrinology 1996; 137(7):3140-3143.
97. Aronson M. Hypothesis: involution of the thymus with aging—programmed and beneficial. Thymus 1991; 18(1):7-13.
98. George AJ, Ritter MA. Thymic involution with ageing: obsolescence or good housekeeping? Immunol Today 1996; 17(6):267-272.
99. Sfikakis PP, Kostomitsopoulos N, Kittas C et al. Tamoxifen exerts testosterone-dependent and independent effects on thymic involution. Int J Immunopharmacol 1998; 20(6):305-312.
100. Olsen NJ, Kovacs WJ. Effects of androgens on T and B lymphocyte development. Immunol Res 2001; 23(2-3):281-288.
101. Olsen NJ, Viselli SM, Fan J et al. Androgens accelerate thymocyte apoptosis. Endocrinology 1998; 139(2):748-752.
102. Leposavic G, Obradovic S, Kosec D et al. In vivo modulation of the distribution of thymocyte subsets by female sex steroid hormones. Int Immunopharmacol 2001; 1(1):1-12.

Chapter 8

Autoimmune Diseases, Aging and the CD4$^+$ Lymphocyte:
Why Does Insulin-Dependent Diabetes Mellitus Start in Youth, but Rheumatoid Arthritis Mostly at Older Age?

Jacek M. Witkowski*

Introduction

Autoimmune diseases, still sometimes called 'autoaggression', result from the impaired and inappropriate reaction of the immune system to self antigens and cause cell and tissue damage and acute or chronic inflammatory processes. However, many profoundly different pathologies are collected together under the common designation 'autoimmune diseases'. From the biogerontological viewpoint, the important point, of course, is whether any of the autoimmune diseases occurs more frequently in the elderly than in young people. It is established knowledge that the incidence of autoimmunity increases in the elderly. However, only certain defined autoimmune diseases, of which the most characteristic (and commonest) example would be rheumatoid arthritis (RA), follow this pattern. Reciprocally, other autoimmune disease with age-related prevalence, such as insulin-dependent diabetes mellitus (IDDM), are typical for children and young adults. It has to be stressed that the age-dependence of both these examples (and others) is clearly not absolute. There are many examples of RA occurring in young adults or even teenagers; in fact, some studies are describing semi-separate 'diseases' of juvenile rheumatoid arthritis, early onset RA and late onset RA, with the latter being the most frequent.[1,1-7] On the other hand, IDDM is observed also in the middle-aged and elderly.[8-10] This suggests that although the underlying mechanism(s) of these and other autoimmune diseases is inappropriate recognition of self, inducing pathological immune reactions, the mechanisms leading to the immune attack are different.

Autoantigens in IDDM and RA

A major question here is whether autoimmune diseases that are more frequent in the elderly are a cause or a consequence of (aging-dependent?) changes in (certain) type(s) of the cells involved in this immune reactivity? In other words, is the aging process itself changing the immune system or parts thereof, so as to make autoimmune attack more likely? And if so, why is this true for RA and a few other autoimmune diseases, but not for all of them? Let us consider some known similarities and dissimilarities between IDDM—as a key example of an autoimmune disease typical for young individuals—and RA—which is prevalent rather in the middle-aged and elderly. Both diseases involve immune-mediated destruction of target tissues; in the case of IDDM, the beta cell

*Jacek M. Witkowski—Department of Pathophysiology, Medical University of Gdansk, Poland. Email: jawit@amg.gda.pl

Immunosenescence, edited by Graham Pawelec. ©2007 Landes Bioscience and Springer Science+Business Media.

of pancreatic islets and in RA, although the target antigens are still undefined, both the cartilage of affected joints as well as the adjacent bone. Early manifestations of RA include *synovitis* or inflammation of the synovium (the connective tissue lining the joint cavity and providing synovial fluid lubricating liquid for the joint). The latter process involves proliferation and accumulation of different cell types, including macrophages, subpopulations of lymphocytes and fibroblasts—the archetypal cells of the connective tissue. In the case of the IDDM, target antigen discovery was focused on protein(s) of the pancreatic β cells, such as preproinsulin, or the precursor protein for insulin itself,[11-14] glutamic acid decarboxylase GAD65,[11,15-22] tyrosyl phosphatases IA-2 and phogrin,[23-25] and carboxypeptidase E(H);[26] the latter targeted only by autoantibodies and not specific activated T-cells, which are quite abundant for the other autoantigens, for instance for GAD65.[11,14,21,27-29] In the case of RA, it would be logical to look for possible inducer (auto)antigens either in the connective tissue cells or the extracellular matrix. Some such candidate autoantigens have been proposed, including certain cartilage proteins, collagen type II, proteoglycans and calpastatin.[30,30-38] With the exception of the latter, all of them belong to the components of the extracellular matrix of the connective tissue (mostly cartilage) and are synthesized and secreted by fibroblasts and their derived chondrocytes (cartilage cells); calpastatin, on the other hand, is an endogenous inhibitor of a neutral cysteine (SH-) protease calpain, which in various forms is present in practically all cell types.[39-44] Thus, their epitopic fragments could be presented as autoantigens on the surface of fibroblasts, chondrocytes and, in the case of calpastatin, also other cells and potentially targeted by the immune cells for elimination.

Thus it seems that the type of (auto)antigen(s) involved does not provide any indications for understanding the difference in the usual time of onset for IDDM v.s. RA. Aging is associated with the accumulation of so-called advanced glycation endproducts, AGE.[45-48] It could therefore by hypothesized that AGE-dependent modification of proteins, as well as AGE products themselves, could serve as antigens preferentially in the elderly. In fact, some AGE products (e.g., pentosidine and N(epsilon)-carboxymethyllysine) have been found to be present in relatively high amounts in the inflamed synovium of RA patients;[49-52] others may modify the structure of IgM and IgG antibodies,[53-55] possibly transmuting them into something 'alien' to be recognized by immune cells and culminating in the production of rheumatoid factor.[53,54] On the other hand, the same AGE products can be found not only in the minor fraction of poorly-controlled IDDM patients who suffer from the disease for many years, in a sense 'aging with their disease',[56] but also in prepubertal IDDM children.[57,58] Thus, AGE product accumulation in the elderly and their biomolecule-modifying activity cannot be considered a cause for different timing of autoimmune reactivity in these diseases.

If not the structure, could it be the level of autoantigen expression that is the problem? Increased secretion of a protein could—in theory—lead to its increased presentation to the immune system and so stochastically lead to increased frequency of impaired reaction. However, data on the possible age-dependent changes in the level of expression or production of most of the abovementioned autoantigens are virtually nonexistent. Only calpastatin has been reported to be decreased in the red blood cells of aged individuals, possibly due to its excessive cleavage by calpain.[59] Whether cleaved calpastatin fragments could more easily induce autoimmune responses remains unknown and even if such a possibility were confirmed, this finding would not explain why RA (for which calpastatin might be an autoantigen) develops later in life than IDDM.

Influence of Infection

Autoimmune diseases are sometimes associated with prior infection or another provision of the individual with foreign antigenic proteins which—in genetically-susceptible individuals—trigger the autoimmune mechanism. This seems to be true for IDDM, where Coxsackie B and adenovirus infections are listed among the important events preceding the onset of disease. It has been demonstrated that a homology exists between viral proteins from these species and GAD65, suggesting that in this case autoaggression would be a result of molecular mimicry.[27,60,61] The relation between prior infection and RA is not so well-documented, but it was demonstrated long ago that Coxsackie virus infection is also frequent in cases of early onset (juvenile) rheumatoid arthritis.[62,63] This suggests a potentially accelerating role of such an infection in the development of this disease as well. On the other hand, many bacterial infections (including Chlamydia, Proteus, Streptococcus and E. coli) are variably presented as increasing the chance to manifest RA; altogether, these infections seem to be present in 10-20% of RA patients, which makes them less likely as a cause of the disease. Again, there is no published suggestion about how such an infection would affect the immune system to start attacking its own cartilage and bone.

In the light of findings that cytomegalovirus (CMV) antibodies and specific $CD8^+$ lymphocytes are increasingly frequent in elderly people, it is tempting to speculate that CMV infection has a role in precipitating RA symptoms. Indeed, evidence of CMV (and EBV) infection (as either the virus DNA or antiviral antibodies) was found more frequently in RA patients than in age-matched healthy controls.[64-67] Whether the infection is among the causes of the disease (inducing the aberrant immune reactivity of chronically-infected individuals to self antigens) or among its consequences (derailed immune system of the RA patients being unable to clear the viruses), as well as any possible explanation of the relation between chronic CMV infection (in a majority of the population starting at an early age) and a disease manifested at later age remains to be investigated. It is noteworthy, however, that preceding CMV infection and similarity between the virus and islet β-cell antigens was demonstrated for IDDM as well.[60]

Gene-Disease Associations

Another question that comes to mind in association with the problem of differentiation between (usually) early onset of IDDM and relatively late onset of RA is possible age-related genetic diversification of the patients. It is well known that both diseases are associated with increased frequencies of specific (and partially overlapping) haplotypes of HLA Class II; namely, variants of DR4 and DRB1 for RA and DR3, DR4 and DQB1 for IDDM.[68-73] On the other hand, the role of the HLA system in aging and an association between the aging process and the MHC polymorphisms of an individual is part of the general background supporting the "immunological theory of aging" proposed many years ago.[74] Interestingly, even excluding the juvenile form, RA can be subdivided into two types of manifestation, differing in clinical features (including type and number of joints involved, presence of extra-joint symptoms) and in time of clinical onset: early- and late-onset RA, EORA and LORA respectively.[5,75,76] It has been shown that EORA is significantly more frequent in the bearers of DR4 haplotypes.[77] So, is there a difference in the aging pattern between the bearers of HLA types associated with either disease? Multiple reports show that this is in fact the case.[74,78,79] Reciprocally, the relative frequency of HLA-DR7 is increased, while DR4 and DR3 are decreased, in the healthy elderly;[79] as the two latter are more frequent in the IDDM and RA groups (suffering from potentially life-shortening diseases), they seem to be a potent common denominator and similar for the two autoimmune diseases. Therefore, different onset times also cannot be explained in this manner.

Immune Aging

Having excluded genetics (HLA) and environment (the source of antigens) as factors that variably affect the immune system in IDDM and RA leading to different onset times of these two diseases, a remaining possibility is that IDDM is an abnormal reaction of otherwise normal lymphocytes, while the same cells in RA patients are affected by aging processes. As both diseases involve overproduction of more or less specific autoantibodies, the B lymphocyte is probably not culpable here, which points towards the different properties of helper or $CD4^+$ T-cells in RA and IDDM. These can be considered in two ways: first, early and late-onset disease may differ in the composition of major $CD4^+$ cell subpopulations, namely in the proportion of naïve to memory cells and of known functional Th subclasses. One such difference could be the domination of one of the functional subclasses of the helper T-cells. Thus, IDDM is characterized by the functional domination (reflected by cytokine production) of Th1 cells (secreting IL 2, IFN-γ and TNF-α),[80-85] while RA was initially considered rather as a Th2-dominated disease, resulting in oversecretion of

IL 4, IL 6, IL 10 etc. However, both Th1 and Th2 dysbalance may be prevalent at different stages of the disease.[86-91] Does this difference take us closer to understanding why IDDM tends to occur early but RA late in life? Probably not, as we can see domination of either type in various other (auto)immune diseases with no correlation of any single Th type with the age of disease onset.

Does normal aging change the proportions of Th1/2 cells? Numerous reports support this notion, but although experiments on young and old mice have usually led to the conclusion that aging is associated with a Th1 to Th2 shift, the matter seems much less consistent in humans and both types of response seem to dominate depending on experimental setup, type of stimulation etc.[92-97] Thus, aging-related changes in the dominant cytokine response pattern cannot be proposed as the cause of late onset of RA (albeit as mentioned earlier, IDDM is associated with a single type of response—the Th1).

Aging is associated with the accumulation of memory T-cells and reduction of those T-cells with a naïve phenotype. The result of this population shift is reduced immune responses to new antigenic challenges with relatively stable responses to recall antigens, albeit the latter is only temporary, probably due to stepwise exhaustion ('using-up') of the frequently-stimulated lymphocyte clones. In general, this and other features exhibited by T-cells of elderly people results in lower effectiveness of the immune response, whether measured by the ability to proliferate to antigenic or mitogen challenge, or by cytokine synthesis. In the light of these observations, comparison of the phenotypes of $CD4^+$ lymphocytes in IDDM and in RA does show a significant difference. While IDDM was initially thought to be a disease in which naïve, activated $CD4^+$ lymphocytes dominated in the circulation, it was recently shown to involve the accumulation of memory T-cells as well.[98-102] On the other hand, RA can be considered more like 'normal' aging: here, the dominant $CD4^+$ lymphocytes have the memory phenotype $CD4^+CD45RO^+$.[103-106] This in fact is one of the first observations pointing towards the possibility of the behavior of T-cells of RA patients resembling (or causing?) aging of the immune T-cells.

Are there any other similarities between T-cells of RA patients and T-cells of 'physiologically', or healthy, aging individuals and are they typical for RA but not IDDM? The in vitro proliferative response of T-cells from the elderly and of $CD4^+$ lymphocytes in particular, is poor compared to cells from young individuals, whether measured using ^3H-thymidine incorporation, or calculating population doublings, or assessing the number of available productive precursor cells giving viable progeny.[107,108] This feature itself, when translated to in vivo situations, makes the response of the $CD4^+$ cells of an elderly individual to an antigenic challenge less effective and at the end of life ineffective. It seems strange to look for similar symptoms in an autoimmune disease, where the immune response appears surplus and uncontrolled. Yet it has been known for a relatively long time that $CD4^+$ lymphocytes of RA patients respond as poorly to stimulation in vitro as those of healthy elderly, although the decrease in proliferative capacities of RA cells occurs, of course, earlier in life.[6,103,109-113] Similar, albeit much more poorly documented, observations have been made for the T-cells of IDDM children.[114]

Telomere Lengths

$CD4^+$ lymphocytes of elderly individuals exhibit significantly shorter telomeres than those of young healthy individuals (reviewed *inter alia* in refs. 115-117) This supposedly reflects their 'proliferative history' (i.e., the number of times such cells divided since their generation in the thymus). At least in vitro, similar to fibroblasts, T-cells also fail to proliferate indefinitely and succumb at around the "Hayflick limit"—the maximal number of cellular division before the onset of senescence when the cells cannot divide any more.[118] Shortening telomeres are thought to be part of the mechanism that prevents further division of these old cells. Telomeres of $CD4^+$ lymphocytes of RA patients are shorter than those of their healthy, age-matched counterparts and resemble the length of telomeres usually seen in the lymphocytes of much older healthy people.[113,119,120] This constitutes another feature by which T-cells of RA patients can be considered 'prematurely aged'. However and perhaps not so surprisingly, considering that some T-cells had to undergo multiple divisions when stimulated by β-cell autoantigens or viral 'mimicry' antigens, there are reports showing telomere shorten in T-cells of IDDM patients as well.[121]

Recent Thymic Emigrants

Another observation along the same lines concerns the number of so-called sjTRECs in the T-cell population. The sjTRECs are TCR rearrangement excision circles, or small circular molecules of DNA excised from the T-cell receptor genes of maturing T-cell at the time of their rearrangement.[122] These small DNA circles remain in the cell until it leaves the thymus and can be detected and quantified by molecular methods; thus, they mark the "early emigrants" from the thymus.[122-125] With time, when the thymic output of new T-cells decreases and peripheral proliferation of T-lymphocytes 'dilutes' the thymic emigrants, the levels of TRECs drop.[126-129] The same decrease in the numbers of early thymic emigrant T-cells measured via their TREC content occurs in RA patients; and again, it occurs significantly earlier in life for the patients than for their healthy age-matched counterparts.[130-132] Of course, lowered thymic output in the case of both healthy aging and RA individuals may contribute to the reduced proportion of naïve T-cells. Unlike telomere length, there are virtually no reports on the numbers of TRECs (or early thymic emigrants) among the $CD4^+$ lymphocytes of IDDM patients and reports from animal experiments are inconclusive.[133,134] This might suggest that IDDM is not related to the loss of naïve T-cells; however, it may simply reflect the fact that strong T-cell activation observed in IDDM patients still occurs even in the young organism with a functional thymus, that is providing enough new naïve T-cells.

T-Cell Activation

Appropriate immune reactions by $CD4^+$ lymphocytes depend on signal transduction between the TCR-CD3 complex and genes involved in proliferation and cytokine synthesis on the one hand and on appropriate costimulatory signals on the other. For both $CD4^+$ and majority of $CD8^+$ lymphocytes, the latter are provided by the ligation of the CD28 molecule by CD80/86, the specific ligands present on antigen presenting cells. It is well established that age-associated changes of CD28 expression on CD8 cells and to some extent on CD4 cells can result in decreases to below the detection level, transforming them into $CD28^-$ cells and that the proportion of these cells increases with age.[108,135-138] Whether these $CD28^-$ lymphocytes are inert ballast waiting for elimination, or whether they perform some function in the immune system remains unresolved. However, similar accumulation of $CD4^+$ $CD28^-$ cells is reported for RA patients.[111,139-144] In addition, not only the number of $CD4^+$ lymphocytes that have lost CD28 increases, but also the amount (number of molecules) of CD28 on those $CD4^+$ cells that do retain the $CD28^+$ phenotype is significantly decreased.[139,141,145,146] Considering the importance of costimulatory signaling via CD28 molecules it is tempting to speculate that reduction of their numbers on the surface of patients' $CD4^+$ cells might have functional consequences. In support of this, we have shown recently that decreased numbers of CD28 molecules on the surface of $CD4^+$ lymphocytes of healthy elderly is correlated with the increased time required by these cells to initiate first mitosis after contact with a stimulator (immobilized anti-CD3 antibody).[147] Our yet unpublished observations show similar relation for the $CD4^+$ lymphocytes of RA patients.

Regulation of CD28 expression has been associated with the action of TNF on $CD4^+$ lymphocytes. From the mechanistic point of view it is easily understandable that during a chronic inflammatory condition like RA, known to increase the levels of secreted TNF, the expression of CD28 should be decreased; relationships between TNF and CD28 have been recently confirmed by the finding that the CD28 expression level in anti-TNF treated patients normalizes.[139] This observation, interesting in itself and possibly of practical value, has also some connotations with the process of $CD4^+$ cell aging. As mentioned above, the proportion of $CD4^+CD28^-$ cells increases with age and the levels of CD28 molecule on those cells retaining it do decrease, also in apparently healthy individuals. Can this be related in some way to the sub-clinical inflammatory status of these individuals? It was observed that at least some apparently healthy individuals exhibit elevated levels of circulating proinflammatory cytokines, such as TNF and IL 6.[6,93,108,148] These cytokines (especially TNF) could be involved in the downregulation of CD28 and impairment of $CD4^+$ lymphocyte function in 'apparently healthy' elderly people, conforming to the hypothesis of inflammaging.[149-151] The behavior of CD28 seems to represent a second major difference between the CD4$^+$ cells in RA and in IDDM. In the course of the latter, no major changes in the expression of CD28 were reported; in fact, a recent study concluded that there are no differences in expression levels and proportions of CD28 between IDDM patients, NIDDM and healthy individuals.[152]

T Regulatory Cells

Finally, autoimmune diseases would seem to be the 'natural' situation where impaired action of suppressor or regulatory T-cells (Tregs) is implicated. Nonetheless, available data for IDDM do not support any major changes in the proportion and function of CD4$^+$CD25$^+$ Treg cells compared to normal controls, although there are also some studies showing the opposite.[153-157] In contrast, in RA the numbers of Tregs defined as CD4$^+$ CD25$^+$ cells seem to be either normal, or elevated both in the peripheral blood and in the synovial fluid.[158-161] Analysis of the proportions of Tregs in the healthy elderly also shows normal-to-increased values[145,162] another similarity with RA. Thus, an autoimmune disease does not necessarily involve the paralysis of regulatory T-cells.

T-Cell Antigen Receptor Repertoire

A further parallel between the state of CD4$^+$ lymphocytes in the healthy elderly and in RA is the reduced repertoire of their T-cell receptors (TCR), observed for both cohorts as increased frequencies of cells belonging to specific clones (characterized by narrow specificities of their TCR β-chains).[103,111,113,128,163] A reduced TCR repertoire may lead to inability of the immune system of either healthy elderly or a RA patient to recognize an antigen and to mount effective responses against it. Comparison of the TCR repertoire in IDDM patients with healthy controls yields more variable results, from no difference to significant contraction analogous to that observed in the RA or in healthy aged people.[164-167]

Conclusion

There is no single cause of late onset of rheumatoid arthritis that would differentiate it from type 1 diabetes mellitus. On the other hand, certain features of the CD4$^+$ lymphocytes of RA patients indicate the possibility of premature aging of these cells. Mechanistically speaking, this postulated premature aging of the RA CD4$^+$ lymphocytes would lead to overall exhaustion of their reactivity, but with some reactivity (of specific clones?) remaining strong or even enhanced. How the process of T-cell aging would translate into the symptoms of an autoimmune disease remains a mystery; however, known dynamics of the disease (its appearing in a proportion of relatively young people) suggest that the aging process in T-cells occurs with individualized speed: it is slow for non-RA individuals, earlier for those who exhibit disease symptoms late in life and faster in those exhibiting them earlier in life. Full understanding of the relationships between aging of the CD4$^+$ lymphocytes and the onset of RA remains elusive. Interestingly, accelerated aging may not be a characteristics only of RA patients' CD4$^+$ lymphocytes; it was shown that certain features like telomere shortening concern not only the lymphocytes, but also the myeloid bone marrow cells, indicating the possibility of more generalized cellular aging in RA patients and possibly a defect of the stem cells.[119] Also for osteoblasts, it was found that in RA patients they show accelerated aging manifesting itself as lower proliferation, shorter telomeres and other senescent cell markers.[168,169] In the light of these observations, RA—unlike IDDM where the CD4$^+$ cells seem to be 'doing their job', i.e., reacting to (unfortunately one's own) antigen(s)—stands out as a consequence of a more generalized process of accelerated aging.

References

1. Flato B, Lien G, Smerdel A et al. Prognostic factors in juvenile rheumatoid arthritis: a case-control study revealing early predictors and outcome after 14.9 years. J Rheumatol 2003; 30:386.
2. Ansell BM. Juvenile rheumatoid arthritis, juvenile chronic arthritis and juvenile spondyloarthropathies. Curr Opin Rheumatol 1992; 4:706.
3. Tsokos GC, Inghirami G, Pillemer SR et al. Immunoregulatory aberrations in patients with polyarticular juvenile rheumatoid arthritis. Clin Immunol Immunopathol 1988; 47:62.
4. Le Parc JM. [Inflammatory arthritis of the elderly]. Rev Prat 2005; 55:2115.

5. Calvo-Alen J, Corrales A, Sanchez-Andrada S et al. Outcome of late-onset rheumatoid arthritis. Clin Rheumatol 2005; 24:485.
6. Fulop T Jr, Larbi A et al. Ageing, autoimmunity and arthritis: Perturbations of TCR signal transduction pathways with ageing—a biochemical paradigm for the ageing immune system. Arthritis Res Ther 2003; 5:290.
7. Mazneva LM, Vasil'eva EV, Timofeeva EB et al. [Clinico-immunological study of rheumatoid arthritis in middle-aged and elderly patients]. Ter Arkh 1987; 59:56.
8. Harris MI, Robbins DC. Prevalence of adult-onset IDDM in the US population. Diabetes Care 1994; 17:1337.
9. Mizota M, Uchigata Y, Moriyama S et al. Age-dependent association of HLA-A24 in Japanese IDDM patients. Diabetologia 1996; 39:371.
10. Sanjeevi CB, Gambelunghe G, Falorni A et al. Genetics of latent autoimmune diabetes in adults. Ann N Y Acad Sci 2002; 958:107.
11. Ott PA, Dittrich MT, Herzog BA et al. T-cells recognize multiple GAD65 and proinsulin epitopes in human type 1 diabetes, suggesting determinant spreading. J Clin Immunol 2004; 24:327.
12. Durinovic-Bello I, Boehm BO, Ziegler AG. Predominantly recognized proinsulin T helper cell epitopes in individuals with and without islet cell autoimmunity. J Autoimmun 2002; 18:55.
13. Semana G, Gausling R, Jackson RA et al. T-cell autoreactivity to proinsulin epitopes in diabetic patients and healthy subjects. J Autoimmun 1999; 12:259.
14. Rudy G, Stone N, Harrison LC et al. Similar peptides from two beta cell autoantigens, proinsulin and glutamic acid decarboxylase, stimulate T-cells of individuals at risk for insulin-dependent diabetes. Mol Med 1995; 1:625.
15. Rathmann S, Rajasalu T, Rosinger S et al. Preproinsulin-specific $CD8^+$ T-cells secrete IFNgamma in human type 1 diabetes. Ann N Y Acad Sci 2004; 1037:22-5.:22.
16. Kim SK, Tarbell KV, Sanna M et al. Prevention of type I diabetes transfer by glutamic acid decarboxylase 65 peptide 206-220-specific T-cells. Proc Natl Acad Sci USA 2004; 101:14204.
17. Reijonen H, Mallone R, Heninger AK et al. GAD65-specific $CD4^+$ T-cells with high antigen avidity are prevalent in peripheral blood of patients with type 1 diabetes. Diabetes 2004; 53:1987.
18. Ou D, Metzger DL, Wang X et al. beta-cell antigen-specific CD56(+) NKT-cells from type 1 diabetic patients: autoaggressive effector T-cells damage human CD56(+) beta cells by HLA-restricted and nonHLA-restricted pathways. Hum Immunol 2002; 63:256.
19. Quinn A, McInerney MF, Sercarz EE. MHC class I-restricted determinants on the glutamic acid decarboxylase 65 molecule induce spontaneous CTL activity. J Immunol 2001; 167:1748.
20. Ou D, Mitchell LA, Metzger DL et al. Cross-reactive rubella virus and glutamic acid decarboxylase (65 and 67) protein determinants recognised by T-cells of patients with type I diabetes mellitus. Diabetologia 2000; 43:750.
21. Roep, BO. T-cell responses to autoantigens in IDDM. The search for the Holy Grail. Diabetes 1996; 45:1147.
22. Petersen JS, Hejnaes KR, Moody A et al. Detection of GAD65 antibodies in diabetes and other autoimmune diseases using a simple radioligand assay. Diabetes 1994; 43:459.
23. Bottazzo GF, Bosi E, Cull CA et al. IA-2 antibody prevalence and risk assessment of early insulin requirement in subjects presenting with type 2 diabetes (UKPDS 71). Diabetologia 2005; 48:703.
24. Seissler J, de Sonnaville JJ, Morgenthaler NG et al. Immunological heterogeneity in type I diabetes: presence of distinct autoantibody patterns in patients with acute onset and slowly progressive disease. Diabetologia 1998; 41:891.
25. Kawasaki E, Eisenbarth GS, Wasmeier C et al. Autoantibodies to protein tyrosine phosphatase-like proteins in type I diabetes. Overlapping specificities to phogrin and ICA512/IA-2. Diabetes 1996; 45:1344.
26. Aguilar-Diosdado M, Parkinson D, Corbett JA et al. Potential autoantigens in IDDM. Expression of carboxypeptidase-H and insulin but not glutamate decarboxylase on the beta-cell surface. Diabetes 1994; 43:418.
27. Schloot NC, Willemen SJ, Duinkerken G et al. Molecular mimicry in type 1 diabetes mellitus revisited: T-cell clones to GAD65 peptides with sequence homology to Coxsackie or proinsulin peptides do not crossreact with homologous counterpart. Hum Immunol 2001; 62:299.
28. Liu J, Purdy LE, Rabinovitch S et al. Major DQ8-restricted T-cell epitopes for human GAD65 mapped using human CD4, DQA1*0301, DQB1*0302 transgenic IA(null) NOD mice. Diabetes 1999; 48:469.
29. Durinovic-Bello I, Hummel M, Ziegler AG. Cellular immune response to diverse islet cell antigens in IDDM. Diabetes 1996; 45:795.
30. Hueber W, Kidd BA, Tomooka BH et al. Antigen microarray profiling of autoantibodies in rheumatoid arthritis. Arthritis Rheum 2005; 52:2645.

31. Reines BP. Is rheumatoid arthritis premature osteoarthritis with fetal-like healing? Autoimmun Rev 2004; 3:305.
32. Glant TT, Buzas EI, Finnegan A et al. Critical roles of glycosaminoglycan side chains of cartilage proteoglycan (aggrecan) in antigen recognition and presentation. J Immunol 1998; 160:3812.
33. Poole AR, Witter J, Roberts N et al. Inflammation and cartilage metabolism in rheumatoid arthritis. Studies of the blood markers hyaluronic acid, orosomucoid and keratan sulfate. Arthritis Rheum 1990; 33:790.
34. Deberg M, Labasse A, Christgau S et al. New serum biochemical markers (Coll 2-1 and Coll 2-1 NO2) for studying oxidative-related type II collagen network degradation in patients with osteoarthritis and rheumatoid arthritis. Osteoarthritis Cartilage 2005; 13:258.
35. Cho ML, Yoon CH, Hwang SY et al. Effector function of type II collagen-stimulated T-cells from rheumatoid arthritis patients: cross-talk between T-cells and synovial fibroblasts. Arthritis Rheum 2004; 50:776.
36. Myers LK, Higgins GC, Finkel TH et al. Juvenile arthritis and autoimmunity to type II collagen. Arthritis Rheum 2001; 44:1775.
37. Kotaniemi A, Isomaki H, Hakala M et al. Increased type I collagen degradation in early rheumatoid arthritis. J Rheumatol 1994; 21:1593.
38. Lark MW, Bayne EK, Flanagan J et al. Aggrecan degradation in human cartilage. Evidence for both matrix metalloproteinase and aggrecanase activity in normal, osteoarthritic and rheumatoid joints. J Clin Invest 1997; 100:93.
39. Branca D. Calpain-related diseases. Biochem Biophys Res Commun 2004; 322:1098.
40. Reverter D, Sorimachi H, Bode W. The structure of calcium-free human m-calpain: implications for calcium activation and function. Trends Cardiovasc Med 2001; 11:222.
41. Huang Y, Wang KK. The calpain family and human disease. Trends Mol Med 2001; 7:355.
42. Iwaki-Egawa S, Matsuno H, Yudoh K et al. High diagnostic value of anticalpastatin autoantibodies in rheumatoid arthritis detected by ELISA using human erythrocyte calpastatin as antigen. J Rheumatol 2004; 31:17.
43. Lackner KJ, Schlosser U, Lang B et al. Autoantibodies against human calpastatin in rheumatoid arthritis: epitope mapping and analysis of patient sera. Br J Rheumatol 1998; 37:1164.
44. Menard HA, el-Amine M. The calpain-calpastatin system in rheumatoid arthritis. Immunol Today 17:545.
45. Suji G, Sivakami S. Glucose, glycation and aging. Biogerontology 2004; 5:365.
46. Yan SF, Ramasamy R, Naka Y et al. Glycation, inflammation and RAGE: a scaffold for the macrovascular complications of diabetes and beyond. Circ Res 2003; 93:1159.
47. Meli M, Frey J, Perier C. Native protein glycoxidation and aging. J Nutr Health Aging 2003; 7:263.
48. Boulanger E, Dequiedt P, Wautier JL. [Advanced glycosylation end products (AGE): new toxins?]. Nephrologie 2002; 23:351.
49. Drinda S, Franke S, Canet CC et al. Identification of the advanced glycation end products N(epsilon)-carboxymethyllysine in the synovial tissue of patients with rheumatoid arthritis. Ann Rheum Dis 2002; 61:488.
50. Chen JR, Takahashi M, Suzuki M et al. Pentosidine in synovial fluid in osteoarthritis and rheumatoid arthritis: relationship with disease activity in rheumatoid arthritis. J Rheumatol 1998; 25:2440.
51. Miyata T, Ishiguro N, Yasuda Y et al. Increased pentosidine, an advanced glycation end product, in plasma and synovial fluid from patients with rheumatoid arthritis and its relation with inflammatory markers. Biochem Biophys Res Commun 1998; 244:45.
52. Rodriguez-Garcia J, Requena JR, Rodriguez-Segade S. Increased concentrations of serum pentosidine in rheumatoid arthritis. Clin Chem 1998; 44:250.
53. Newkirk MM, Goldbach-Mansky R, Lee J et al. Advanced glycation end-product (AGE)-damaged IgG and IgM autoantibodies to IgG-AGE in patients with early synovitis. Arthritis Res Ther 2003; 5:R82-R90.
54. Lucey MD, Newkirk MM, Neville C et al. Association between IgM response to IgG damaged by glyoxidation and disease activity in rheumatoid arthritis. J Rheumatol 2000; 27:319.
55. Ligier S, Fortin PR, Newkirk MM. A new antibody in rheumatoid arthritis targeting glycated IgG: IgM anti-IgG-AGE. Br J Rheumatol 1998; 37:1307.
56. Sell DR, Lapolla A, Odetti P et al. Pentosidine formation in skin correlates with severity of complications in individuals with long-standing IDDM. Diabetes 1992; 41:1286.
57. Chiarelli F, MM. de, Mezzetti A et al. Advanced glycation end products in children and adolescents with diabetes: relation to glycemic control and early microvascular complications. J Pediatr 1999; 134:486.
58. Berg TJ, Clausen JT, Torjesen PA et al. The advanced glycation end product Nepsilon-(carboxymethyl)lysine is increased in serum from children and adolescents with type 1 diabetes. Diabetes Care 1998; 21:1997.

59. Schwarz-Benmeir N, Glaser T, Barnoy S et al. Calpastatin in erythrocytes of young and old individuals. Biochem J 1994; 304:365.
60. Roep BO, Hiemstra HS, Schloot NC et al. Molecular mimicry in type 1 diabetes: immune cross-reactivity between islet autoantigen and human cytomegalovirus but not Coxsackie virus. Ann N Y Acad Sci 2002; 958:163-5.:163.
61. Bach JM, Otto H, Jung G et al. Identification of mimicry peptides based on sequential motifs of epitopes derived from 65-kDa glutamic acid decarboxylase. Eur J Immunol 1998; 28:1902.
62. Lozovksaia LS, Soboleva VD, Iakovleva AA. [Etiological connection between juvenile rheumatoid arthritis and the chronic form of coxsackie virus infection]. Vopr Virusol 1996; 41:122.
63. Soboleva VD, Lozovskaia LS, Mitchenko AF. [Viremia in children with rheumatoid arthritis]. Vopr Virusol 1983; 207.
64. Petrov AV. [Frequency of different infectious agents persistence in mononuclear leukocytes of blood and synovial fluid in patients with rheumatoid arthritis]. Lik Sprava 2005; 28.
65. Alvarez-Lafuente R, Fernandez-Gutierrez B, de MS et al. Potential relationship between herpes viruses and rheumatoid arthritis: analysis with quantitative real time polymerase chain reaction. Ann Rheum Dis 2005; 64:1357.
66. Petrov AV, Dudar' LV, Malyi KD. [Persistence of various infective agents in blood mononuclear leukocytes in a debut of rheumatoid arthritis]. Ter Arkh 2004; 76:32.
67. Mehraein Y, Lennerz C, Ehlhardt S et al. Latent Epstein-Barr virus (EBV) infection and cytomegalovirus (CMV) infection in synovial tissue of autoimmune chronic arthritis determined by RNA- and DNA-in situ hybridization. Mod Pathol 2004; 17:781.
68. Awata T, Hagura R, Urakami T et al. Age-dependent HLA genetic heterogeneity of IDDM in Japanese patients. Diabetologia 1995; 38:748.
69. Demaine AG, Hibberd ML, Mangles D et al. A new marker in the HLA class I region is associated with the age at onset of IDDM. Diabetologia 1995; 38:623.
70. Ziegler AG, Rabl W, Albert E et al. [Insulin autoantibodies and islet cell antibodies in recently appearing diabetes mellitus type I. Association with age of manifestation and HLA phenotype]. Dtsch Med Wochenschr 1991; 116:1737.
71. Watanabe I, Taneichi K, Shibaki H. [Associations of HLA-DR 4, DR 2, DRw 9 with clinical findings in rheumatoid arthritis]. Ryumachi 1989; 29:39.
72. Gran JT, Husby G, Thorsby E. The association between rheumatoid arthritis and the HLA antigen DR4. Ann Rheum Dis 1983; 42:292.
73. Schernthaner G, Ludwig H, Eibl M et al. [HL-A system and diabetes mellitus]. Schweiz Med Wochenschr 1976; 106:514.
74. Caruso C, Candore G, Colonna RG et al. HLA, aging and longevity: a critical reappraisal. Hum Immunol 2000; 61:942.
75. Narayanan K, Rajendran CP, Porkodi R et al. Late onset rheumatoid arthritis—a clinical and laboratory study. J Assoc Physicians India 2001; 49:311.
76. Yaretzky A, Feldman J, Alterman P et al. [Rheumatoid arthritis in the elderly]. Harefuah 1997; 132:10-72.
77. Pease CT, Bhakta BB, Devlin J et al. Does the age of onset of rheumatoid arthritis influence phenotype?: a prospective study of outcome and prognostic factors. Rheumatology (Oxford) 1999; 38:228.
78. Ricci G, Colombo C, Ghiazza B et al. Association between longevity and allelic forms of human leukocyte antigens (HLA): population study of aged Italian human subjects. Arch Immunol Ther Exp (Warsz) 1998; 46:31.
79. Papasteriades C, Boki K, Pappa H et al. HLA phenotypes in healthy aged subjects. Gerontology 1997; 43:176.
80. Ott PA, Herzog BA, Quast S et al. Islet-cell antigen-reactive T-cells show different expansion rates and Th1/Th2 differentiation in type 1 diabetic patients and healthy controls. Clin Immunol 2005; 115:102.
81. Karlsson Faresjo MG, Ernerudh J, Ludvigsson J. Cytokine profile in children during the first 3 months after the diagnosis of type 1 diabetes. Scand J Immunol 2004; 59:517.
82. Kero J, Gissler M, Hemminki E et al. Could TH1 and TH2 diseases coexist? Evaluation of asthma incidence in children with coeliac disease, type 1 diabetes, or rheumatoid arthritis: a register study. J Allergy Clin Immunol 2001; 108:781.
83. Azar ST, Tamim H, Beyhum HN et al. Type I (insulin-dependent) diabetes is a Th1- and Th2-mediated autoimmune disease. Clin Diagn Lab Immunol 1999; 6:306.
84. Kallmann BA, Huther M, Tubes M et al. Systemic bias of cytokine production toward cell-mediated immune regulation in IDDM and toward humoral immunity in Graves' disease. Diabetes 1997; 46:237.
85. Shimada A, Charlton B, Rohane P et al. Immune regulation in type 1 diabetes. J Autoimmun 1996; 9:263.

86. Kidd P. Th1/Th2 balance: the hypothesis, its limitations and implications for health and disease. Altern Med Rev 2003; 8:223.
87. Gerli R, Bistoni O, Russano A et al. In vivo activated T-cells in rheumatoid synovitis. Analysis of Th1- and Th2-type cytokine production at clonal level in different stages of disease. Clin Exp Immunol 2002; 129:549.
88. van Roon JA, Bijlsma JW, Lafeber FP. Suppression of inflammation and joint destruction in rheumatoid arthritis may require a concerted action of Th2 cytokines. Curr Opin Investig Drugs 2002; 3:1011.
89. Berner B, Akca D, Jung T et al. Analysis of Th1 and Th2 cytokines expressing $CD4^+$ and $CD8^+$ T-cells in rheumatoid arthritis by flow cytometry. J Rheumatol 2000; 27:1128.
90. Verhoef CM, van Roon JA, Vianen ME et al. Lymphocyte stimulation by CD3-CD28 enables detection of low T-cell interferon-gamma and interleukin-4 production in rheumatoid arthritis. Scand J Immunol 1999; 50:427.
91. van Roon JA, Verhoef CM, van Roy JL et al. Decrease in peripheral type 1 over type 2 T-cell cytokine production in patients with rheumatoid arthritis correlates with an increase in severity of disease. Ann Rheum Dis 1997; 56:656.
92. Gasparoni A, Ciardelli L, Avanzini A et al. Age-related changes in intracellular TH1/TH2 cytokine production, immunoproliferative T-lymphocyte response and natural killer cell activity in newborns, children and adults. Biol Neonate 2003; 84:297.
93. Gardner EM, Murasko DM. Age-related changes in Type 1 and Type 2 cytokine production in humans. Biogerontology 2002; 3:271.
94. Sandmand M, Bruunsgaard H, Kemp K et al. Is ageing associated with a shift in the balance between Type 1 and Type 2 cytokines in humans? Clin Exp Immunol 2002; 127:107.
95. Sakata-Kaneko S, Wakatsuki Y, Matsunaga Y et al. Altered Th1/Th2 commitment in human $CD4^+$ T-cells with ageing. Clin Exp Immunol 2000; 120:267.
96. Karanfilov CI, Liu B, Fox CC et al. Age-related defects in Th1 and Th2 cytokine production by human T-cells can be dissociated from altered frequencies of $CD45RA^+$ and $CD45RO^+$ T-cell subsets. Mech Ageing Dev 1999; 109:97.
97. Shearer GM. Th1/Th2 changes in aging. Mech Ageing Dev 1997; 94:1.
98. Kretowski A, Mysliwiec J, Turowski D et al. Analysis of recently activated, memory and naive lymphocyte T subsets in the peripheral blood of patients with Graves' disease and insulin-dependent diabetes mellitus. Rocz Akad Med Bialymst 1999; 44:226-34.:226.
99. Petersen LD, Duinkerken G, Bruining GJ et al. Increased numbers of in vivo activated T-cells in patients with recent onset insulin-dependent diabetes mellitus. J Autoimmun 1996; 9:731.
100. Peakman M, Alviggi L, Hussain MJ et al. Increased expression of T-cell markers of immunological memory associated with protection from type I diabetes. A study of identical twins. Diabetes 1994; 43:712.
101. Peakman M, Warnock T, Vats A et al. Lymphocyte subset abnormalities, autoantibodies and their relationship with HLA DR types in children with type 1 (insulin-dependent) diabetes and their first degree relatives. Diabetologia 1994; 37:155.
102. Ilonen J, Surcel HM, Kaar ML. Abnormalities within CD4 and CD8 T-lymphocytes subsets in type 1 (insulin-dependent) diabetes. Clin Exp Immunol 1991; 85:278.
103. Goronzy JJ, Weyand CM. Aging autoimmunity and arthritis: T-cell senescence and contraction of T-cell repertoire diversity—catalysts of autoimmunity and chronic inflammation. Arthritis Res Ther 2003; 5:225.
104. Nanki T, Lipsky PE. Cytokine, activation marker and chemokine receptor expression by individual $CD4^{(+)}$ memory T-cells in rheumatoid arthritis synovium. Arthritis Res 2000; 2:415.
105. Nielsen H, Petersen AA, Skjodt H et al. Blood levels of $CD11b^+$ memory T-lymphocytes are selectively upregulated in patients with active rheumatoid arthritis. APMIS 1999; 107:1124.
106. Thomas R, McIlraith M, Davis LS et al. Rheumatoid synovium is enriched in CD45RBdim mature memory T-cells that are potent helpers for B-cell differentiation. Arthritis Rheum 1992; 35:1455.
107. Fulop T, Larbi A, Wikby A et al. Dysregulation of T-cell function in the elderly : scientific basis and clinical implications. Drugs Aging 2005; 22:589.
108. Pawelec G, Effros RB, Caruso C et al. T-cells and aging (update february 1999). Front Biosci 1999; 4:D216-69.:D216-D269.
109. Wagner U, Schulze-Koops H. [T-lymphocytes—do they control rheumatic immune responses?]. Z Rheumatol 2005; 64:377.
110. Fournier C. Where do T-cells stand in rheumatoid arthritis? Joint Bone Spine 2005; 72:527.
111. Goronzy JJ, Weyand CM. T-cell regulation in rheumatoid arthritis. Curr Opin Rheumatol 2004; 16:212.
112. Weyand CM, Fulbright JW, Goronzy JJ. Immunosenescence, autoimmunity and rheumatoid arthritis. Exp Gerontol 2003; 38:833.

113. Koetz K, Bryl E, Spickschen K et al. T-cell homeostasis in patients with rheumatoid arthritis. Proc Natl Acad Sci USA 2000; 97:9203.
114. Nervi S, tlan-Gepner C, Fossat C et al. Constitutive impaired TCR/CD3-mediated activation of T-cells in IDDM patients co-exist with normal co-stimulation pathways. J Autoimmun 1999; 13:247.
115. Saretzki G, Von ZT. Replicative aging, telomeres and oxidative stress. Ann NY Acad Sci 2002; 959:24.
116. Malaguarnera L, Ferlito L, Imbesi RM et al. Immunosenescence: a review. Arch Gerontol Geriatr 2001; 32:1.
117. Engelhardt M, Martens UM. The implication of telomerase activity and telomere stability for replicative aging and cellular immortality (Review). Oncol Rep 1998; 5:1043.
118. Perillo NL, Walford RL, Newman MA et al. Human T-lymphocytes possess a limited in vitro life span. Exp Gerontol 1989; 24:177.
119. Schonland SO, Lopez C, Widmann T et al. Premature telomeric loss in rheumatoid arthritis is genetically determined and involves both myeloid and lymphoid cell lineages. Proc Natl Acad Sci USA 2003; 100:13471.
120. Goronzy JJ, Fujii H, Weyand CM. Telomeres, immune aging and autoimmunity. Exp Gerontol 2006; 41:246.
121. Jeanclos E, Krolewski A, Skurnick J et al. Shortened telomere length in white blood cells of patients with IDDM. Diabetes 1998; 47:482.
122. Al-Harthi L, Marchetti G, Steffens CM et al. Detection of T-cell receptor circles (TRECs) as biomarkers for de novo T-cell synthesis using a quantitative polymerase chain reaction-enzyme linked immunosorbent assay (PCR-ELISA). J Immunol Methods 2000; 237:187.
123. Ye P, Kirschner DE. Measuring emigration of human thymocytes by T-cell receptor excision circles. Crit Rev Immunol 2002; 22:483.
124. Hazenberg MD, Verschuren MC, Hamann D et al. T-cell receptor excision circles as markers for recent thymic emigrants: basic aspects, technical approach and guidelines for interpretation. J Mol Med 2001; 79:631.
125. McFarland RD, Douek DC, Koup RA et al. Identification of a human recent thymic emigrant phenotype. Proc Natl Acad Sci USA 2000; 97:4215.
126. van den Dool C, de Boer RJ. The effects of age, thymectomy and HIV Infection on alpha and beta TCR excision circles in naive T-cells. J Immunol 2006; 177:4391.
127. Nasi M, Troiano L, Lugli E et al. Thymic output and functionality of the IL-7/IL-7 receptor system in centenarians: implications for the neolymphogenesis at the limit of human life. Aging Cell 2006; 5:167.
128. Naylor K, Li G, Vallejo AN et al. The influence of age on T-cell generation and TCR diversity. J Immunol 2005; 174:7446.
129. Nobile M, Correa R, Borghans JA et al. De novo T-cell generation in patients at different ages and stages of HIV-1 disease. Blood 2004; 104:470.
130. Thewissen M, Linsen L, Somers V et al. Premature immunosenescence in rheumatoid arthritis and multiple sclerosis patients. Ann NY Acad Sci 2005; 1051:255-62.:255.
131. Ponchel F, Morgan AW, Bingham SJ et al. Dysregulated lymphocyte proliferation and differentiation in patients with rheumatoid arthritis. Blood 2002; 100:4550.
132. Koetz K, Bryl E, Spickschen K et al. T-cell homeostasis in patients with rheumatoid arthritis. Proc Natl Acad Sci USA 2000; 97:9203.
133. Ramanathan S, Norwich K, Poussier P. Antigen activation rescues recent thymic emigrants from programmed cell death in the BB rat. J Immunol 1998; 160:5757.
134. Zadeh HH, Greiner DL, Wu DY et al. Abnormalities in the export and fate of recent thymic emigrants in diabetes-prone BB/W rats. Autoimmunity 1996; 24:35.
135. Effros RB. Loss of CD28 expression on T-lymphocytes: a marker of replicative senescence. Dev Comp Immunol 1997; 21:471.
136. Vallejo AN, Weyand CM, Goronzy JJ. Functional disruption of the CD28 gene transcriptional initiator in senescent T-cells. J Biol Chem 2001; 276:2565.
137. Vallejo AN, Brandes JC, Weyand CM et al. Modulation of CD28 expression: distinct regulatory pathways during activation and replicative senescence. J Immunol 1999; 162:6572.
138. Vallejo AN, Nestel AR, Schirmer M et al. Aging-related deficiency of CD28 expression in $CD4^+$ T-cells is associated with the loss of gene-specific nuclear factor binding activity. J Biol Chem 1998; 273:8119.
139. Bryl E, Vallejo AN, Matteson EL et al. Modulation of CD28 expression with anti-tumor necrosis factor alpha therapy in rheumatoid arthritis. Arthritis Rheum 2005; 52:2996.
140. Lewis DE, Merched-Sauvage M, Goronzy JJ et al. Tumor necrosis factor-alpha and CD80 modulate CD28 expression through a similar mechanism of T-cell receptor-independent inhibition of transcription. J Biol Chem 2004; 279:29130.

141. Bryl E, Vallejo AN, Weyand CM et al. Down-regulation of CD28 expression by TNF-alpha. J Immunol 2001; 167:3231.
142. Namekawa T, Wagner UG, Goronzy JJ et al. Functional subsets of CD4 T-cells in rheumatoid synovitis. Arthritis Rheum 1998; 41:2108.
143. Weyand CM, Klimiuk PA, Goronzy JJ. Heterogeneity of rheumatoid arthritis: from phenotypes to genotypes. Springer Semin Immunopathol 1998; 20:5.
144. Martens PB, Goronzy JJ, Schaid D et al. Expansion of unusual $CD4^+$ T-cells in severe rheumatoid arthritis. Arthritis Rheum 1997; 40:1106.
145. Bryl E, Witkowski JM. Decreased proliferative capability of $CD4(^+)$ cells of elderly people is associated with faster loss of activation-related antigens and accumulation of regulatory T-cells. Exp Gerontol 2004; 39:587.
146. Vallejo AN, Bryl E, Klarskov K et al. Molecular basis for the loss of CD28 expression in senescent T-cells. J Biol Chem 2002; 277:46940.
147. Witkowski JM, Bryl E. Paradoxical age-related cell cycle quickening of human $CD4(^+)$ lymphocytes: a role for cyclin D1 and calpain. Exp Gerontol 2004; 39:577.
148. Rink L, Cakman I, Kirchner H. Altered cytokine production in the elderly. Mech Ageing Dev 1998; 102:199.
149. Prelog M. Aging of the immune system: a risk factor for autoimmunity? Autoimmun Rev 2006; 5:136.
150. Boren E, Gershwin ME. Inflamm-aging: autoimmunity and the immune-risk phenotype. Autoimmun Rev 2004; 3:401.
151. Vallejo AN, Weyand CM, Goronzy JJ. T-cell senescence: a culprit of immune abnormalities in chronic inflammation and persistent infection. Trends Mol Med 2004; 10:119.
152. Tsutsumi Y, Jie X, Ihara K et al. Phenotypic and genetic analyses of T-cell-mediated immunoregulation in patients with Type 1 diabetes. Diabet Med 2006; 23:1145.
153. Yang Z, Zhou Z, Huang G et al. The $CD4(^+)$ regulatory T-cells is decreased in adults with latent autoimmune diabetes. Diabetes Res Clin Pract 2006.
154. Bisikirska BC, Herold KC. Regulatory T-cells and type 1 diabetes. Curr Diab Rep 2005; 5:104.
155. Juedes AE, von Herrath MG. Regulatory T-cells in type 1 diabetes. Diabetes Metab Res Rev 2004; 20:446.
156. Homann D, von HM. Regulatory T-cells and type 1 diabetes. Clin Immunol 2004; 112:202.
157. Arreaza GA, Sharif S, Cameron MJ et al. Role of regulatory T-cells in the pathogenesis of autoimmune diabetes. Curr Dir Autoimmun 2001; 4:308-32.:308.
158. Minami R, Sakai K, Miyamura T et al. [The role of $CD4^+CD25^+$ regulatory T-cells in patients with Rheumatoid Arthritis]. Nihon Rinsho Meneki Gakkai Kaishi 2006; 29:37.
159. Ruprecht CR, Gattorno M, Ferlito F et al. Coexpression of CD25 and CD27 identifies $FoxP3^+$ regulatory T-cells in inflamed synovia. J Exp Med 2005; 201:1793.
160. Leipe J, Skapenko A, Lipsky PE et al. Regulatory T-cells in rheumatoid arthritis. Arthritis Res Ther 2005; 7:93.
161. Mottonen M, Heikkinen J, Mustonen L et al. $CD4^+$ $CD25^+$ T-cells with the phenotypic and functional characteristics of regulatory T-cells are enriched in the synovial fluid of patients with rheumatoid arthritis. Clin Exp Immunol 2005; 140:360.
162. Trzonkowski P, Szmit E, Mysliwska J et al. $CD4^+CD25^+$ T regulatory cells inhibit cytotoxic activity of CTL and NK cells in humans-impact of immunosenescence. Clin Immunol 2006; 119:307.
163. Goronzy JJ, Weyand CM. T-cell development and receptor diversity during aging. Curr Opin Immunol 2005; 17:468.
164. Manfras BJ, Claudi-Boehm S, Kreienberg R et al. T-cell receptor repertoire and function in umbilical cord blood lymphocytes from newborns of type 1 diabetic mothers. Acta Diabetol 2004; 41:167.
165. Luppi P, Zanone MM, Hyoty H et al. Restricted TCR V beta gene expression and enterovirus infection in type I diabetes: a pilot study. Diabetologia 2000; 43:1484.
166. Santamaria P, Lewis C, Jessurun J et al. Skewed T-cell receptor usage and junctional heterogeneity among isletitis alpha beta and gamma delta T-cells in human IDDM [corrected]. Diabetes 1994; 43:599.
167. Malhotra U, Spielman R, Concannon P. Variability in T-cell receptor V beta gene usage in human peripheral blood lymphocytes. Studies of identical twins, siblings and insulin-dependent diabetes mellitus patients. J Immunol 1992; 149:1802.
168. Yudoh K, Matsuno H, Kimura T. [Relationship between periarticular osteoporosis and osteoblast senescence in patients with rheumatoid arthritis]. Clin Calcium 2001; 11:612.
169. Yudoh K, Matsuno H, Osada R et al. Decreased cellular activity and replicative capacity of osteoblastic cells isolated from the periarticular bone of rheumatoid arthritis patients compared with osteoarthritis patients. Arthritis Rheum 2000; 43:2178.

Chapter 9

Role of Chemokines and Chemokine Receptors in Diseases of Ageing

Erminia Mariani,* Adriana Rita Mariani and Andrea Facchini

Abstract

Chemokines play an important role in orchestrating leukocyte recruitment and activation during inflammation. Given the ubiquity of chemokines involved in inflammatory tissue destruction, it is not surprising that they contribute to numerous human pathologies. Epidemiological studies have suggested that chronic low-grade inflammation is related to several diseases of ageing with an inflammatory pathogenesis (such as atherosclerosis, type 2 diabetes, osteoarthritis and Alzheimer's disease) and to increased mortality risk. In this chapter, we will briefly review the properties of chemokines and their receptors and highlight the roles of these chemoattractants in the above selected diseases of ageing.

Introduction

Inflammation is fundamentally an acute protective response occurring in the vascularized connective tissue in response to any insult. In acute situations, or at low levels, it has a relatively short duration, deals with the abnormality and promotes healing; when uncontrolled or chronically sustained at high levels, it has a longer duration and may be potentially harmful, damaging viable host tissues and possibly underlying the pathogenesis of many diseases.

A critical function of inflammation is the delivery of leukocytes to the site of injury which is achieved by increased local blood flow, structural changes in the micro-vessels to permit leukocyte migration and their accumulation in the focus of lesion (Fig. 1).[1,2] Chemokines play an important role in orchestrating leukocyte recruitment and activation during this inflammation (for reviews, see refs. 3-8). Nonetheless, very few data are available on profiles of chemokines in healthy ageing, considering the increasing importance that these molecules are gaining with regard to the regulation of immune responses. Alterations in the production of chemokines or in their recognition by cells of the immune system may be responsible for at least some changes observed in immune responses with ageing.[9] A progressive age-related increase of plasma concentrations of some chemokines has been observed by our group (unpublished data and personal communications) and by other investigators in healthy elderly subjects.[10-13] However, the increase of these inflammatory molecules is still far from the levels evident during acute inflammation, thus indicating that ageing is associated with a low-grade basal inflammation. The possible cause of this increase may be an "in vivo" preferential activation of circulating mononuclear cells occurring in healthy aged subjects, in agreement with the spontaneous production of chemokines that we[14] and others[15] demonstrated in vitro and with the progressive age-related increase of circulating monocytes that these subjects displayed (our unpublished observations).

*Corresponding Author: Erminia Mariani—Laboratorio di Immunologia e Genetica, Istituto di Ricerca Codivilla-Putti, IOR, Via di Barbiano 1/10, 40136 Bologna, Italy.
Email: marianie@alma.unibo.it

Immunosenescence, edited by Graham Pawelec.

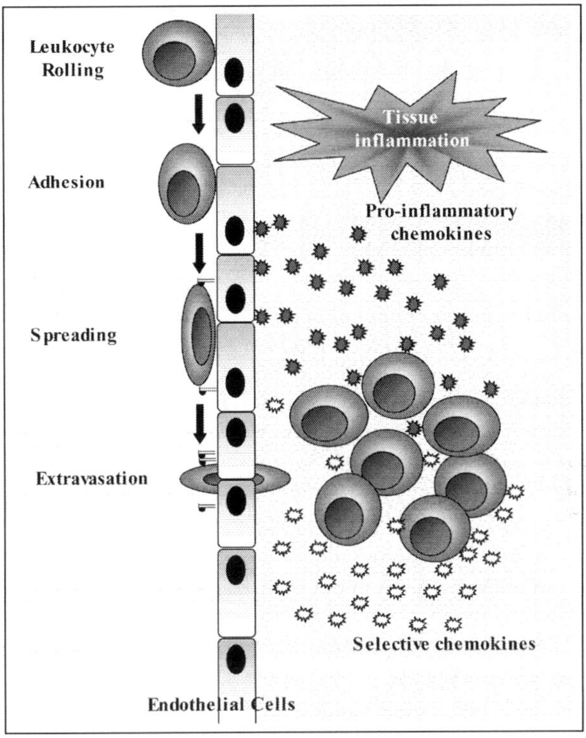

Figure 1. Mechanism of chemokine-mediated recruitment of leukocytes. Leukocyte recruitment is a multi-step process usually occurring in the post capillary vascular system. Leukocytes that circulate in the bloodstream constantly monitor abnormal signals from the endothelial cell by marginating and rolling on surface selectins of endothelial cells. Chemokines (synthesized by the endothelial cells, upon activation by pro-inflammatory cytokines or produced by tissue cells and subsequently transported across the endothelium), are secreted and bind to glycosoaminoglycans on the endothelial cell surface. The activation of the cognate receptor mediates integrin activation, flow arrest, adhesion of the leukocyte to the adhesion molecules expressed on the endothelial cell surface, followed by extravasation by diapedesis, across the endothelial cell barrier. The resultant chemokine gradient provides a directional signal that the cells may use to navigate towards the site of inflammation, where the high concentration of chemokines desensitises the receptors. The cells can be activated to secrete pro-inflammatory mediators and exert their effects.

Given the ubiquity of chemokine involvement in inflammatory tissue destruction, it is not surprising that numerous medical fields are co-opted[16,17] Epidemiological studies have suggested that the chronic low-grade inflammation is related to several diseases of ageing with an inflammatory pathogenesis and to increased mortality risk.[18] In this chapter, we will briefly review the properties of chemokines and their receptors and highlight the roles of these chemoattractants in selected diseases of ageing.

The Chemokine System

Chemokines are a superfamily of small heparin-binding peptides of low molecular weight that control the trafficking of specific cell subpopulations both in physiological and pathological processes. In the last few years, it has been demonstrated that these mediators play a role in embryonic development, hematopoiesis, angiogenesis, host defence, inflammation, immunity, AIDS and cancer.[19-21]

Chemokines

According to the arrangement of positionally conserved cysteine residues near the amino terminus, chemokines are classified into four families: CC, CXC, CX3C and C.[22] A systematic nomenclature has been proposed in the past years[23] but it is not yet widely adopted. Therefore, in order to avoid confusion, the historical name will be used in this chapter. The largest family consists of CC chemokines, including at least 28 ligands (Table 1). The second family consists of CXC chemokines, including 16 ligands (Table 2). Fractalkine is the only member of the third, CX3C, family of chemokines. Its domain is fused to a mucin-like stalk, forming a cell adhesion receptor capable of arresting cells under physiologic flow conditions. Finally, the fourth family (C chemokines) includes lymphotactin (Table 2).

Chemokine Receptors

Each family of chemokines interacts with a reciprocal family of seven transmembrane domain receptors coupled to trimeric G proteins (GPCR)[23] which activate multiple intracellular signaling pathways that eventually lead to cytoskeletal rearrangements and cell mobilization. In addition to their role in cell recruitment, chemokines may induce leukocyte activation and control lymphocyte differentiation and effector function.[19,22,23] At present, ten receptors for CC, six receptors for CXC, one for C and one for CX3C chemokines have been identified (Table 3). They do not interact as a single receptor/ligand pair but often act with promiscuity of binding and redundancy. A comparison of the properties exhibited by different ligands of a common receptor suggests the existence of control mechanisms to limit redundancy. A new degree of complexity has emerged with the discovery that chemokines can also act as receptor antagonists. All these mechanisms seem to operate to increase the selectivity of cell recruitment.

Functional Families of Chemokines

According to a recent classification that uses physiological characteristics, chemokines can be divided into two main functional families: inflammatory (alternatively, inducible) and homeostatic (alternatively, constitutive, housekeeping or lymphoid) (Tables 1 and 2). This distinction is not absolute and some members cannot be assigned clearly to either one of the two functional categories and therefore are identified as "dual-function" chemokines.[4,6-8]

The inflammatory chemokines are induced by pathogens and proinflammatory stimuli in both resident tissue cells and leukocytes[19,22,23] and recruit leukocytes in response to physiological stress. These chemokines play a role in innate immunity and in inflammation. The homeostatic chemokines, on the other hand, are constitutively expressed in separate microenvironments within lymphoid tissues, skin and mucosa. They are involved in basal leukocyte trafficking and homing, as well as in development.[19,22,23]

Atherosclerosis

Atherosclerosis is a chronic disease with an inflammatory pathogenesis, developing in response to damage of the vessel wall.[24,25] The infiltration of mononuclear cells into the intima, the proliferation of smooth muscle cells and the deposition of extracellular matrix represent the main vascular modifications. Therefore, endothelial dysfunction (possibly determined, for example, by hypertension, cigarette smoking, increased plasma levels of cholesterol and/or homocysteine, diabetes) is indicated as the first step in atherosclerosis, establishing a reduced vasodilatation as well as a proinflammatory and prothrombotic condition.[24,26,27]

Both in animal models and humans, different chemokines including IL-8, IP-10, I-309 and receptor CXCR2 have been identified in atherosclerotic lesions, but the most promising chemokine for a pathogenetic role in atherosclerosis is MCP-1 and its CCR2 receptor.[8,17] In fact, mice in which either MCP-1 or CCR2 were genetically deleted, presented less lipid and lower macrophage deposition and smaller atherosclerotic lesions than mice genetically susceptible to atherosclerosis.[28-30] In contrast, mice with a deficiency of CCR5 which is not activated by MCP-1 remained vulnerable to atherosclerosis.[31] Furthermore, lipid loaded foam cells, derived from macrophages, which characterize early atherosclerotic plaques, express MCP-1; oxidized low density

Table 1. CC chemokine ligands

Systematic Nomenclature	Historical Designation of Ligand	Producing Cells	Functional Families
CCL1	TCA3/I-309	Mo, T, MastC, MiC, As	Dual function
CCL2	MCP-1/MCAF/TDFC	Mo, F, EndoC, EpC, N, MastC, G, MesC, DC	Inflammatory
CCL3	MIP-1α/LD78α	Mo, N, Eo, F, MastC, G, MesC	Inflammatory
CCL4	MIP-1β/HC21	Mo, N, Eo, F, MastC, Ba, NK	Inflammatory
CCL5	RANTES	T, Mo, F, MesC	Inflammatory
CCL6	C10	Mo, Eo, MiC	Inflammatory
CCL7	MCP-3/MARC	Plt, Mo, MastC, F, EndoC, EpC	Inflammatory
CCL8	MCP-2	Mo, F	Inflammatory
CCL9/10	MIP-1γ	Mo, DC, lung, liver, thymus, pancreas	Inflammatory
CCL11	Eotaxin-1	EndoC, EpC, Eo, lung	Inflammatory
CCL12	MCP-5	Mo, lymph node	Inflammatory
CCL13	MCP-4/CKβ10	DC, EpC, lung, thymus, intestine	Inflammatory
CCL14	HCC-1	Bone marrow, gut, spleen, liver, SmC	Inflammatory
CCL15	HCC-2/Lkn/MIP-1δ	DC, Mo, T, B, NK	Homeostatic
CCL16	HCC-4/LEC/LCC-1	Mo	Homeostatic
CCL17	TARC	DC, Mo, EpC, F, SmC	Dual function
CCL18	DC-KC1/PARC/MIP-4	DC, Mo	Homeostatic
CCL19	MIP-3β/ELC/Exodus-3	N, lymph node, spleen, thymus, intestine	Homeostatic
CCL20	MIP-3α/LARC/Exodus-1	Mo, T, N, EndoC, liver, lung, thymus, placenta, appendix	Dual function
CCL21	SLC/6Ckine/Exodus-2	EndoC, lymph node	Homeostatic
CCL22	MDC/STCP-1	DC, Mo, B, T, NK, EpC	Dual function
CCL23	MPIF-1/CKβ8/ CKβ8-1	DC, Mo, lung, liver	Inflammatory
CCL24	Eotaxin-2/CKβ6/MPIF-2	Mo, T, lung, spleen, thymus, liver	Inflammatory
CCL25	TECK	DC, EpC, EndoC, gut	Dual function
CCL26	Eotaxin-3	EndoC, heart, ovary	Inflammatory
CCL27	CTACK/ILC	K, placenta, skin	Inflammatory
CCL28	MEC	EpC, EndoC	Inflammatory

Abbreviations: As: astrocyte, Ba: basophil, DC: dendritic cell, Eo: eosinophil, EC: endothelial cell, EpC: epithelial cell, F: fibroblast, G: glioblastoma, K: keratinocyte, Mo: monocyte/macrophage, MC: mast cell, MeC: mesangial cell, MiC: microglia cell, N: neutrophil, NK: natural killer cell, Plt: platelet, SmC: smooth muscle cell, T: T lympocyte.

lipoprotein (ox-LDL) induces the production of this chemokine in endothelial and smooth-muscle cells, indicating a linker role for MCP-1 between ox-LDL and the recruitment of foam cells to the vessel wall.[8,32,33] Increased circulating levels of MCP-1, correlating with LDL cholesterol, have been observed in patients with cardiovascular risk factors such as hyperlipemia.[34] Patients with coronary heart disease (CHD) presented increased expression of MCP-1 and patients with acute coronary syndromes have higher serum MCP-1 levels than those with stable angina.[35,36] MCP-1 has also been found in diseased human carotid arteries.[8,17,37]

Together with MCP-1, also elevated serum levels of IL-8 and IP-10 have been reported to precede CHD, in agreement with clinical and preclinical studies that suggested that these mediators

Table 2. CXC, C and CX3C chemokine ligands

Systematic Nomenclature	Historical Designation of Ligand	Producing Cells	Functional Families
CXCL1	GROα/MGSA-β	Mo, N, EndoC, F, Mel	Inflammatory
CXCL2	GROβ/MGSA-β	Mo, N, EndoC, F, Mel	Inflammatory
CXCL3	GROγ/MGSA-γ	Mo, N, EndoC, F, Mel	Inflammatory
CXCL4	PF4	Plt, Mk	Inflammatory
CXCL5	ENA-78	EndoC, Plt, Eo	Inflammatory
CXCL6	GCP-2	EndoC, F, Mo	Inflammatory
CXCL7	NAP-2/CTAP-III	Plt, EndoC, Mo (thymus)	Inflammatory
CXCL8	IL-8/NAP-1/MDNCF/ MIP-2	EndoC, N, Plt, As, G, MesC,Ba, NK	Inflammatory
CXCL9	MIG	Mo, N	Dual function
CXCL10	IP-10/CRG-2	Mo, K, N, F, EndoC, As, G	Dual function
CXCL11	I-TAC/β-R1/H174/IP-9	As, Mo, N	Dual function
CXCL12	SDF-1αβ/PBSF	EndoC, EpC, lung	Homeostatic
CXCL13	BLC/BCA-1	EndoC, Mo, DC, lymphnode, spleen	Homeostatic
CXCL14	BRAK/Bolekine	F, B, Mo	Homeostatic
CXCL15	Lungkine	EndoC (lung)	Inflammatory
CXCL16		B, Mo, DC	Dual function
XCL1	Lymphotactin a/SCM-1α/ATAC	T, MastC, NK	Inflammatory
XCL2	Lymphotactin b/SCM-1β	T, NK, spleen	Inflammatory
CX3CL1	Fractalkine/neurotactin	APC, EndoC, DC, Ne, T	Inflammatory

APC: antigen presenting cell, As: astrocyte, B: B lymphocyte, Ba: basophil, DC: dendritic cell, Eo: eosinophil, EndoC: endothelial cell, EpC: epithelial cell, F: fibroblast, G: glioblastoma, K: keratinocyte, Mo: monocyte/macrophage, MastC: mast cell, MesC: mesangial cell, MiC: microglia cell, Mk: megakariocyte, Mel: melanoma cells, N: neutrophil, Ne: neuron, NK: natural killer cell, Plt: platelet, T: T lymphocyte.

might be involved in atheroma formation and myocardial infarction risk.[38-40] A polymorphism in the promoter of MCP-1 (−2518A/G) is associated with increased transcription of the MCP-1 gene and homozygous patients were found to be at higher risk for CHD.[41] In contrast, the relationship of another polymorphism of CCR2- is not clear.[42]

Some evidence associates MCP-1 and infectious agents with restenosis and atherosclerosis. Following Chlamydia pneumoniae infection, endothelial cells express MCP-1,[43] while smooth muscle cells infected by cytomegalovirus, express the viral US28 receptor, which is responsive to MCP-1, RANTES and Fractalkine chemokines.[44]

Epidemiological studies have recently implicated Fractalkine and its receptor (CX3CR1) in human atherosclerosis.[45] Fractalkine is a chemokine expressed by inflamed endothelium. It induces leukocyte adhesion and migration, through interaction with its receptor CX3CR1, which, similar to CXCR2 (IL-8R), has also been implicated in the early formation of atheroma.[46] CX3CR1 deficiency in apolipoprotein $E^{-/-}$ mice decreased the formation of atherosclerotic lesions[47] with an evident reduction in macrophage accumulation.

The prevalence and severity of CHD is lower in people heterozygous or homozygous for CX3CR1 polymorphism[48] which provides protection against calcified atherosclerotic lesions and is associated with a lower risk of heart attack/unstable angina.[49]

Recently, it was demonstrated that monocyte-specific recruitment after angioplasty or stent implantation in an animal model was selectively blocked by targeting the MCP-1 receptor CCR2,[50]

Table 3. Chemokine receptors

Receptor	Expression	Ligands
CCR1	DC, T, Mo, N, NK, B, MastC, As	CCL3, 5–9, 14, 15, 16, 23
CCR2	Mo, Eo, N	CCL2, 7, 8, 12, 13
CCR3	Eo, Ba, MastC, N, T, EndoC	CCL5, 7, 8, 11, 13, 15, 24, 26
CCR4	DC, Th2, NK, Mo, Plt, Ba	CCL17, 22
CCR5	Mo, T, DC	CCL3, 4, 5, 8, 14
CCR6	DC, T, NK, B	CCL20
CCR7	T, NK, B, DC, Mo	CCL19, 21
CCR8	T, Mo, Ty	CCL1, 16, 17
CCR9	T, B, Ty	CCL25
CCR10	T, B	CCL2, 7, 26, 27, 28
CXCR1	N, Mo, T, NK, Ba, MastC, EndoC	CXCL1, 7, 8
CXCR2	N, MastC, Mo, T, NK, As, EndoC, Ne	CXCL1–3, 5–8
CXCR3	T, NK, EndoC	CXCL9–11
CXCR4	T, Mo, DC, NK, EndoC, EpC, B, Eo	CXCL12
CXCR5	B, T, Mo	CXCL13
CXCR6	T	CXCL16
XCR1	T, MastC, NK, Ne	CL1–2
CX3CR1	Mo, N, NK, T, MiC, MastC	CX3CL1

As: Astrocyte, B: B lymphocyte, Ba: basophil, DC: Dendritic cell, Eo: Eosinophil, EndoC: Endothelial cell, Mo: Monocyte/macrophage, MastC: Mast cell, MiC: Microglia cell, N: Neutrophil, Ne: neuron, NK: Natural killer cell, Plt: Platelet, T: T lymphocyte, Ty: thymocyte

thus reducing the neointimal hyperplasia. In addition to these inflammatory chemokines, also decreased plasma levels of SDF-1 (a homeostatic chemokine), have been described in patients with angina. SDF-1 mediates anti-inflammatory and matrix stabilizing effects, contributing to plaque stabilization.[51]

Type 2 Diabetes

Type 2 diabetes (T2D) is a complex disease comprising both environmental and genetic factors.[52] Insulin resistance is a key feature in the pathogenesis of T2D and may precede by 10-20 years the onset of hyperglycemia and the clinical manifestation of the disease. Inflammation may predispose to a prediabetic state by increasing insulin resistance, since prediabetic subjects have increased plasma levels of inflammatory proteins without primary defects of pancreas β-cell functions.[53] Excessive amounts of adipose tissue are associated with the development of Type 2 diabetes, an obesity-related disorder.[54] Adipocytes, mainly viewed as fat stores, do have a metabolically active role, secreting a family of cytokines referred as to adipokines[55,56] that exacerbate insulin resistance by desensitising insulin receptors. Recently, chemokines came into the focus of diabetes research, since some studies found that in mouse models and in humans, obesity was associated with infiltration of macrophages into adipose tissue[57,58] Adipose tissue from obese mice exhibited a significant upregulation of immune genes including chemokines.[57,58] Adipocytes express CCR2[59] which when activated by its ligand MCP-1 induces the expression of inflammatory genes and impaired uptake of insulin-dependent glucose. Furthermore, adipocytes synthesize MCP-1, creating conditions for a positive autocrine-feedback loop and potent signals for the recruitment of macrophages.[59] MCP-1 is more expressed in obese mice and is insulin responsive, as demonstrated by the its secretion induced by insulin both in vitro in adipocytes and in obese ob/ob mice in vivo.[60] Obese mice with a deficiency of CCR2 have improved insulin resistance, a finding which provides support for a potentially important role of MCP-1 in the metabolic syndrome.[61]

High glucose levels can also increase MCP-1 secretion by endothelial cells[62] and monocytes.[63] As both insulin and glucose appear to influence MCP-1 secretion, this interaction might be important in T2D which is characterised by elevated glucose and insulin levels. In addition to MCP-1, MIP-1α, IL-8 and IP-10 are also up-regulated and released by adipocytes [56] and may be involved in obesity in animal models. It is important to note that IL-8 expression in adipose tissue and in endothelial cells was found to be positively regulated by glucose.[64,65]

Elevated systemic levels of MCP-1, IL-8 and IP-10 are associated with incident T2D[66,67] and increased expression levels of monocyte CCR2 correlated with glucose control.[68] In addition, the MONICA/KORA study with a follow-up of more than 10 years, demonstrated that whereas IL-8 levels were elevated in T2D patients only, systemic concentrations of RANTES were higher in individuals with T2D and strongly associated with impaired glucose tolerance (IGT),[67] in agreement with the increase of systemic IL-8 levels observed in obese subjects with IGT during an oral glucose tolerance test.[65] The finding that levels of RANTES, but not IL-8, are already significantly increased in subjects with IGT argues for a different role of these chemokines in the development of type 2 diabetes.

It has been suggested that CCR5-mediated recruitment of monocytes and the differentiation of these cells into macrophages in the glomeruli may be associated with the onset and progression of diabetic nephropathy.[69] Renal MCP-1 expression was found in tubulo-interstitial lesions[70] and MCP-1 concentrations were associated with the degree of proliferative retinopathy[71] indicating a potential influence of MCP-1 in the pathogenesis of microangiopathic complications in diabetes. In addition, the diabetic state stimulates the expression of MCP-1 and RANTES by mesangial cells.[72] Finally, RANTES (−28C/G) and CCR5 (59029A/G) gene promoter polymorphisms are independently associated with diabetic nephropathy, suggesting that the RANTES −28G and CCR5 59029A genotypes may be independent risk factors and may have an additive effect on nephropathy.[73]

Osteoarthritis

Osteoarthritis (OA) is a degenerative disease leading to an alteration of metabolic processes and destruction of articular cartilage. The OA disease process affects the entire joint structure, including the synovial membrane, bone, ligaments and periarticular muscles.[74,75] Despite its widespread occurrence in the aged population, the pathogenesis of OA remains largely unknown. In OA, the normal balance between synthesis and degradation of the cartilage matrix is biased toward degradation and it has been shown that cytokines (such as IL-1 and TNF-α, the known catabolic cytokines for cartilage metabolism) as well as functional changes of chondrocytes themselves (behaving like activated macrophages and releasing shared inflammatory mediators) play major roles in the process of deterioration by inducing expression of proteinases such as those of the matrix metalloproteinase (MMP) family.[76-78] In OA, traditionally considered a non-inflammatory arthropathy (since cartilage lacks blood and lymphatic vessels and neural tissue), inflammation has nonetheless been well-documented[79,80] and accumulating evidence suggests the involvement of chemokines and their receptors in the disease process. Fibroblast-like synoviocytes from patients with OA can produce IL8, MCP-1, MIP-1α and RANTES both in vivo and in vitro.[81-83] Furthermore, synovial fluid, blood vessels and cells lining the synovial membrane are all found to contain IL-8, which has been suggested to promote chondrocyte hyperthrophic changes associated with early disease.[84]

Abundant expression of MCP-1, mainly in the intimal lining layer and RANTES diffusely in synovial tissue and synovial fluid, has also been reported.[85,86] MCP-2 and MCP-3, two ligands structurally similar to MCP-1, which influence migration especially of lymphocytes and monocytes, showed a marked expression in the synovial tissue mainly in the intimal lining layer. In addition, the expression of HCC-1, HCC-2 and HCC-4, was also observed.[86] Although most of the described chemokines and receptors appear to be expressed at higher levels also in other inflammatory arthropathies, MIP-1β is found at significantly greater levels in the synovial fluid of patients with OA. MIP-1β, which is a ligand for CCR5, may be responsible for a substantial fraction of chemotactic activity for monocytes, in OA synovial fluid, probably reflecting production by different cell types, since it is poorly produced by unstimulated OA chondrocytes.[87]

Chondrocytes are reported to express mRNA for MCP-1, MIP-1α and RANTES. These chemokines are also present intracellularly and are produced following chondrocyte stimulation. CCR2 and CCR5 receptors were also observed.[88,89] Although the mechanism underlying MCP-1 production by OA chondrocytes is still unclear, fragments of hyaluronan produced from damaged OA cartilage might contribute to MCP-1 production by chondrocytes, as observed in other tissues.[90,91] On the other hand, MCP-1 and RANTES participate in degradation of many components of cartilage (including aggrecans, type II, IX, X, XI collagen, laminin and fibronection) through the regulation of MMP expression, suppressing proteoglycan synthesis and also increasing proteoglycan release.[78] Chondrocytes also express Eotaxin-1. The trigger of eotaxin-1 by pro-inflammatory cytokines further results in enhanced expression of its own receptors (CCR3, CCR5) and of MMP, suggesting that it may play an important role in cartilage degradation in OA.[92] Elevated levels of MCP-1, RANTES and Eotaxin-1 were also observed in patient plasma,[92] while synovial fluid of OA patients contained increased levels of SDF-1, another chemokine able to induce chondrocytes to release MMP.[93] Furthermore, GRO-α is present in the joint fluid and, together with its receptor, is up-regulated in OA cartilage. GRO-α induces articular chondrocyte hyperthrophy and calcification, suggesting a link between inflammation and altered differentiation of articular chondrocytes.[85] It can also activate apoptotic pathways, inducing chondrocyte death and progressive cell depletion.[94] CCR1 and CCR5 are abundantly expressed in synovial tissue and by a large number of synovial macrophages, indicating up regulation of these receptors and/or accumulation of CCR1- and CCR5-positive cells in the inflamed synovial tissue.[95] The expression of CCR1, CCR2, CCR3, CCR5, CXCR1, CXCR2 and CXCR3 is observed on the surface of a limited number of OA chondrocytes but in relatively large numbers inside the cell, in particular CCR3 and CCR5 and CXCR1 and CXCR2 are up-regulated.[75,85,96,97] Furthermore, the expression of CCR5 is modulated by stimulation with RANTES, suggesting an autocrine/paracrine pathway in which the ligand up-regulates the expression of its own receptor.

The development of joint cartilage degeneration in OA is followed by modifications of subchondral bone undergoing a higher metabolism and abnormal production of pro-inflammatory mediators by stromal cells and osteoblasts. Expression of GRO-α, IL-8, MCP-1, MIP-1α and β and RANTES by osteoblasts and stromal cells is increased both in vitro and in bone biopsies from OA patients.[98-100] Furthermore, the bone reabsorbing activity of osteoclast precursors was promoted by the interaction of SDF-1 with its own receptor CXCR4 present on these cells.[101]

Alzheimer's Disease

Alzheimer's disease (AD) is a progressive age-related neurodegenerative disorder that is the most common form of dementia affecting people 65 years and older.[102] The pathologic features of AD are the presence of senile plaques and neurofibrillary tangles in the brain. Senile plaques are extracellular beta-amyloid protein (Aβ) deposits arising from dysregulated metabolism of beta amyloid precursor protein (βAPP), while neurofibrillary tangles are intraneuronal structures composed of tau protein.[103]

A disturbed balance between the production and the degradation of Aβ can trigger chronic inflammatory processes in microglial cells and astrocytes.[104,105] Microglial cells are the most important cells of the innate immune system in the brain. They play the role of cerebral macrophages and recruit and stimulate astrocytes. They can be activated by factors such as brain trauma, ischaemia or neurodegeneration.[105] Exposure of microglial cells to Aβ causes their activation and leads to the production of pro-inflammatory cytokines and chemokines (IL-8, MIP-1α and MCP-1).[106-108] In agreement with these findings, the stimulation of peripheral circulating macrophages with Aβ induces a similar pro-inflammatory response (the production of IL-8, MCP-1, MIP-1α and MIP-1β) and migration across a human blood-brain barrier model, possibly leading to an increased inflammatory burden.[109-111] MCP-1 was demonstrated by immunohistochemistry only in mature senile plaques and in reactive microglia of brain tissues from patients with AD, suggesting that MCP-1—related inflammatory events induced by reactive microglia contribute to the maturation of senile plaques.[112]

Astrocytes, the most common cells in the brain, can be also activated by Aβ peptides to synthesize various pro-inflammatory molecules similar to those produced by microglia.[113] In APP SW transgenic mice, reactive astrocytes were found in close proximity to fibrillary and diffuse Aβ deposits.[114] Electron microscopy revealed that Aβ was also present in astrocyte processes.[115] While previous studies suggested that astrocytes may play a role in Aβ processing, their main function is thought to be associated with the release of pro-inflammatory products. In particular, reactive astrocytes have been shown to secrete pro-inflammatory mediators such as MCP-1 and RANTES in response to stimulation with Aβ42.[110] In turn, these chemokines attract microglial cells which further express pro-inflammatory products, contributing to additional neuronal damage. Enhanced expression of MCP-1 was found after brain damage, suggesting a role in astrocytosis, a characteristic of the AD brain, which represents either a reaction to degrade the increasing amounts of toxic Aβ peptides or an effort to replace dying neurons by astrocyte proliferation.[116]

Recently, MCP-1, IP-10 and IL-8 levels have been evaluated in cerebrospinal fluid. Both MCP-1 and IL-8 levels were higher in patients with amnestic mild cognitive impairment (MCI) and patients with AD, whereas IP-10 levels were increased only in patients with MCI and mild but not severe AD. The presence of inflammatory molecules is likely to be a very early event in AD pathogenesis, preceding the clinical onset of the disease, as demonstrated by subjects with MCI who developed AD over time. IP-10 is specifically increased in MCI and seems to decrease with the progression of AD, whereas MCP-1 and IL-8 are up-regulated also in the late stages of the disease, suggesting a role in phases in which neurodegeneration is prevalent.[117,118] Cerebrospinal fluid MCP-1 levels were higher than in blood also in the presence of an intact blood-brain barrier.[119]

CCR1 is an early and specific marker of AD and it appears to be part of the neuroimmune response to Aβ42-positive neuritic plaques.[120] CCR3, CCR5 and CXCR2 were found elevated in AD brain. In particular, CCR3 and CCR5 were observed on reactive microglia and in senile plaques,[121] while CXCR2 is prevalent in distrophic neuritis.[122]

Polymorphisms at the gene encoding RANTES (–403A/G) and receptors CCR2 (V64I) and CCR5 (Delta32) are not associated with risk or clinical outcome.[123,124] Other studies demonstrated an absence of homozygosity for the polymorphism CCR2-64I, suggesting a protective effect of the mutated allele on the occurrence of AD.[125] Conflicting results have been obtained for polymorphisms at the gene encoding MCP-1 (–2518A/G), reported not to be associated[123,124] or considered an independent risk factor for AD in an Italian population.[126] In addition, a progressive significant increase of MCP-1 serum levels in AD patients carrying one or two G mutated alleles suggested a contribution of this polymorphism to increased inflammatory process occurring in AD.[127]

Future Prospects

Over the past several years, many reports from different scientific disciplines have demonstrated that the field of chemokine activity extends far beyond their chemoattractant properties; a growing mass of evidence now indicates their crucial contributions to a variety of diseases. Since current knowledge suggests that blocking interactions of chemokine ligands with their cognate receptors is a suitable approach to treat these various diseases, at least some of these molecules are potentially interesting targets for biological interventions. The use of small specific inhibitor molecules is becoming an attractive way to target these. In practice, however, chemokine receptors have proven difficult to antagonize, perhaps because of the large surface of interaction with the chemokine ligand. Nonetheless, antagonists of a number of chemokine receptors are in phase 1-2 trials for different clinical indications (such as joint, neurological, viral, pulmonary, intestinal diseases).

Clinical trials to evaluate whether the blockade of CCR2 can diminish insulin resistance in Type 2 Diabetes are in progress (Phase 1 trial, sponsored by Incyte).[8] There is also considerable interest in the use of CCR2 antagonists for the treatment of atherosclerosis. CCR1 also appears to have a role in other joint diseases. Although other receptors and ligands are involved as well, blockade of CCR1 and CCR5 may be a potentially effective therapeutic approach to reduce synovial inflammation in a variety of arthritides.[8] Future clinical trials will demonstrate whether targeting this family is of therapeutic value and which receptors are the best targets for each pathology.

Acknowledgements

This work was partially supported by grants from Bologna University (60% fund), Ricerca Corrente IOR, Italian Health Ministry fund and was performed under the aegis of the EU ImAginE project (QLK6-CT-1999-02031) and more recently the ZINCAGE project (FOOD-CT-2003-506850).

References

1. Licastro F, Candore G, Lio D et al. Innate immunity and inflammation in ageing: a key understanding age-related diseases. Immun Ageing 2005; 2:8.
2. Sarkar D, Fisher PB. Molecular mechanisms of ageing-associated inflammation. Cancer Lett 2006; 236:13-23.
3. Luster AD. Chemokines—chemotactic cytokines that mediate inflammation. N Engl J Med 1998; 338:436-445.
4. Moser B, Loetscher P. Lymphocyte traffic control by chemokines. Nature Immunol 2001; 2:123-128.
5. Ono SJ, Nakamura T, Miyazaki D et al. Chemokines: role in leukocyte development, trafficking and effector function: J Allergy Clin Immunol 2003; 111:1185-1199.
6. Moser B, Wolf M, Walz et al. Chemokines: multiple levels of leukocyte migration control. Trends Immunol 2004; 25:75-84.
7. Moser B, Willimann. Chemokines: role in inflammtion and immune surveillance. Ann Rheum Dis 2004; 63 suppl II:ii84-ii89.
8. Charo IF, Ransohoff RM. The many roles of chemokines and chemokine receptors in inflammation. N Engl J Med 2006; 354:610-621.
9. Pawelec G, Barnett Y, Forsey R et al. T-cells and ageing, 2002. Front Biosci 2002; 7:d1056-d1183.
10. Inadera H, Egashira K, Takemoto M et al. Increase in circulating levels of monocyte chemoattractant protein-1 with aging. J Interferon Cytokine Res 1999; 19:1179-1182.
11. Gerli R, Monti D, Bistoni O et al. Chemokines, sTNF-Rs and sCD30 serum levels in healthy aged people and centenarians. Mech Ageing Dev 2000; 12:37-46.
12. Antonelli A, Rotondi M, Fallahi P et al. Increase of CXC chemokine CXCL10 and CC chemokine CCL2 serum levels in normal ageing. Cytokine 2006; 34:32-38.
13. Wieczorowska-Tobis K, Niemir ZI, Podkowka R et al. Can an increased level of circulating IL-8 be a predictor of human longevity? Med Sci Monit 2006; 12:CR118-CR121.
14. Pulsatelli L, Meliconi R, Mazzetti I et al. Chemokine production by peripheral blood mononuclear cells in elderly subjects. Mech Ageing Dev 2000; 12:89-100.
15. Gabriel P, Cakman I, Rink L. Overproduction of monokines by leukocytes after stimulation with lipopolysaccaride in the elderly. Exp Gerontol 2002; 37:235-247.
16. Gerard C, Rollins BJ: Chemokines and disease. Nature Immunol 2001; 2:108-115.
17. Dong VM, McDermott DH, Abdi R. Chemokines and diseases. Eur J Dermatol 2003; 13:224-230.
18. Wick G, Jansen-Durr P, Berger P et al. Diseases of aging. Vaccine 2000; 18:1567-1583.
19. Rossi D, Zlotinick A. The biology of chemokines and their receptors. Annu Rev Immunol 2000; 18:217-242.
20. Zlotnick A, Yoshie O. Chemokines: a new classification system and their role in immunity. Immunity 2000; 12:121-127.
21. Homey B, Muller A, Zlotnick A. Chemokines: agents for the immunotherapy of cancer? Nat Rev Immunol 2002; 2:175-184.
22. Baggiolini M. Chemokines in pathology and medicine. J Intern Med 2001; 250:91-104.
23. Murphy PM, Baggiolini M, Charo IF et al. International union of pharmacology. XII. Nomenclature for chemokine receptors. Pharmacol Rev 2000; 52:145-176.
24. Ross R. Atherosclerosis—an inflammatory disease. N Engl J Med 1999; 340:115-126.
25. Francisco G, Hernandez C Simò R. Serum markers of vascular inflammation in dyslipemia. Clin Chim Acta 2006; 369:1-16.
26. Davignon J, Ganz P. Role of endothelial dysfunction in atherosclerosis. Circulation 2004; 109: III27-32.
27. Vaapatalo H, Mervaala E. Clinically important factors influencing endothelial function. Med Sci Monit 2001; 7:1075-1085.
28. Gosling J, Slaymaker S, Gu L et al. MCP-1 deficiency reduces susceptibility to atherosclerosis in mice that overexpress human apolipoprotein B. J Clin Invest 1999; 103:773-778.
29. Gu L, Okada Y, Clinton SK et al. Absence of monocyte chemoattractant protein-1 reduces atherosclerosis in low density lipoprotein receptor-deficient mice. Mol Cell 1998; 2:275-281.
30. Boring L, Gosling J, Cleary M et al. Decreased lesion formation in CCR2-/- mice reveals a role for chemokines in the initiation of atherosclerosis. Nature 1998; 394:894-897.

31. Kuziel WA, Dawson TC, Quinones MK et al. CCR5 deficiency is not protective in the early stages of atherogenesis in apoE Knockout mice. Atherosclerosis 2003; 167:25-32.
32. Nelken NA, Coughlin SR, Gordon D et al. Monocyte chemoattractant protein-1 in human atheroma plaques. J Clin Invest 1991; 88:1121-1127.
33. Yu X, Dluz S, Graves DT et al. Elevated expression of monocyte chemoattractant protein-1 by vascular smooth muscle cells in hypercholesterolemic primates. Proc Natl Acad Sci USA 1992; 89:6953-6957.
34. Kowalski J, Okopien B, Madej A et al. Levels of sICAM-1, sVCAM-1 and MCP-1 in patients with hyperlipoproteinemia IIa and IIb. Int J Clin Pharmacol Ther 2001; 39:48-52.
35. Matsumori A, Furukawa Y, Hashimoto T et al. Plasma levels of the monocyte chemotactic and activating factor/monocyte chemoattractant protein-1 are elevated in patients with acute myocardial infarction. J Mol Cell Cardiol 1997; 29:419-423.
36. Nishiyama K, Ogawa H, Yasue H et al. Simultaneous elevation of the levels of circulating monocyte chemoattractant protein-1 and tissue factor in acute coronary syndrome. Jpn Circ J 1998; 62:710-712.
37. Larsson PT, Hallertstam S, Rosfors S et al. Circulating markers of inflammation are related to carotid artery atherosclerosis. Int Angiol 2005; 24:43-51.
38. Wang N, Tabas I, Winchester R et al. Interleukin 8 is induced by cholesterol loading of macrophages and expressed by macrophage foam cells in human atheroma. J Biol Chem 1996; 271:8837-8842.
39. Rothenbacher D, Muller-Scholze S, Herder C et al. Differential expression of chemokines, risk of stable coronary heart disease and correlation with established cardiovascular risk markers. Arterioscler Thromb Vasc Biol 2006; 26:194-199.
40. Herder C, Baumert J, Thorand B et al. Chemokines and Incident coronary heart disease: results from the MONICA/KORA Augsburg case-cohort study, 1984-2002. Arterioscler Thromb Vasc Biol 2006; 26.
41. Szalai C, Duba J, Prohaszka Z et al. Involvement of polymorphism in the chemokine system in the susceptibility for coronary artery disease (CAD): coincidence of elevated Lp(a) and MCP-1 –2518G/G genotype in CAD patients. Atherosclerosis 2001; 158:233-239.
42. Gonzalez P, Alvarez R, Batalla A et al. Genetic variation at the chemokine receptors CCR5/CCR2 in myocardial infarction. Genes Immun 2001; 2:191-195.
43. Molestina RE, Dean D, Miller RD et al. Characterization of a strain of Chlamydia pneumoniae isolated from a coronary atheroma by analysis of the amp 1 gene and biological activity in human endothelial cells. Infect Immun 1998; 66:1370-1376.
44. Streblow DN, Sodenberg-Naucler C, Vieira J et al. The human cytomegalovirus chemokine receptor U28 mediates vascular smooth muscle cell migration. Cell 1999; 99:511-520.
45. Lesnik P, Haskell CA, Charo IF. Decreased atherosclerosis in CX3CR1-/- mice reveals a role for fraktalkine in atherogenesis, J Clin Invest 2003; 111:333-340.
46. Boisvert WA, Santiago R, Curtiss LK et al. A leukocyte homologue of the IL-8 receptor CXCR-2 mediates the accumulation of macrophages in atherosclerotic lesions of LDL receptor deficient mice. J Clin Invest 1998; 101:353-363.
47. Combadiere C, Potteaux S, Gao JL et al. Decreased atherosclerotic lesion formation in CX3CR1/apolipoprotein E double Knockout mice. Circulation 2003; 107:1009-1016.
48. McDermott DH, Halcox JP, Schenke WH et al. Association between polymorphism in the chemokine receptor CX3CR1 and coronary vascular endothelial dysfunction and atherosclerosis. Circ Res 2001; 89:401-407.
49. Moatti D, Faure S, Fumeron F et al. Polymorphism in the fractalkine receptor CX3CR1 as a genetic risk factor for coronary artery disease. Blood 2001; 97:1925-1928.
50. Horvath C, Welt FG, Nedelman M et al. Targeting CCR2 or CD18 inhibits experimental in-stent restenosis in primates: inhibitory potential depends on the type of injury and leukocyte targeted. Circ Res 2002; 90:488-494
51. Damas JK, Waehre T, Yndestad A et al. SDF-1 alpha in unstable angina: potential anti-inflammatory and matrix stabilizing effects. Circulation 2002; 106:36-42.
52. Prentki M, Nolan CJ. Islet β cell failure in type 2 diabetes. J Clin Invest 2006; 116:1802-1812.
53. Festa A, Hanley AJ, Tracy RP et al. Inflammation in the prediabetic state is related to increased insulin resistance rather than decreased insulin secretion. Circulation 2003; 108:1822-1830.
54. Kadowaki T, Yamauchi T, Kubota N et al. Adiponectin and adiponectin receptors in insulin resistance, diabetes and the metabolic syndrome. J Clin Invest 2006; 116:1784-1792.
55. Friedman JM. The function of leptin in nutrition, weight and physiology. Nutr Rev 2002; 60:S1-S14.
56. Juge-Aubry CE, Henrichot E, Meier CA. Adipose tissue: a regulator of inflammation. Best Practice Res Clin Endocrinol Metab 2005; 19:547-566.
57. Weisberg SP, McCann D, Desai M et al. Obesity is associated with macrophage accumulation in adipose tissue. J Clin Invest 2003; 112:1796-1808.
58. Ku H, Barnes GT, Yang Q et al. Chronic inflammation in fat plays a crucial role in the development of obesity-related insulin resistance. J Clin Invest 2003; 112:1821-1830.

59. Gerhardt CC, Romero IA, Cancello R et al. Chemokines control fat accumulation and leptin secretion by cyltured human adipocytes. Mol Cell Endocrinl 2001; 175:81-92.
60. Sartipy P, Loskutoff DJ. Monocyte chemoattractant protein-1 in obesity and insulin resistance. Proc Nat Acad Sci USA 2003; 100:7265-7270.
61. Weisberg SP, Hunter D, Huber R et al. CCR2 modulates inflammatory and metabolic effects on high-fat feeding. J Clin Invest 2006; 116 :115-124.
62. Takaishi M, Taniguchi T, Takahashi A et al. High glucose accelerate MCP-1 production via p38 MAPK in vascular endothelial cells. Biochem Biophys Res commun 2003; 305:122-128.
63. Shanmigam N, Reddy MA, Guha M et al. High glucose-induced expression of proinflammatory cytokine and chemokine genes in monocytic cells. Diabetes 2003; 52:1256-1264.
64. He G, Bruun JM, Lihn AS et al. Stimulation of PAI-I and adipokines by glucose in human adipose tissue in vitro. Biochem Biophys Res Comm 2003; 310:878-883.
65. Straczkowski M, Kowalska I, Nilolajuk A et al. Plasma interleukin 8 concentration in obese subjects with impaired glucose tolerance. Cardiovasc Diabetol 2003; 2:5.
66. Zietz B, Buchler C, Herfarth H et al. Caucasion patients with type diabetes mellitus have elevated levels of monocyte chemoattractant protein-1 that are not influenced by the –2518A/G promoter polymorphism. Diabetes Obes Metab 2005; 7:570-578.
67. Herder C, Baumert J, Thorand B et al. Chemokines as risk factor for type 2 diabetes: results from the MONICA/Kora Ausburg study, 1984-2002. Diabetologia 2006; 49:921-929.
68. Mine S, Okada Y, Tanikawa T et al. Increased expression levels of monocyte CCR2 and monocyte chemoattractant protein-1 in patients with diabetes mellitus. Biochem Biophys Res Comm 2006; 344:780-785.
69. Chow F, Ozols DJ, Nikolic-Paterson RC et al. Macrophages in mouse type 2 diabetic nephropathy: correlation with diabetic state and progressive renal injury. Kidney Int 2004; 65:1007-1012.
70. Wada T, Furuiki K, Sakai N et al. Up-regulation of monocyte chemoattractant protein-1 in tubulointerstitial lesions of human diabetic nephropathy. Kidney Int 2000; 58:1492-1499.
71. Mitamura Y, Takeuki S, Matsuda A et al. Monocyte chemotactic protein-1 in the vitreous of patients with proliferative diabetic retinopathy. Ophthalmologica 2001; 215:415-418.
72. Schlodorff D. Nelson PJ, Luckow J et al. Chemokine and renal disease. Kidney Int. 1997; 51: 610-621.
73. Mokubo A, Tanaka Y, Nakajiama K et al. Chemotactic cytokine receptor 5 (CCR5) gene promoter polymorphism (59029A/G) is associated with diabetic nephropathy in Japanese patients type 2 diabetes: a 10-year longitudinal study. Diabetes Res Clin Pract 2006; 73:89-94.
74. Schurman DJ, Smith RL. Osteoarthritis. Clin Orthop Rel Res 2004; 427S:S183-S189.
75. Yuan GH, Masuko-Hongo K, Sakata M et al. The role of C-C chemokines and their receptors in Osteoarthritis. Arthritis Rheum 2001; 44:1056-1070.
76. Westacott CI, Sharif M. Cytokines in osteoarthritis: mediators or markers of joint destruction? Semin Arthritis Rheum 1996; 25:254.272.
77. Shimmeri M, Masuda K, Kikuchi T et al. Production of cytokines by chondrocytes and its role in proteoglycan degradation. J Rheumatol Suppl 1991; 27:89-91.
78. Borzì RM, Mazzetti I, Marcu K et al. Chemokines in cartilage degradation. Clin Orthop Rel Res 2004; 427S:S53-S61.
79. Smith MD, Triantafillou S, Parker A et al. Synovial membrane inflammation and cytokine production in patients with early osteoarthritis. J Rheumatol 1997; 24:365-371.
80. Haywood L, McWilliams DF, Pearson CI et al. Inflammation and angiogenesis in osteoarthritis. Arthritis Rheum 2003; 48:2173-2177.
81. Villinger PM, Terkeltaub R, Lotz M. Production of monocyte chemoattractant protein-1 by inflammed synovial tissue and cultured synoviocytes. J Immunol 1992; 149:722-727.
82. Seitz M, Loetscher P, Dewald B et al. Production of interleukin-1 receptor antagonist, inflammatory chemotactic proteins and prostaglandin E by rheumatoid and osteoarthritic synoviocytes-regulation by IFN-gamma and IL-4. J Immunol 1994; 152:2060-2065.
83. Volin MV, Shah MR, Takuira M et al. RANTES expression and contribution to monocyte chemotaxis in arthritis. Clin Immunol Immunopathol 1998; 89:44-53.
84. Merz D, Liu R, Johnson K et al. IL-8/CXCL8 and growth-related oncogene alpha/CXCL1 induce chondrocyte hypertrophic differentiation. J Immunol 2003; 171:4406-4415.
85. Conti P, Reale M, Barbacane RC et al. Differential production of RANTES and MCP-1 in synovial fluid from the inflammed human knee. Immunol Lett 2002; 80:105-111.
86. Haringman JJ, Smeets TJM, Reinders-Blankert P et al. Chemokine and chemokine receptors expression in paired peripheral blood mononuclear cells and synovial tissue of patients with rheumatoid arthritis, osteoarthritis and reactive arthritis. Ann Rheum Dis 2006; 65:294-300.

87. Koch AE, Kunkel SL, Shah MR et al. Macrophage inflammatory protein –1 beta: a C-C chemokine in osteoarthritis. Clin Immunol Immunopathol 1995; 77:307-314.
88. Borzì RM, Mazzetti I, Macor S et al. Flow cytimetric analysis of intracellular chemokines in chondrocytes in vivo: costitutive expression and enhancement in osteoarthritis and rheumatoid arthritis. FEBS Lett 1999; 455:238-242.
89. Pulsatelli L, Dolzani P, Piacentini A et al. Chemokine production by human chondrocytes. J Rheumatol 1999; 26:1991-2000.
90. McKee CM, Penno MB, Cowman M et al. Hyaluronan (HA) fragments induce chemokine gene expression in alveolar machropages: The role of HA size and CD44. J Clin Invest 1996; 98:2403-2413.
91. Beck-Simmer B, Oertli B, Pasch T et al. Hyaluronan induces monocyte chemoattractant protein-1 (MCP-1) expression in renal tubular epithelial cells. J Am Soc Nephrol 1998; 9:2283-2290.
92. Hsu YH, Hsieh MS, Liang YC et al. Production of the chemokine eotaxin-1 in osteoarthritis and its role in cartilage degradation. J Cell Biochem 2004; 93:929-939.
93. Kanbe K, Takagishi K, Chen Q. Stimulation of matrix metalloproteinase 3 release from human chondrocytes by the interaction of stromal cell-derived factors 1 and CXC chemokine receptor 4. Arthritis Rheum 2002; 46:130-137.
94. Borzì RM, Mazzetti I, Magagnoli G et al. Growth-related oncogene alpha induction of apoptosis in osteoarthritis chondrocytes. Arthritis Rheum 2002; 46:3201-3211.
95. Haringman JJ, Ludikuize J, Tak PP. Chemokines in joint disease: the key to inflammation? Ann Rheum Dis 2004; 63:1186-1194.
96. Silvestri T, Meliconi R, Pulsatelli L et al. Down modulation of chemokine receptor cartilage expression in inflammatory arthritis. Rheumatology 2000; 42:14-18.
97. Borzì RM, Mazzetti I, Cattini L et al. Human chondrocytes express functional chemokine receptors and release matrix degrading enzymes in response to C-X-C and C-C chemokines. Arthritis Rheum 2000; 43:1734-1741.
98. Lisignoli G, Toneguzzi S, Pozzi C et al. Proinflammaotry cytokine and chemokine production and expression by human osteoblasts isolated from patients with rheumatoid arthritis and osteoarthritis. J Rheumatol 1999; 26:791-799.
99. Lisignoli G, Toneguzzi S, Pozzi C et al. Chemokine expression by subchondral bone marrow stromal cells isolated from osteoarthritis (OA, and rheumatoid arthritis (RA) patients. Clin Exp Immunol 1999; 116:371-378.
100. Lisignoli G, Toneguzzi S, Grassi F et al. Different chemokines are expressed in human arthritis bone biopsies: IFN-γ and IL-6 differently modulate IL-8, MCP-1 and RANTES production by arthritic osteoblasts. Cytokine 2002; 20:231-238.
101. Grassi F, Cristino S, Toneguzzi S et al. CXCL12 chemokine upregulates bone resorption and MMP-9 release by human osteoclasts: CXCL12 levels are increased in synovial and bone tissue of rheumatoid arthritis patients. J Cell Physiol 2004; 199:244-251.
102. Ravaglia G, Forti P, Maioli F et al. Incidence and etiology of dementia in a large elderly Italian population. Neurology 2005; 64:1525-1530.
103. Tuppo EE, Arias HR. The role of inflammation in Alzheimer's disease. Int J Biochem Cell Biol 2005; 37:289-305
104. Akiyama H, Barger S, Barnim S et al. Inflammation and Alzheimer's disease. Neurobiol Ageing 2000; 21:383-421.
105. Blasko I, Stampfer-Kountchev M, Robatscher P et al. How chronic inflammation can affect the brain and support the development of Alzheimer's disease in old age: the role of microglia and astrocytes. Aging Cell 2004; 3:169-176.
106. Rogers J, Lue LF. Microglial chemotaxis, activation and phagocytosis of amyloid β-peptide as linked phenomena in Alzheimer's disease. Neurochem Int 2001; 39:333-340.
107. Lue LF, Brigham EF, Yang LB et al. Inflammatory repertoire of Alzheimer's disease and nondemented elderly microglia in vitro. Glia 2001; 35:72-79.
108. Franciosi S, Choi HB, Su K et al. IL-8 enhancement of amyloid-beta (Abeta 1-42)-induced expression and production of pro-inflammatory cytokines and COX-2 in cultured human microglia. J Neuroimmunol 2005; 159:66-74.
109. Fiala M, Zhang L, Gan X et al. Amyloid-beta induces chemokine secretion and monocyte migration across a human blood-brain barrier model. Mol Med 1998; 4:480-489.
110. Smits HA, Rijsmus A, van Loon JH et al. Amyloid-beta-induced chemokine production in primary human macrophages and astrocytes. J Neuroimmunol. 2002; 127:160,168.
111. McGeer EG, McGeer PL. Inflammatory processes in Alzheimer's disease. Prog Neuo-Psychopharmacol Biol PsY 2003; 27:741-749.
112. Ishizuka K, Kimura T, Igatayi R et al. Identification of monocyte chemoattractant protein-1 in senile plaques and reactive microglia of Alzheimer's disease. Psychiatry Clin Neurosci 1997; 1:135-138.

113. Savchenko VL, McKanna JA, Nikonenko IR et al. Microglia and astrocytes in the adult rat brain: comparative immunocytochemical analysis demonstrates the efficacy of lipocortin 1 immunoreactivity. Neuroscience 2000; 96:195-203.
114. Benzing WC, Wujek JR, Ward EK et al. Evidence for glial-mediated inflammation in aged APP(SW) transgenic mice. Neurobiol Aging 1999; 20:581-589.
115. Kurt MA, Davies DC, Kidd M. B-amyloid immunoreactivity in astrocytes in Alzheiner's disease brain biopsies: an electron microscope study. Exp Neurol 1999; 158:221-228.
116. Little AR, Benkovic SA, Miller DB et al. Chemically induced neuronal damage and gliosis : enhanced expression of the pro-inflammatory chemokine, monocyte chemoattractant protein (MCP-1), without a corresponding increase in proinflammatory cytokines. Neuroscience 2002; 115:307-320.
117. Galimberti D, Fenoglio C, Lovati C et al. Serum MCP-1 levels are increased in mild cognitive impairment and mild Alzheimer's disease. Neurobiol Aging 2006; 27:262-269.
118. Galimberti D, Schoonenboom N, Scheltens P et al. Intrathecal chemokine syntesis in mild cognitive inpairment and Alzheimer's disease. Arch Neurol 2006; 63:538-543.
119. Galimberti D, Schoonenboom N, Scarpini E et al. Chemokines in serum and cerebrospinal fluid of Alzheimer's disease patients. Ann Neurol 2003; 53:547-548.
120. Halks-Miller M, Schroeder ML, Haroutunian V et al. CCRl is an early and specific marker of Alzheimer's. Ann Neurol 2003; 54:638-646.
121. Xia MQ, Qin SX, Wu LJ et al. Immunohistochemical study of the beta-chemokine receptors CCR3 and CCR5 and their ligand in normal and Alzheimer's disease brains. Am J Pathol 1998; 153:31-37.
122. Hesselgesser J, Horuk R. Chemokine and chemokine receptor expression in the central nervous system. J Neurovirol 1999; 5:13-26.
123. Huerta C, Alverez V, Mata IF et al. Chemokines (RANTES and MCP-1) and chemokine receptors (CCR2 and CCR5) gene polymorphisms in Alzheimer's and Parkinson's disease. Neurosci Lett 2004; 370:151-154.
124. Combarros O, Infante J, Liorca J et al. No evidence for association of the monocyte chemoattractant protein-1 (–2518) gene polymorphism and Alzheimer's disease. Neurosci Lett 2004; 360:25-28.
125. Galimberti D, Fenoglio C, Lovati C et al. CCR2-64I polymorphism and CCR5Delta32 deletion in patients with Alzheimer's disease. J Neurol Sci 2004; 225:79-83.
126. Pola R, Flex A, Gaetani E et al. Monocyte chemoattractant protein-1 (MCP-1) gene polymorphism and risk of Alzheimer's disease in Italians. Exp Gerontol. 2004; 39:1249-1252.
127. Fenoglio C, Galimberti D, Lovati C et al. MCP-1 in Alzheimer's disease patients: A-2518G polymorphism and serum levels. Neurobiol Aging 2004; 25:1169-1173.

Chapter 10

The Efficacy of Vaccines to Prevent Infectious Diseases in the Elderly

Dietmar Herndler-Brandstetter and Beatrix Grubeck-Loebenstein*

Abstract

Infectious diseases still represent a major challenge to human progress and survival. Especially elderly persons are more frequently and severely affected by infectious diseases and they display distinct features with respect to clinical presentation and treatment. Although vaccinations are considered a vital medical procedure for preventing morbidity and mortality caused by infectious diseases, the protective effect of vaccinations is abrogated in elderly persons. This is due to a decline in the functions of the immune system referred to as immunosenescence. The first part of this chapter will therefore summarize the status quo of the efficacy of vaccines in preventing morbidity and mortality caused by typical infectious diseases in the elderly, such as influenza, pneumonia and tuberculosis. The second part will then elucidate the underlying age-related mechanisms which may contribute to the decreased efficacy of vaccines. Based on the complex mechanisms involved in immunosenescence, strategies will be outlined which may be successful in enhancing protective immune responses following vaccination in elderly persons.

Introduction

With respect to the current demographic development in many countries, including the European Union and the United States of America, infectious diseases in geriatric patients are becoming an increasingly important issue. Infections in elderly persons are not only more frequent and more severe, but they also have distinct features regarding clinical presentation, microbial epidemiology and treatment. Urinary tract infections, lower respiratory tract infections, skin and soft tissue infections, infective endocarditis, bacterial meningitis, tuberculosis and herpes zoster appear to have a higher prevalence in elderly persons. In developed countries like the United States, pneumonia, influenza and septicemia are ranked among the ten major causes of deaths in people aged 65 years and older.[1] The reasons for the increased susceptibility to infectious diseases include epidemiological elements, immunosenescence, malnutrition and age-dependent anatomical alterations.

Infectious diseases still represent a major challenge to human progress and survival as they are responsible for about 20% of all deaths in the world. This is not only related to microbial and viral factors but also to social and environmental determinants, such as social upheaval, urbanization, air travel, natural disasters and climate change.[2] Newly emerging infectious diseases include acquired immune deficiency syndrome (AIDS), hepatitis C, several hemorrhagic fevers, severe acute respiratory syndrome (SARS) and avian influenza. The resurgence of several other infectious diseases is supported by the increased occurrence of multiple drug-resistant microorganisms such as *Staphylococcus aureus, Mycobacterium tuberculosis, Escherichia coli* and *Streptococcus pneumoniae.*

*Corresponding Author: Beatrix Grubeck-Loebenstein—Institute for Biomedical Aging Research, Austrian Academy of Sciences, Rennweg 10, 6020 Innsbruck, Austria. Email: beatrix.grubeck-loebenstein@oeaw.ac.at

Immunosenescence, edited by Graham Pawelec. ©2007 Landes Bioscience and Springer Science+Business Media.

Altogether, this represents an enormous economic burden on health care systems all over the world. For instance, the annual costs of medical care for treating infectious diseases in the United States alone is about $120 billion and for treating antimicrobial-resistant infections it may be as high as $5 billion.[3]

A great success story was the implementation of large-scale vaccination strategies that led to the eradication of smallpox in 1980[4] and to a drastic reduction of poliomyelitis, tetanus, diphtheria, measles, pertussis and meningitis. Presently, vaccinations are still considered the most cost-effective medical procedure for preventing morbidity and mortality caused by infectious diseases. 26 different infectious diseases can be prevented by vaccinations and 61 vaccines are being developed according to a 2004 survey by the Pharmaceutical Research and Manufacturers of America.[5] The new candidate vaccines are intended to provide protection against diseases caused by rotavirus, herpes zoster and papilloma virus and will be available from 2007 onwards (Table 1). But also improved vaccines against influenza, pneumonia and tuberculosis are currently being tested in clinical trials (Table 1). This chapter now outlines the relevance of vaccines to fight infectious diseases in old age and how age-related changes within the immune system contribute to the decreased efficacy of vaccines. It also discusses the progress made in the development of vaccines with improved immunogenicity in elderly persons.

The Role of Vaccines in Fighting Infectious Diseases in Old Age

Outbreaks of deadly infectious diseases such as Ebola, Marburg, SARS or the H5N1 avian influenza regularly alert the world, whereas there is not much public attention paid to infectious diseases that cause substantial morbidity and mortality among the elderly population. For instance, influenza, invasive *Streptococcus pneumoniae* infection, urinary tract and skin infections have a higher prevalence in elderly persons.[6] Old individuals may also fail to respond sufficiently to therapy and frequently suffer from opportunistic infections, recurrent infections with the same pathogen or reactivation of latent diseases, such as those caused by *Mycobacterium tuberculosis* or the Varicella zoster virus. There are no vaccines available for many infectious pathogens that are frequent in elderly subjects and existing vaccines are underused and often do not assure such an effective protection as in young subjects. The following paragraphs will highlight the most important infectious diseases which threaten the elderly population and will provide information on epidemiology, vaccine availability and efficacy, vaccination coverage and general health authority recommendations.

Influenza

Influenza is a highly contagious, acute viral respiratory disease that causes significant morbidity and mortality. The annual outbreaks affect approximately 5-20% of the population worldwide with 3-5 million cases of severe illness and up to one million deaths each year. Especially elderly people and persons that are chronically ill or otherwise immunocompromised are at enhanced risk. For example, during influenza epidemics, Barker and Mullooly reported two deaths per 100,000 healthy people below 65 years of age compared with 797 per 100,000 in those over 65 with two or more high-risk conditions.[7] In contrast to measles, smallpox and poliomyelitis, influenza is caused by viruses that undergo continuous antigenic variation and possess an animal reservoir. Therefore, we are recognizing annual epidemics that have been interrupted by three pandemics (Spanish influenza, H1N1, 1918-1919; Asian influenza, H2N2, 1957-1958 and Hong Kong influenza, H3N2, 1968), caused by new influenza virus strains with increased virulence.

Influenza viruses are enveloped viruses containing eight single-stranded RNA segments which encode for viral proteins, such as hemagglutinin (HA), neuraminidase (NA), matrix protein (M1) and nucleoprotein (NP) (Fig. 1). Influenza viruses belong to the family Orthomyxoviridae and are divided into three genera, influenza virus A, B and C, based on antigenic differences in two of their structural proteins, M and NP. Disease symptoms caused by Influenza C are rare whereas Influenza B often causes sporadic outbreaks, especially in residential communities like nursing homes. Influenza A viruses are further divided into subtypes according to the antigenicity of their

Table 1. The developmental status of vaccines against some human pathogens

Disease or Pathogen	Product Name	Company	Type of Vaccine	Developmental Status
Cytomegalovirus	-	Chiron	subunit	phase II
	-	Sanofi Pasteur	live-attenuated	phase II
	-	Vical	nucleic acid	phase I
Herpes zoster (Shingles)	Zostavax	Merck and Co	live-attenuated	preregistration
	-	GSK	subunit	phase I
Human papillomavirus	Cervarix	GSK, MedImmune	virus-like particle L1 + adjuvant AS04	preregistration
	Gardasil	Merck and Co	virus-like particle L1	preregistration
Influenza	CAIV-T	MedImmune, Wyeth	live-attenuated	licensed in the US
	FluMist	MedImmune	nasal, live-attenuated	phase III
	-	Chiron	cell-culture based	phase III (Europe)
	FluBLOK	Protein Sciences	subunit, cell-culture based	phase III
	FluINsure	ID Biomedical	nasal, subunit	phase III
	Fluviral	ID Biomedical	split-virus	phase III
	-	GSK	subunit	phase I, II
	-	Protein Sciences	subunit (H5)	phase I
Pneumonia	-	Wyeth	9-valent conjugate	phase III
	Streptorix	GSK	11-valent conjugate	phase III
	StreptAvax	ID Biomedical	subunit	phase II
	-	GSK	subunit	phase I
Rotavirus	Rotateq	Merck and Co	live-attenuated	preregistration
	Rotarix	GSK, Avant Ther.	oral, live-attenuated	preregistration
Tuberculosis	rBCG30	Aeras Global TB Vaccine Foundation	live-attenuated	phase I
	-	GSK	subunit + adjuvant AS02A	phase I
	-	Corixa	subunit	phase I
Varicella, Mumps,		GSK	live-attenuated	phase III
Measles, Rubella		Merck and Co	live-attenuated	phase III

GSK, GlaxoSmithKline; adapted from ref. 88.

major envelope glycoproteins, HA and NA. With at least 15 different hemagglutinin and 9 different neuraminidase subtypes, there is considerable antigenic variation among influenza viruses. The human influenza viruses are currently limited to three hemagglutinin (H1, H2 and H3) and two neuraminidase subtypes (N1 and N2), whereas birds are the predominant hosts for the other subtype strains. HA initiates viral infection by binding to sialic acid residues on the carbohydrates of glycoproteins present on epithelial cells of the respiratory tract. Therefore, high-affinity IgA and IgG antibodies against HA may prevent infection from influenza virus. In contrast, NA cleaves the sialic acid from viral and cellular proteins to promote the release of newly synthesized influenza viruses from the infected host's plasma membrane.

Although antiviral drugs with moderate efficacy are available, active immunization represents the most vital element in the prophylaxis of influenza disease. However, the frequently occurring

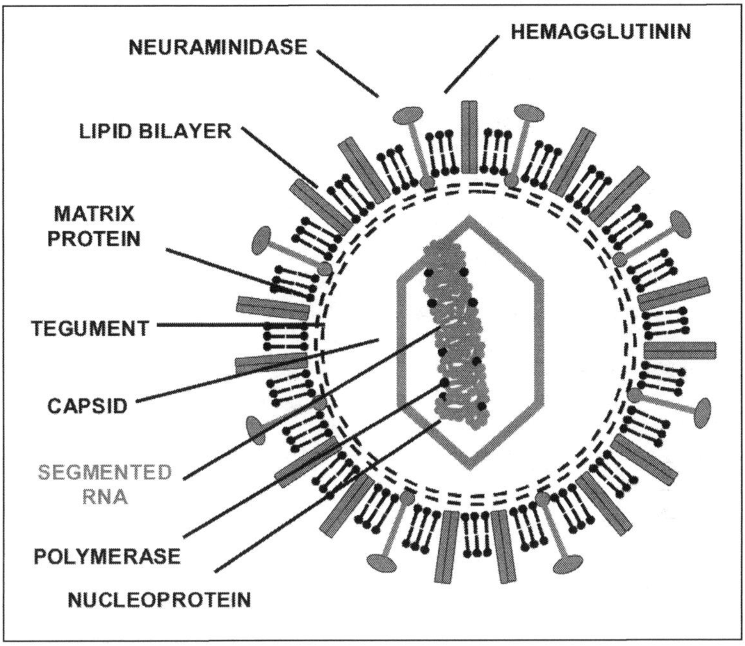

Figure 1. Schematic representation of the influenza virus structure. See text for more details.

antigenic drift requires an annual modification of the vaccine components according to the recommendations of the WHO. Therefore, vaccination has to be repeated annually to ensure protection against the circulating influenza strains. But vaccination coverage differs largely within European countries. In 2002, the rate of vaccine distribution was highest in Spain, Belgium, The Netherlands, United Kingdom and Germany (between 18.1 and 20.3%) and lowest in Poland, Czech Republic, Lithuania and Latvia (between 1.9 and 7.1%). Canada and the United States had the highest rate of vaccine distribution, being 32.8 and 28.9%, respectively. Remarkably, 70% of US citizens aged 65 and above have been vaccinated against influenza.[8] Although there are several vaccines available, the efficacy of many vaccines in preventing influenza disease in elderly persons is only around 56%.[9] Especially very old and frail persons show a decreased response to influenza vaccines.[10] The reduced vaccine efficacy is due to low levels of IgA and IgG antibodies, delayed peak antibody titers and shortened maintenance of titers after vaccination. Nevertheless, immunization in elderly people has been shown to be safe, cost-effective and associated with reduced rates of hospitalization and influenza-related deaths.[11,12] In particular, the efficacy of influenza vaccination to reduce mortality in elderly people is greater after repeated annual vaccination than after first administration.[13] Presently, influenza vaccines can be classified in split-virus, subunit, virosomal and live-attenuated vaccines (Table 2). Split-virus vaccines are used since the 1980s, are cheap and offer a good protection for children above 6 months of age and adults. Recently, subunit vaccines with new adjuvants have been developed (Fluad®, Addigrip®) that show an increased immunogenicity, a favorable safety profile and may be more suitable for the vaccination of elderly persons.[14] Additionally, invariant antigens, such as M1 and NP, may also play an important role in protection and could be used in vaccines to induce long-lasting immunity to a variety of different influenza strains. Another strategy to enhance immunogenicity may the use of virosomes that are nontoxic, biodegradable lipid-based antigen-presentation systems.[15] Virosomal influenza vaccines, such as Inflexal V®, Influvac Plus® and Invivac® have been on the market in several European countries for a number of years.

Table 2. An assortment of vaccines available in the European Union

Disease	Registered Name	Company	Vaccine Composition	Type of Vaccine
Diphtheria,	dT reduct	Sanofi Pasteur	dT	toxoid
Tetanus,	Tdpur	Chiron	dT	toxoid
Pertussis	Polio Salk	Sanofi Pasteur	IPV	inactivated
and/or	Boostrix	GSK	dTaP	subunit (aP)
Poliomyelitis	Repevax	Sanofi Pasteur	dTaP-IPV	-
	Hexavac	Sanofi Pasteur	dTaP-HiB-Polio-HepB	-
	Infanrix-HiB+IPV	GSK	dTaP-HiB-IPV	-
	Infanrix Hexa	GSK		-
Hepatitis A	Avaxim	Sanofi Pasteur	*	inactivated
	Epaxal	Berna	*	virosome
	Havrix	GSK	*	inactivated
	Vaqta	Merck and Co	*	inactivated
	Twinrix	GSK	HepA+B	inactivated (HepA)
Influenza	Addigrip	Sanofi Pasteur	*	subunit + adjuvant
	Begrivac	Chiron	*	split-virus
	Fluad	Chiron	*	subunit + adjuvant
	Fluarix	GSK	*	split-virus
	Fluvirin	Chiron	*	subunit
	Fluzone	Sanofi Pasteur	*	split-virus
	Inflexal V	Berna	*	virosome
	Influvac Plus	Solvay	*	virosome
	Invivac	Solvay	*	virosome
	Sandovac	Sanofi Pasteur	*	subunit
	Vaxigrip	Sanofi Pasteur	*	split-virus
Pneumonia	Prevnar	Wyeth	*	7-valent conjugate
	Pneumo 23 "Merieux"	Sanofi Pasteur	*	23-valent conjugate
	Pneumovax 23	Merck and Co	*	23-valent conjugate
Tick-borne	Encepur	Chiron	*	inactivated whole virus
encephalitis	FSME immune	Baxter	*	inactivated whole virus
Varicella	Varilrix	GSK	*	live-attenuated
(Chickenpox)	Varivax II	Merck and Co	MMR-V	live-attenuated
Yellow fever	Stamaril	Sanofi Pasteur	*	live-attenuated
	YF-Vax	Sanofi Pasteur	*	live-attenuated
	Arilvax	Chiron	*	live-attenuated

aP, acellular pertussis; d, diphtheria; GSK, GlaxoSmithKline; HepA+B, hepatitis A+B; HiB, haemophilus influenzae B; IPV, inactivated polio virus; MMR-V, measles-mumps-rubella-varicella; T, tetanus; *, vaccine that protects only against the pathogen indicated under "Disease"; adapted from ref. 88.

They display a high immunogenicity and a similar safety profile in elderly persons compared with inactivated influenza vaccines.[16-18] Furthermore, new live-attenuated and subunit influenza vaccines are currently in clinical trials that promise to have an increased efficacy of protection (Table 1). Especially live-attenuated influenza vaccines are believed to elicit strong T-cell responses and should be able to enhance antibody levels after vaccination. Importantly, before administration of live-attenuated vaccines to elderly or immunocompromised persons, an acceptable safety profile has to be demonstrated. Irrespective of the improvement of influenza vaccines for elderly persons, vaccination of children is clinically effective[19] and high vaccination coverage among pupils has demonstrated to induce herd immunity and to reduce mortality in older adults.

Pneumonia

Streptococcus pneumoniae is an important cause of invasive clinical manifestations such as bacterial pneumonia, meningitis and septicemia, particularly in young children and the elderly. 80 to 90% of deaths associated with *Streptococcus pneumoniae* infection occur in people aged 60 years and above. Several further groups at higher risk of invasive pneumococcal disease have been defined, including individuals with splenic dysfunction, immunosuppression, chronic pulmonary or cardiac disease, diabetes mellitus and chronic liver disease. Generally, antibiotic therapy has to be initiated as soon as possible to reduce the risk of complications due to pneumonia, meningitis or sepsis. Nonetheless, 50% of all deaths occur within the first 48 hours despite adequate antibiotic therapy. This may be due to the increased occurrence of multiple drug-resistant pneumococcal strains. In 2002, the proportion of penicillin-resistant *Streptococcus pneumoniae* was reported to be 53% in France and more than 25% in Israel, Poland, Romania and Spain.[21] After recovering from pneumococcal infection, people are not necessarily immune, because there are about 90 different serotypes and immunity will be guaranteed only to the strain that has caused the infection. Currently, pneumococcal vaccines are available that include up to 23 strains which are responsible to cause disease in almost 90% of all cases (Table 2). These vaccines offer protection against invasive pneumococcal disease in 65% of the general elderly population [22,23] whereas in elderly persons with high risk factors, the protective effect of vaccination seems to be only moderate. Although many European countries recommend the administration of pneumococcal vaccines to all those >65 years of age, vaccination coverage among the elderly population is very low. This may be due to the high costs of the vaccine and its unsatisfying efficacy in elderly people. But more immunogenic vaccines are currently in different phases of clinical trials (Table 1) and promise to be more efficient in old age. Additionally, implementing pneumococcal vaccination for children may decrease the incidence of pneumococcal disease in the elderly by reducing transmission and possibly accomplishing herd immunity.[24]

Tuberculosis

Each year, about 8 million people are infected worldwide with the tubercle bacillus *Mycobacterium tuberculosis* and 1.6 million of them die. The EU25 has a tuberculosis (TB) burden of more than 50,000 new cases per year, with the highest incidences in Latvia, Lithuania and Estonia (50-100 cases/100,000). The risk of developing a disease following TB infection is about 5-10% during lifetime and individuals above 65 years of age have a four-fold increased risk of developing TB than the average population.[25] TB is also frequently diagnosed with delay due to an atypical manifestation in old age. This may lead to an increased morbidity and mortality and to a spreading of the disease, in particular within institutionalized elderly persons.[26] Further difficulties include the increased emergence of new, multiple drug-resistant strains with higher transmissibility,[27] the poor efficacy of the current bacille Calmette Guérin (BCG) vaccine in protecting adults and elderly people from pulmonary infection[28] and the increased risk of TB co-infection in HIV positive patients.[29] However, in the past few years, several TB vaccine candidates have entered phase I clinical trials, including adjuvanted subunit vaccines as well as improved live recombinant strains of the current BCG vaccine (Table 1). All these vaccine candidates are supposed to induce an effective and sustainable cellular immune response which is thought to be crucial to protect the host from an intracellular pathogen such as *Mycobacterium tuberculosis*.[30]

Herpes Zoster

Primary infection with the Varicella zoster virus (VZV) causes chickenpox which is usually a mild disease in childhood. The virus then persists in a latent form in sensory ganglia until its reactivation which results in the clinical manifestation of herpes zoster (shingles). Between 13 and 26% of persons with herpes zoster develop complications, such as postherpetic neuralgia.[31] Postherpetic neuralgia also increases with age with a prevalence of 50% in people aged 70 years and above. The incidence and severity of herpes zoster increase with age, because VZV reactivation is associated with a progressive decline in cell-mediated immunity to VZV.[32,33] Routine vaccination of children using a tetravalent vaccine that protects against measles, mumps, rubella and varicella will soon be available (Table 1) and may reduce the incidence of chickenpox as well as the reactivation of VZV in later life. Since 1995, a live-attenuated Oka strain VZV vaccine is on the market that has shown clinical efficacy in preventing children from chickenpox.[34] However, the currently available VZV vaccines have not been proven to adequately boost T-cell responses in older adults and to prevent reactivation of herpes zoster. Recently, a vaccine that may prevent herpes zoster virus reactivation has been submitted for registration. This live-attenuated VZV vaccine has been developed to prevent reactivation of herpes zoster in the elderly.[35,36] This is of particular importance, because the elderly population has not been vaccinated against but may have been frequently infected by VZV. For instance, more than 90% of adults in the United States have had chickenpox. As a consequence, it is estimated that up to 800,000 people in the United States suffer from shingles each year and the incidence is expected to increase as the population ages. Thus, reactivation of herpes zoster and its clinical manifestations represents a serious health burden to the growing elderly population and could be counteracted by potent vaccines.

Cytomegalovirus

The cytomegalovirus (CMV), a B-herpesvirus, has also been shown to persist throughout life until its reactivation as a result of immune suppression or deficiency. CMV infection is quite common and affects 60-100% of the adult population, depending on the area. The CMV is transmitted via person-to-person contacts but immunocompetent subjects mostly do not recognize infection as it causes no or few unspecific symptoms. However, a CMV infection represents a severe health problem in immunocompromised persons (e.g., due to immunosuppressive disease, chemotherapy or transplantation) or in a fetus as a result of congenital infection. Research results over the past decade suggest that CMV favors an accelerated aging of the immune system as CMV infection is chronic and the organism is forced to continuously prevent virus reactivation.[37-39] Despite the high frequency of CMV-specific $CD8^+$ T-cells, the virus usually can not be eliminated by the immune system. This is because the virus has evolved several mechanisms to escape the host's immune defense.[40] For instance, CMV encodes for a type of proteins called immunoevasins that modulate the presentation of viral peptides or directly suppress cellular immune responses. Hence, the accumulation of CMV-specific T-cells substantially constricts the diversity of the T-cell repertoire[41] and leads to the production of proinflammatory cytokines, such as gamma interferon and tumor necrosis factor alpha.[37] This imbalance in the cytokine production profile may not only promote the pathogenesis of age-related diseases[42] but leads to a decreased production of antibodies following influenza vaccination in elderly persons.[43] A few antiviral substances including ganciclovir, valganciclovir, foscarnet and cidofovir are available to prevent CMV infection in immunocompromised patients. But antiviral therapy is limited by its severe adverse reactions, such as neutropenia, nephrotoxicity, hypocalcemia and seizures. Another strategy is the adoptive transfer of donor-derived CMV-specific $CD4^+$ and $CD8^+$ T-cells that may restore the host's immunity against CMV.[44] Despite the need of a safe and potent vaccine that prevents CMV disease, no vaccine candidate has yet entered the market. A few vaccines against CMV are currently in phase I/II clinical trials (Table 1). Active immunization against CMV could reduce the incidence of neonatal infections as well as complications in immunocompromised persons and may prevent CMV-associated premature aging of the immune system when applied early in life.

Pertussis

Pertussis (whooping cough) is a highly contagious respiratory system infection caused by the bacterium *Bordetella pertussis* and rarely by *B. parapertussis, B. bronchiseptica* or other pathogens.

Each year, more than 20 million cases of pertussis are reported worldwide, 90% of which occur in developing countries, with an estimated 200,000 to 300,000 fatalities. The implementation of routine childhood vaccination against pertussis has reduced the high mortality rate among children. Although most infants are being immunized against pertussis in industrialized countries, immunity usually fades during adolescence. Consequently, a significant rise in pertussis incidence has been noticed in adolescents and adults.[45] However, the reported pertussis cases in adults and elderly people are likely to be underestimated because symptoms of disease may be characterless and make clinical diagnosis difficult. Among nonvaccinated elderly people the attack rate of pertussis is high (53%) and up to 10% of elderly persons may die from intracranial bleeding while they are symptomatic for pertussis.[46] Regular booster immunizations should thus be considered for adults and elderly persons, which is indispensable to remain protected from disease.

Tetanus and Diphtheria

Tetanus is acquired via environmental exposure to the spores of *Clostridium tetani*, which are present in soil worldwide. The disease is caused by a potent neurotoxin produced by the bacterium in dead tissue, e.g., dirty wounds. Diphtheria is a bacterial disease caused by *Corynebacterium diphtheriae* and is transmitted from person to person through close physical contact. The public health burden of both diseases has been low in developed countries due to routine immunization. However, outbreaks of diphtheria have been reported in the independent states of the former Soviet Union, Algeria, China, Iraq, Sudan, Thailand and other countries. Thus, maintaining high vaccination coverage is important to prevent the outbreak of new diphtheria epidemics. Although vaccines that prevent from tetanus and diphtheria have been used for routine immunization for a long time all over the world, few studies exist that document their efficacy in elderly people. The vaccination coverage among elderly subjects is decreasing in several European countries and up to 40% of appropriately vaccinated elderly persons do not have protective tetanus-specific antibody concentrations.[47-49] Therefore, public health authorities of some European countries have recommended five instead of ten year booster vaccination intervals for people over 60 years of age. Additionally, strategies should be developed to draw public attention to the problem of immunizations in the elderly, to inform general practitioners and to increase vaccination acceptance.

Travel Vaccines

The increasing mobility of elderly persons recognized worldwide is accompanied by an enhanced risk to encounter new antigens. This may be of concern because elderly persons possess a limited T-cell repertoire that may not guarantee full responsiveness to a wide variety of new antigens (see below for details). Nevertheless, in vitro experiments have demonstrated that naive T-cells from elderly persons can still be stimulated by neoantigens, at least to the recombinant Etr protein of TBE virus and rabies virus.[50] Based on an assessment of the risks for travel-related diseases, including the destination, the type of journey and the duration, vaccination is recommended to protect from typhoid and yellow fever, hepatitis A and B, Japanese encephalitis, tick-borne encephalitis (TBE) or rabies. But elderly persons should also check whether they have followed the recommended booster intervals of routine immunizations, e.g., against tetanus, diphtheria, poliomyelitis, measles or influenza.

TBE is caused by a virus that is primarily transmitted to humans by infected ticks. There are three genetically closely related subtypes of the TBE virus known (European, Siberian and Far Eastern subtype). TBE is among the most dangerous neuro-infectious diseases in Europe and Asia and is responsible for up to 12,000 cases of TBE annually, most of them occurring in Russia, Czech Republic and the Baltic states.[51] Up to 30% of adults with clinically confirmed TBE infection develop meningitis or meningoencephalitis and the lethality of TBE in Europe is up to 1%. Yet, there is no specific therapy available and, therefore, active immunization with inactivated whole virus provides the only efficient protection from TBE disease (Table 2).[52] Importantly, more and more TBE cases are reported in people over 50 years of age and vaccination coverage in this population is lower than average. Therefore, future strategies should increase the vaccination coverage among elderly persons and assure that they stick to regular booster intervals. Anyhow, regular boosters should be given throughout life as this may favor the maintenance of long-lasting humoral immunity against TBE[53] and may decrease the risk of immunization failures in the elderly.

Hepatitis A is an acute disease of the liver caused by the hepatitis A virus (HAV), a nonenveloped virus belonging to the Picornaviridae family. Each year, an estimated 1.5 million cases of hepatitis A occur worldwide. HAV infection induces life-long immunity and is usually asymptomatic in young children, whereas adults frequently experience symptomatic disease. HAV is acquired directly from infected persons by close contact or by the consumption of contaminated drinking water, vegetables, fruits or shells. HAV vaccination is recommended when traveling to tropic and subtropic countries that have an increased risk of infection. For instance, the risk of hepatitis A infection of persons traveling to developing countries was estimated to be 3 to 20 cases per 1000 persons per month of stay, varying with destination, living conditions and age.[54] Improved sanitary standards in developed countries have reduced the opportunity for environmental exposure to HAV and have lowered the overall incidence of infection. Paradoxically, susceptibility to the virus increased because of the decrease in natural immunity. Consequently, less than 20% of persons born after 1945 have a natural immunity against HAV.[54] In contrast to hepatitis B and C that may lead to the manifestation of a chronic infection, clinical illness after hepatitis A infection is usually mild in young individuals. But increasing age represents an enhanced risk of severe infection and mortality rates are about 2% for persons over 40 and 4% for those over 60 years of age.[54] Several vaccines against hepatitis A are available (Table 2) and a study of 773 adults showed that immunogenicity and safety profiles between Twinrix® and Havrix® are comparable.[56] But there is some evidence of lower antibody titers with advanced age. For instance, the seroconversion rates 8 months after two doses of Havrix® were found to be 85% and 60% for adults < 35 years and >35 years, respectively.[57] After the recommended immunization schedule with Twinrix®,seroprotection was 92% and 63% for adults <40 years and >60 years, respectively.[58] Therefore, it may be useful to measure HAV antibodies in elderly persons, as in the case of vaccination failure, boosters have shown to be effective.[58] It is further recommended that the vaccine is given at least 3 to 4 weeks before travel due to a slower onset of the antibody response in elderly individuals.[59]

Another travel vaccine is directed against yellow fever (YF), which is endemic in tropic regions of Africa and South America. YF is transmitted by the bite of infective Aedes aegypti and other mosquitoes that bite during daylight hours in regions below 2500 meters of altitude. Most infections lead to an acute illness characterized by fever, muscular pain, headache, anorexia, nausea and/or vomiting, often with bradycardia. After a few days, about 15% of patients progress to a second phase, with resurgence of fever, development of jaundice, abdominal pain and haemorrhagic manifestations. Half of these persons die 10-14 days after the onset of illness. The WHO estimates that a total of 200,000 cases of YF occur each year, with about 30,000 deaths.[60] YF also represents a significant risk to more than 3 million travelers that visit YF-endemic areas each year. Neonates and elderly individuals demonstrate the highest mortality when infected by the YF virus. As there is no specific antiviral treatment against YF available yet, vaccination is the only way to protect persons from YF disease. The currently available vaccine contains a live-attenuated 17D strain virus (Table 2) and has been shown to be safe and highly potent.[61] However, due to the increased use in international travelers, it has become evident that advanced age might be a risk factor for serious adverse effects and even death.[62] Compared with persons aged 25-44 years, individuals aged <75 had an 18-fold greater risk to experience serious adverse events after vaccination. The rate for systemic illness requiring hospitalization or leading to death after YF vaccination was reported to be 3.5 per 100,000 among people 65 to 75 years of age and 9.1 per 100,000 for people more than 75 years. Furthermore, there are no studies available that demonstrated the efficacy of the YF vaccine in elderly persons. Accordingly, recommendations and manufacturing standards have been modified to increase vaccine safety in elderly persons. Although the benefit-risk ratio still favors the vaccination of people at high risk for infection and outlines the vaccine's fundamental role in disease prevention and control, efforts to improve safety and to ensure vaccine efficacy in elderly persons are of urgent need.

How Does Immunosenescence Influence Vaccine Efficacy?

The term immunosenescence refers to a complex remodeling of the immune system in old age and may contribute significantly to morbidity and mortality in the elderly. Thymic involution, telomere shortening, T-cell signal transduction changes, alterations in the interaction of the innate and adaptive immune response, impaired DNA repair and antioxidant mechanisms as well as persistent antigenic stress may all be factors contributing to immunosenescence. Although perturbations of innate immune system components have been described, much of the decrease in immunoresponsiveness seen in elderly people is associated with changes in T-cell responses. This is due to the continuous loss of functional thymic tissue with increasing age.[63,64] The thymus, the central lymphoid organ, is responsible for the maturation and selection of so-called naive T-cells that regenerate the peripheral T-cell pool and retain the capability of the immune system to respond to a variety of different pathogens. In old age, the number of naive T-cells decreases while the number of antigen-experienced T-cells increases.[65,66] These antigen-experienced T-cells include a substantial proportion of senescent memory-effector T-cells that accumulate in elderly persons. Senescent memory-effector T-cells display phenotypic (loss of costimulatory molecules such as CD28 and CD40L) as well as functional changes (altered cytokine production profile, decreased proliferative response, shortened telomeres, increased resistance to programmed cell-death and restricted T-cell diversity).[67,68] Of particular importance, the senescent $CD8^+CD28^-$ memory-effector T-cell population predominantly produces the pro-inflammatory cytokine gamma interferon (IFNγ), but does not produce interleukin 2 (IL-2) and the anti-inflammatory, B-cell stimulating cytokine IL-4.[43] Recent data also support the hypothesis that chronic infection with the cytomegalovirus, a β-herpesvirus, may lead to a decrease in the size of the naive and early memory $CD8^+$ T-cell pool, but to an increase in the number of dysfunctional, IFNγ-producing $CD8^+CD28^-$ memory-effector T-cells (Fig. 2).[37] One clinical consequence of the accumulation of $CD8^+CD28^-$ T-cells is an impaired generation of protective antibody levels after vaccination.[43,69] Furthermore, the age-dependent increase in the level of pro-inflammatory cytokines may lead to ubiquitous chronic inflammatory responses in old age[42] and may therefore support the development of age-related chronic diseases, such as atherosclerosis,[70] rheumatoid arthritis[71] and Alzheimer's disease.[72,73]

Although individuals maintain a relatively constant total number of peripheral B-cells during aging, each B-cell subset comprises severe perturbations in size, dynamics and repertoire. The alterations affecting the B-cell subsets are due to a decreased generation of B-cell precursors, such as early lymphoid precursors and pro-B-cells. Cell-intrinsic as well as micro-environmental disturbances are both likely to contribute to the decreased output of pro-B-cells. Furthermore, alterations in environmental factors also impair overall V(D)J recombinase activity among pro-B-cells[74] which accounts for the limited B-cell repertoire frequently detected in elderly persons.[75] Although no decrease in overall serum immunoglobulin levels have been observed during aging, the antibodies generated in old age are of lower affinity due to a shift in antibody isotypes from IgG to IgM.[76] Of particular importance, B-cells from elderly individuals are stimulated 70% less efficiently by follicular dendritic cells than B-cells from young subjects,[77] suggesting loss of B-cell function, in part due to the decreased expression of costimulatory molecules, such as CD40 or CD27.[78] Impaired T-cell-mediated immunity as well as defects in antigen presentation by antigen presenting cells (APC) also contribute to the decline in B-cell specific functions.[79]

To summarize, environmental factors as well as intrinsic alterations lead to the disturbance of the peripheral B-cell pool characterized by the loss of B-cell costimulatory molecules and loss of B-cell diversity. The cytokine environment as well as T-cell-mediated B-cell stimulation are further important determinants of intact antibody production.[65] Thus, decreased numbers of $CD28^+$ and $CD40L^+$ T-cells and a lack of cytokines such as IL-2 and IL-4, are both likely to endanger normal T-cell/B-cell communication, B-cell growth, differentiation and antibody production in the elderly.

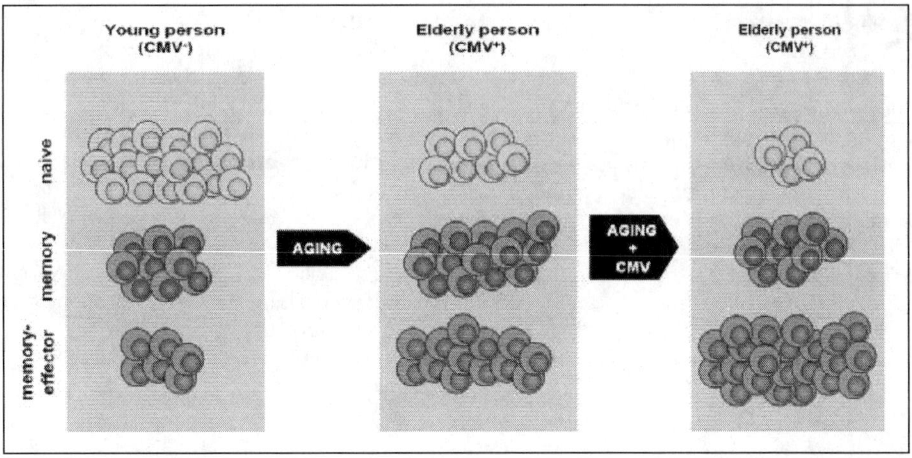

Figure 2. Schematic representation of the effects of age and persistent CMV infection on the peripheral $CD8^+$ T-cell compartment. The number of naive $CD8^+$ T-cells declines with age due to thymic involution whereas antigen-experienced $CD8^+$ T-cells, such as memory and effector-memory cells increase. Persistent CMV infection leads to a further accumulation of senescent and dysfunctional memory-effector $CD8^+$ T-cells. The exact percentage of the various $CD8^+$ T-cell subsets was determined in ref. 37.

How to Improve Vaccine Efficacy in Old Age?

There is a tremendous need to increase the protective effect of vaccines in the elderly. Research of the last decade has provided new insights into the molecular mechanisms of the immune response in old age, which can now be used for the development of potent vaccines. In the past, vaccines were primarily designed to elicit a strong humoral immune response. However, vaccines in elderly persons may be more effective if the stimulate innate immune components and the generation of long-lived memory T-cells. Currently, several strategies are being pursued to increase immunogenicity, to minimize adverse side effects and to increase vaccine acceptance by introducing needle-free injection devices. Proven and promising vaccine technologies are used to design conjugate, subunit, live vector, DNA and live-attenuated vaccines (Table 3).[80] While live-attenuated vaccines (e.g., against varicella, measles or yellow fever) stimulate numerous immune components and display enhanced immunogenicity, conjugate and subunit vaccines (e.g., against influenza) are often supplemented with adjuvants[81] to ensure their protective effect. Generally, adjuvants can be divided into antigen delivery systems (cationic microparticles, proteasomes and virus-like particles) and immune potentiators (e.g., cytokines). These adjuvants may overcome the proposed age-related functional decline of innate immune responses by targeting pattern-recognition receptors, such as the recently identified toll-like receptors or nucleotide-binding oligomerization domain proteins.[82] The enhanced activation of the innate immune system may also improve antigen processing and presentation leading to more potent T and B-cell responses and to sustained immunological memory. Vaccines supplemented with the DNA of cytokines (e.g., IL-2, IL-7, IL-12, IL-15 or IL-21), chemokines or costimulatory molecules may magnify immune responses by generating more and long-lived memory T-cells[83-85] and may overcome immunodominance.[86]

In addition to improve vaccine efficacy, a modification of vaccination strategies for elderly persons has been supported by the results of several vaccination trials. For instance, a decreased response and a shortened duration of protective immunity following booster immunization is a characteristic feature of old age.[87] Thus, several European health authorities have recommended five-year vaccination intervals for tetanus, diphtheria, pertussis and pneumonia. Increased public awareness of regular booster vaccinations in adults should be enforced, as these immunization

Table 3. Attributes of different vaccine types

Category	Advantages	Disadvantages
Conjugate vaccines	High clinical tolerability	Booster immunizations usually required; no CTL responses
Inactivated whole-virus vaccines	High antibody levels	Higher probability of adverse side effects compared to subunit or conjugate vaccines (e.g., replacement of whole-cell pertussis vaccine by acellular pertussis vaccine)
Live-attenuated vaccines	Booster immunizations usually not required; high antibody levels and strong CTL responses	Safety concerns for immunocompromised persons
Live-vector vaccines	Booster immunizations usually not required; high antibody levels and strong CTL responses	Moderate safety concerns for immunocompromised persons
DNA vaccines	Specific manipulation of the cellular immune response; increased efficiency when added to subunit vaccines	Low immunogenicity
Subunit vaccines	High clinical tolerability due to the high purity of single immunogenic peptides only	Moderate immunogenicity can be overcome by supplementation of an adjuvant or DNA
Nonliving antigen delivery systems (e.g., virosomes, liposomes, virus-like particles)	High clinical tolerability; higher antibody levels and moderate CTL responses; delivery of a variety of purified antigens but also DNA	Biostability?

CTL, cytotoxic T-lymphocyte

regimes may be essential to maintain the ability to respond to recall antigens in old age. Recent results also indicate that long-lasting protection but also a good booster effect can be expected even a long time after the last vaccination, when a live-attenuated vaccine (e.g., polio vaccine) is used for primary immunization in early life. New delivery systems that make use of tiny micro-needles or non-injectable application devices (nasal, oral, transcutaneous) may further increase vaccination acceptance, especially in the case of influenza as this vaccination has to be repeated annually.

Conclusions

Infectious diseases in elderly persons are becoming an increasingly important issue. An utmost need represents the development of more immunogenic vaccines for the elderly. The improvement of specific vaccine types regarding immunogenicity and tolerability, the addition of adjuvants, the design of new delivery systems as well as specific immunization regimes should all contribute to an enhanced efficacy of vaccines in elderly persons. Further improvements may comprise the adjustment of vaccination intervals in old age, the increase in vaccine acceptance and vaccination coverage as well as raising people's awareness to stick to the recommended booster vaccination intervals throughout life. In the distant future, vaccines may also play an important role in treating non-infectious diseases such as allergy, autoimmunity, Alzheimer's disease and cancers.

Acknowledgements

The authors wish to acknowledge the support of the Austrian Science Fund and the Austrian Green Cross Society for Preventive Medicine.

References

1. Anderson RN, Smith BL. Deaths: leading causes for 2002. Natl Vital Stat Rep 2005; 53(17):1-89.
2. Weiss RA, McMichael AJ. Social and environmental risk factors in the emergence of infectious diseases. Nat Med 2004; 10(12 Suppl):S70-76.
3. http://www.niaid.nih.gov
4. Fenner F. A successful eradication campaign. Global eradication of smallpox. Rev Infect Dis 1982; 4(5):916-930.
5. Pharmaceutical Research and Manufacturers of America, 2004 Survey: New Medicines in development for Infectious diseases, http://www.phrma.org/newmedicines/resources/2004-04-22.130.pdf 2004.
6. Gavazzi G, Krause KH. Ageing and infection. Lancet Infect Dis 2002; 2(11):659-666.
7. Barker WH, Mullooly JP. Pneumonia and influenza deaths during epidemics: implications for prevention. Arch Intern Med 1982; 142(1):85-89.
8. Influenza and pneumococcal vaccination coverage among persons aged > or = 65 years and persons aged 18-64 years with diabetes or asthma—United States, 2003. MMWR Morb Mortal Wkly Rep 2004; 53(43):1007-1012.
9. Gross PA, Hermogenes AW, Sacks HS et al. The efficacy of influenza vaccine in elderly persons. A meta-analysis and review of the literature. Ann Intern Med 1995; 123(7):518-527.
10. Bouree P. Immunity and immunization in elderly. Pathol Biol (Paris) 2003; 51(10):581-585.
11. Nichol KL, Margolis KL, Wuorenma J et al. The efficacy and cost effectiveness of vaccination against influenza among elderly persons living in the community. N Engl J Med 1994; 331(12):778-784.
12. Mullooly JP, Bennett MD, Hornbrook MC et al. Influenza vaccination programs for elderly persons: cost-effectiveness in a health maintenance organization. Ann Intern Med 1994; 121(12):947-952.
13. Ahmed AE, Nicholson KG, Nguyen-Van-Tam JS. Reduction in mortality associated with influenza vaccine during 1989-90 epidemic. Lancet 1995; 346(8975):591-595.
14. Iob A, Brianti G, Zamparo E et al. Evidence of increased clinical protection of an MF59-adjuvant influenza vaccine compared to a non-adjuvant vaccine among elderly residents of long-term care facilities in Italy. Epidemiol Infect 2005; 133(4):687-693.
15. Gluck R, Mischler R, Finkel B et al. Immunogenicity of new virosome influenza vaccine in elderly people. Lancet 1994; 344(8916):160-163.
16. Conne P, Gauthey L, Vernet P et al. Immunogenicity of trivalent subunit versus virosome-formulated influenza vaccines in geriatric patients. Vaccine 1997; 15(15):1675-1679.
17. Ruf BR, Colberg K, Frick M et al. Open, randomized study to compare the immunogenicity and reactogenicity of an influenza split vaccine with an MF59adjuvanted subunit vaccine and a virosome-based subunit vaccine in elderly. Infection 2004; 32(4):191-198.
18. de Bruijn IA, Nauta J, Cramer WC et al. Clinical experience with inactivated, virosomal influenza vaccine. Vaccine 2005; 23 Suppl 1:S39-49.
19. Ruben FL. Inactivated influenza virus vaccines in children. Clin Infect Dis 2004; 38(5):678-688.
20. Reichert TA, Sugaya N, Fedson DS et al. The Japanese experience with vaccinating schoolchildren against influenza. N Engl J Med 2001; 344(12):889-896.
21. EARSS Annual Report, available at: http://www.earss.rivm.nl/PAGINA/DOC/rep2002/annual-report-2002.pdf. 2002.
22. Melegaro A, Edmunds WJ. The 23-valent pneumococcal polysaccharide vaccine. Part I. Efficacy of PPV in the elderly: a comparison of meta-analyses. Eur J Epidemiol 2004; 19(4):353-363.
23. Wagner C, Popp W, Posch M et al. Impact of pneumococcal vaccination on morbidity and mortality of geriatric patients: a case-controlled study. Gerontology 2003; 49(4):246-250.
24. Lexau CA, Lynfield R, Danila R et al. Changing epidemiology of invasive pneumococcal disease among older adults in the era of pediatric pneumococcal conjugate vaccine. Jama. 2005; 294(16):2043-2051.
25. Dutt AK, Stead WW. Tuberculosis. Clin Geriatr Med 1992; 8(4):761-775.
26. Rajagopalan S, Yoshikawa TT. Tuberculosis in the elderly. Z Gerontol Geriatr 2000; 33(5):374-380.
27. WHO. AntiTB drug resistance in the world. Report no.2. Prevalence and trends. Geneva. Report No.: WHO/CDS/TB/2000.278. 2000.

28. Colditz GA, Brewer TF, Berkey CS et al. Efficacy of BCG vaccine in the prevention of tuberculosis. Meta-analysis of the published literature. Jama 1994; 271(9):698-702.
29. McShane H. Coinfection with HIV and TB: double trouble. Int J STD AIDS 2005; 16(2):95-100; quiz 101.
30. Flynn JL, Chan J. Immunology of tuberculosis. Annu Rev Immunol 2001; 19:93-129.
31. Ragozzino MW, Melton LJ, 3rd, Kurland LT et al. Population-based study of herpes zoster and its sequelae. Medicine (Baltimore) 1982; 61(5):310-316.
32. Arvin AM. Varicella-Zoster virus: pathogenesis, immunity and clinical management in hematopoietic cell transplant recipients. Biol Blood Marrow Transplant 2000; 6(3):219-230.
33. Mehta SK, Cohrs RJ, Forghani B et al. Stress-induced subclinical reactivation of varicella zoster virus in astronauts. J Med Virol 2004; 72(1):174-179.
34. Arvin AM, Gershon AA. Live attenuated varicella vaccine. Annu Rev Microbiol 1996; 50:59-100.
35. Weir E. Vaccination boosts adult immunity to varicella zoster virus. Cmaj 2005; 173(3):249.
36. Oxman MN, Levin MJ, Johnson GR et al. A vaccine to prevent herpes zoster and postherpetic neuralgia in older adults. N Engl J Med 2005; 352(22):2271-2284.
37. Almanzar G, Schwaiger S, Jenewein B et al. Long-term cytomegalovirus infection leads to significant changes in the composition of the $CD8^+$ T-cell repertoire, which may be the basis for an imbalance in the cytokine production profile in elderly persons. J Virol 2005; 79(6):3675-3683.
38. Pawelec G, Akbar A, Caruso C et al. Human immunosenescence: is it infectious? Immunol Rev 2005; 205:257-268.
39. Ouyang Q, Wagner WM, Zheng W et al. Dysfunctional CMV-specific $CD8^+$ T-cells accumulate in the elderly. Exp Gerontol 2004; 39(4):607-613.
40. Holtappels R, Podlech J, Pahl-Seibert MF et al. Cytomegalovirus misleads its host by priming of CD8 T-cells specific for an epitope not presented in infected tissues. J Exp Med 2004; 199(1):131-136.
41. Khan N, Shariff N, Cobbold M et al. Cytomegalovirus seropositivity drives the CD8 T-cell repertoire toward greater clonality in healthy elderly individuals. J Immunol 2002; 169(4):1984-1992.
42. Franceschi C, Bonafe M, Valensin S et al. Inflamm-aging. An evolutionary perspective on immunosenescence. Ann N Y Acad Sci 2000; 908:244-254.
43. Saurwein-Teissl M, Lung TL, Marx F et al. Lack of antibody production following immunization in old age: association with $CD8^+CD28^-$ T-cell clonal expansions and an imbalance in the production of Th1 and Th2 cytokines. J Immunol 2002; 168(11):5893-5899.
44. Riddell SR, Watanabe KS, Goodrich JM et al. Restoration of viral immunity in immunodeficient humans by the adoptive transfer of T-cell clones. Science 1992; 257(5067):238-241.
45. Cherry JD. The epidemiology of pertussis: a comparison of the epidemiology of the disease pertussis with the epidemiology of Bordetella pertussis infection. Pediatrics 2005; 115(5):1422-1427.
46. Mertens PL, Stals FS, Schellekens JF et al. An epidemic of pertussis among elderly people in a religious institution in The Netherlands. Eur J Clin Microbiol Infect Dis 1999; 18(4):242-247.
47. Steger MM, Maczek C, Berger P et al. Vaccination against tetanus in the elderly: do recommended vaccination strategies give sufficient protection. Lancet 1996; 348(9029):762.
48. Grubeck-Loebenstein B, Berger P, Saurwein-Teissl M et al. No immunity for the elderly. Nat Med 1998; 4(8):870.
49. Gergen PJ, McQuillan GM, Kiely M et al. A population-based serologic survey of immunity to tetanus in the United States. N Engl J Med 1995; 332(12):761-766.
50. Gomez I, Marx F, Gould EA et al. T-cells from elderly persons respond to neoantigenic stimulation with an unimpaired IL-2 production and an enhanced differentiation into effector cells. Exp Gerontol 2004; 39(4):597-605.
51. Suss J. Epidemiology and ecology of TBE relevant to the production of effective vaccines. Vaccine 2003; 21 Suppl 1:S19-35.
52. Kunz C. TBE vaccination and the Austrian experience Vaccine. 2003; 21 Suppl 1:S50-55.
53. Rendi-Wagner P, Kundi M, Zent O et al. Immunogenicity and safety of a booster vaccination against tick-borne encephalitis more than 3 years following the last immunisation. Vaccine 2004; 23(4):427-434.
54. Steffen R, Kane MA, Shapiro CN et al. Epidemiology and prevention of hepatitis A in travelers. Jama 1994; 272(11):885-889.
55. Steffen R. Hepatitis A and hepatitis B: risks compared with other vaccine preventable diseases and immunization recommendations. Vaccine 1993; 11(5):518-520.
56. Joines RW, Blatter M, Abraham B et al. A prospective, randomized, comparative US trial of a combination hepatitis A and B vaccine (Twinrix) with corresponding monovalent vaccines (Havrix and Engerix-B) in adults. Vaccine 2001; 19(32):4710-4719.
57. Hopperus Buma AP, van Doornum GJ, Veltink RL et al. Immunogenicity of an inactivated hepatitis A vaccine in Dutch United Nations troops. Vaccine 1997; 15(12-13):1413-1417.

58. Wolters B, Junge U, Dziuba S et al. Immunogenicity of combined hepatitis A and B vaccine in elderly persons. Vaccine 2003; 21(25-26):3623-3628.
59. Reuman PD, Kubilis P, Hurni W et al. The effect of age and weight on the response to formalin inactivated, alum-adjuvanted hepatitis A vaccine in healthy adults. Vaccine 1997; 15(10):1157-1161.
60. WHO. District guidelines for yellow fever surveillance. Geneva, Switzerland. Available at http://www.who.int/emc-documents/yellow_fever/whoepigen9809c.html. 1998.
61. Poland JD, Calisher CH, Monath TP et al. Persistence of neutralizing antibody 30-35 years after immunization with 17D yellow fever vaccine. Bull World Health Organ 1981; 59(6):895-900.
62. Martin M, Weld LH, Tsai TF et al. Advanced age a risk factor for illness temporally associated with yellow fever vaccination. Emerg Infect Dis 2001; 7(6):945-951.
63. Aspinall R andrew D. Thymic involution in aging. J Clin Immunol 2000; 20(4):250-256.
64. George AJ, Ritter MA. Thymic involution with ageing: obsolescence or good housekeeping? Immunol Today 1996; 17(6):267-272.
65. Lazuardi L, Jenewein B, Wolf AM et al. Age-related loss of naive T-cells and dysregulation of T-cell/B-cell interactions in human lymph nodes. Immunology 2005; 114(1):37-43.
66. Zanni F, Vescovini R, Biasini C et al. Marked increase with age of type 1 cytokines within memory and effector/cytotoxic $CD8^+$ T-cells in humans: a contribution to understand the relationship between inflammation and immunosenescence. Exp Gerontol 2003; 38(9):981-987.
67. Grubeck-Loebenstein B, Wick G. The aging of the immune system. Adv Immunol 2002; 80:243-284.
68. Effros RB, Cai Z, Linton PJ. CD8 T-cells and aging. Crit Rev Immunol 2003; 23(1-2):45-64.
69. Goronzy JJ, Fulbright JW, Crowson CS et al. Value of immunological markers in predicting responsiveness to influenza vaccination in elderly individuals. J Virol 2001; 75(24):12182-12187.
70. Shoenfeld Y, Sherer Y, Harats D. Artherosclerosis as an infectious, inflammatory and autoimmune disease. Trends Immunol 2001; 22(6):293-295.
71. Weyand CM, Fulbright JW, Goronzy JJ. Immunosenescence, autoimmunity and rheumatoid arthritis. Exp Gerontol 2003; 38(8):833-841.
72. Blasko I, Grubeck-Loebenstein B. Role of the immune system in the pathogenesis, prevention and treatment of Alzheimer's disease. Drugs Aging 2003; 20(2):101-113.
73. Blasko I, Stampfer-Kountchev M, Robatscher P et al. How chronic inflammation can affect the brain and support the development of Alzheimer's disease in old age: the role of microglia and astrocytes. Aging Cell 2004; 3(4):169-176.
74. Labrie JE, 3rd, Sah AP, Allman DM et al. Bone marrow microenvironmental changes underlie reduced RAG-mediated recombination and B-cell generation in aged mice. J Exp Med 2004; 200(4):411-423.
75. Weksler ME, Szabo P. The effect of age on the B-cell repertoire. J Clin Immunol 2000; 20(4):240-249.
76. Johnson SA, Cambier JC. Ageing, autoimmunity and arthritis: senescence of the B-cell compartment -implications for humoral immunity. Arthritis Res Ther 2004; 6(4):131-139.
77. Aydar Y, Balogh P, Tew JG et al. Age-related depression of FDC accessory functions and CD21 ligand-mediated repair of costimulation. Eur J Immunol 2002; 32(10):2817-2826.
78. Colonna-Romano G, Bulati M, Aquino A et al. B-cells in the aged: CD27, CD5 and CD40 expression. Mech Ageing Dev 2003; 124(4):389-393.
79. McGlauchlen KS, Vogel LA. Ineffective humoral immunity in the elderly. Microbes Infect 2003; 5(13):1279-1284.
80. Levine MM, Sztein MB. Vaccine development strategies for improving immunization: the role of modern immunology. Nat Immunol 2004; 5(5):460-464.
81. Kenney RT, Edelman R. Survey of human-use adjuvants. Expert Rev Vaccines 2003; 2(2):167-188.
82. Pashine A, Valiante NM, Ulmer JB. Targeting the innate immune response with improved vaccine adjuvants. Nat Med 2005; 11(4 Suppl):S63-68.
83. Barouch DH, Santra S, Steenbeke TD et al. Augmentation and suppression of immune responses to an HIV-1 DNA vaccine by plasmid cytokine/Ig administration. J Immunol 1998; 161(4):1875-1882.
84. Kutzler MA, Robinson TM, Chattergoon MA et al. Coimmunization with an optimized IL-15 plasmid results in enhanced function and longevity of CD8 T-cells that are partially independent of CD4 T-cell help. J Immunol 2005; 175(1):112-123.
85. Li Y, Bleakley M, Yee C. IL-21 influences the frequency, phenotype and affinity of the antigen-specific CD8 T-cell response. J Immunol 2005; 175(4):2261-2269.
86. Melchionda F, Fry TJ, Milliron MJ et al. Adjuvant IL7 or IL-15 overcomes immunodominance and improves survival of the $CD8^+$ memory cell pool. J Clin Invest 2005; 115(5):1177-1187.
87. Hainz U, Jenewein B, Asch E et al. Insufficient protection for healthy elderly adults by tetanus and TBE vaccines. Vaccine 2005; 23(25):3232-3235.
88. Herndler-Brandstetter D, Cioca DP, Grubeck-Loebenstein B. Immunizations in the elderly: do they live up to their promise? Wien Med Wochenschr 2006; 156(5-6):130-141.

Chapter 11

Zinc and the Altered Immune System in the Elderly

Hajo Haase and Lothar Rink*

Abstract

Immune function is severely compromised during states of zinc deficiency. A general observation in elderly people is a reduced serum/plasma zinc level. The decline of immune function with aging, so-called immunosenescence, has several parallels with immunological alterations observed during zinc deficiency; thus, it is plausible that both events are associated, implying an impact of altered zinc status on immune function in the elderly. This is supported by several studies showing that zinc supplementation has a beneficial effect on immune function of elderly people and reverses several changes associated with immunosenescence. This chapter therefore argues that zinc deficiency contributes to immunosenescence.

Introduction

Zinc is an essential trace element with multiple functions in enzymatic catalysis, growth and proliferation,[1] and is indispensable for appropriate immune function.[2] The normal serum or plasma zinc level is between 12 and 16 μM and is the source of zinc for leukocytes. Since there is no zinc storage system in the body, zinc deficiency can rapidly lead to impaired zinc supply to immune cells and subsequently compromise immune function.[3]

There is a relationship between aging and an increase in the levels of pro-inflammatory cytokines.[4] Hence, aging is often accompanied by a state of constant low level inflammation and this might interact with, but also be influenced by, zinc levels. During states of inflammation, zinc is redistributed within the body, leading to a reduction of plasma and serum zinc.[5] Several diseases with an immunological background that are known to occur with high frequency in the elderly show a relation to zinc levels. For example, in rheumatoid arthritis a significant reduction of serum zinc levels is observed which are inversely correlated with levels of the pro-inflammatory cytokines tumor necrosis factor (TNF)-α and interleukin (IL)-1β.[6] On the other hand, zinc also has an effect on the disease, since a trial with oral zinc supplementation indicated beneficial effects for the treatment of rheumatoid arthritis.[7] Diabetes is also accompanied by a decrease in zinc levels,[8] and it has been proposed that zinc deficiency might contribute to the onset of diabetes by affecting insulin signal transduction.[9] At least in animal models, zinc supplementation can be beneficial in preventing the onset of diabetes.[10]

While the precise role of zinc in these diseases is still unknown, a correlation between immunosenscence and zinc suggests that the decreasing efficiency of the aging immune system is related to the patient's zinc status and that zinc deficiency contributes to increased infections, malignancies and autoimmune diseases in the elderly.

*Corresponding Author: Lothar Rink—Institute of Immunology, University Hospital, RWTH Aachen University, Pauwelsstrasse 30, 52074 Aachen, Germany. Email: lrink@ukaachen.de

Zinc Deficiency in the Elderly

Poor zinc uptake is common among the elderly and the third national health and nutrition examination survey (NHANES III) showed that persons over 70 years of age had the highest risk for insufficient zinc intake, with nearly 60% of them not having an adequate intake of at least 77% of the recommended daily allowance.[11] This is one potential reason for zinc deficiency, but decreased intestinal absorption which in part depends on the composition of the food, or medications such as diuretics could also cause a negative zinc balance, even when sufficient zinc is ingested. Several diseases that have a higher incidence in the elderly, like diabetes, are also accompanied by increased urinary excretion of zinc.[12] All these factors contribute to the high frequency of an insufficient zinc status in elderly individuals.

It is thus a common finding that serum or plasma zinc levels decrease with age[13,14] (Fig. 1). However, while levels are often significantly lower than in younger individuals, in many cases the subjects are not classified as being zinc deficient, as defined by a serum or plasma concentration below 10.7 μM, corresponding to 70 μg/dL. However, this parameter is not a reliable indicator for zinc status, because marginal zinc deficiency, that may very well have an impact on immune function, can be present in people whose serum or plasma zinc levels are in fact above these reference values. This is illustrated by a study in which plasma zinc did not indicate zinc deficiency in individuals between 50 and 80 years, but who were classified as mildly zinc deficient according to their lymphocyte and granulocyte content. Notably, their serum thymulin activity was reduced, due to limited availability of zinc.[15]

A difference seems to exist between apparently healthy, free living elderly and ill or institutionalized subjects.[16-17] A high occurrence of zinc deficiency was found in two studies with hospitalized elderly patients. In one case[18] 28% were zinc deficient (<10.7 μM plasma zinc), while in the other study[19] even the mean serum zinc of the group was below 10.7 μM and 61% of the patients had a zinc deficiency.

As would be expected from a higher occurrence of zinc deficiency in the sick elderly, a correlation between zinc status and immune function has been observed. The delayed-type hypersensitivity (DTH) reaction to skin antigens was found to be significantly correlated to the plasma zinc concentration and it was suggested that a difference of 1.5 μM plasma zinc is associated with this effect, which corresponds to a small change compared with the physiological range.[20]

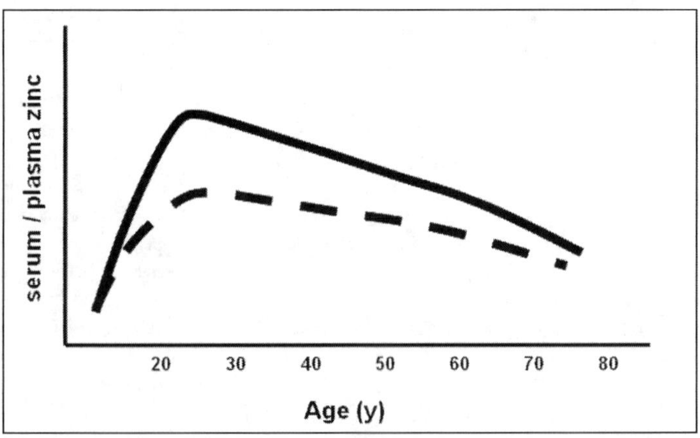

Figure 1. Zinc levels during aging. Age-related changes of the mean zinc levels in plasma or serum of healthy male (—) or female (---) subjects. Both increase during adolescence, leading to a peak that is more pronounced in males, followed by a decline from the third decade of life which can lead to zinc deficiency that can compromise immune function later in life. Due to nutritional imbalance or disease, zinc levels may decline even more rapidly and lie significantly below the mean values in this figure.

When elderly patients undergoing hemodialysis were vaccinated against diphtheria, the fraction which did not respond to the vaccination had significantly lower serum zinc levels than healthy elderly controls of the same age, while the subjects that were vaccinated successfully did not show significant reduction of zinc levels.[21] Another group of elderly with reduced levels of lymphocyte and granulocyte zinc showed diminished IL-2 production upon stimulation with phytohemagglutinin when they were compared to elderly and younger controls which both had significantly higher cellular zinc levels.[22] Finally, in one group of hospitalized patients, not only was there a negative correlation between serum zinc and IgG_2 levels, but zinc deficiency with serum levels below 10.7 μM correlated with significantly higher incidences of congestive cardiopathy, respiratory infections, gastrointestinal diseases and depression.[18]

Comparison of Immunosenescence and the Effects of Zinc Deficiency

As exemplified above, the elderly often have low zinc serum levels; several observations that are commonly made during zinc deficiency are similar to changes in the aging immune system.[23] These are briefly reviewed below.

Innate Immune Response

Phagocytes are affected by both aging and zinc deficiency. While the number of neutrophil granulocytes is generally increased in the elderly, functions like oxidative burst and chemotaxis are reduced.[24] Although zinc deficiency is generally associated with reduced numbers of neutrophils, effects of in vivo zinc deficiency on phagocytosis and oxidative burst were described that also demonstrate an impairment of these functions.[25,26]

NK-cell numbers are increased in the elderly, but the per cell killing ability is reduced.[27,28] Zinc deficiency in vivo also induces a decrease in NK-cell lytic activity, which can be reversed by zinc supplementation.[25,29]

Adaptive Immune Response

Zinc is essential for proliferation, including that of cells of the immune system. Moreover, zinc deficiency is also strongly associated with increased levels of apoptosis in pre B and T-cells,[30] leading to a decline in numbers of mature T and B-cells which is also observed during aging.[27] A hallmark event of increasing age is thymic involution and thymic atrophy is a characteristic of the rare zinc malabsorption syndrome *Acrodermatitis enteropathica*,[3] which leads to severe zinc deficiency due to mutations in the intestinal zinc transport protein hZIP4.[31] In agreement with these effects, T-cell-mediated immunity is particularly affected during aging and zinc deficiency. One effect is on the number of cytotoxic $CD8^+$ T-cells,[27,32] which could (coincidential with the decreased cytolytic activity of NK-cells) lead to the higher incidence of malignancies that is observed with age. Consistent with this, zinc deficiency has been associated with carcinogenesis in the esophagus,[33] zinc status correlates with tumor burden in head and neck cancer,[34] and moderate zinc supplementation can reduce the incidence of prostate cancer.[35]

Cytokines

Several studies indicate that cytokine production changes with age. Reduction of IL-2, soluble IL-2 receptor and interferon (IFN)-γ, accompanied by a rise in the T helper (TH)2 cytokines IL-4 and IL-10, leads to decreased T-cell proliferation and a shift towards TH2.[36,37] These observations correspond well to the TH2 shift observed during zinc deficiency, which is mainly characterized by a reduction of IL-2 and IFN-γ.[32,38,39]

On the other hand, in the elderly, secretion of pro-inflammatory monokines is increased[36] and their cells show an increased production of IL-1β, IL-6, TNF-α and IL-8 after stimulation with lipopolysaccharide in vitro.[40] Zinc deficiency increases the levels of TNF-α, IL-1β and IL-8,[39] and higher concentrations of zinc have been shown to suppress the secretion of pro-inflammatory monokines.[41]

In vitro release of the antiviral cytokine IFN-α in response to viral stimulation of peripheral blood mononuclear cells obtained from elderly subjects was significantly reduced compared to a younger control group, but returned to normal after the cells had been cultured in the presence of 15 μM zinc.[42] This implies a correlation between immunosenescence and zinc deficiency. In the following section, examples of zinc supplementation studies that support this hypothesis are given.

Zinc Supplementation and Immunosenescence

Immune function and zinc levels both decline with age, but either an increased susceptibility to diseases could cause a higher rate of infections and a subsequent loss of zinc, or insufficient nutrition, absorption, or retention of zinc could lead to zinc deficiency that affects immunity and contributes to immunosenescence. Several studies show improvements after zinc supplementation (Table 1), indicating that it is a state of marginal zinc deficiency which increases the susceptibility to infectious diseases. While in general zinc seems to have beneficial effects, there are some studies that do not find any effect at all. A comparison of the different studies is made difficult by the differences of the investigated populations with respect to their health or the zinc supplementation with respect to dose, bioavailability and duration and especially the zinc status prior to supplementation.

T-Cell Numbers and Activity

The numbers of activated T-helper cells and cytotoxic T-cells were significantly increased after supplementation of residents of a public home for older people with 25 mg zinc (as zinc sulfate) per day.[43] In another study,[44] the total number of circulating lymphocytes remained unchanged in a group of elderly that had been supplemented with 100 mg of zinc (as zinc sulfate) for one month; however, the proportion of T-cells was significantly increased by zinc treatment, although this was not accompanied by a change of the response to in vitro stimulation with lymphocyte mitogens.

When DTH was investigated, several studies in which 30 to 100 mg zinc were administered per day found a significant improvement in the reaction to skin antigens,[15,44-46] but this has not been confirmed in all investigations. In a large study,[47] daily supplementation with either 15 mg or 100 mg zinc (as zinc acetate) for 16 months did not have a beneficial effect on DTH nor on the response to different lymphocyte mitogens and only a transient increase in NK-cell activity was observed after three months in the group that had been supplemented with 100 mg zinc. Interestingly the subjects were healthy and their mean serum zinc levels prior to supplementation were normal (approximately 13 μM) and only increased in the 100 mg zinc group during the first 12 months of the study.[47]

Thymulin

One way in which zinc could affect T-cells is thymulin, a hormone secreted by thymic epithelial cells which is essential for differentiation and function of T-cells. A decrease in plasma thymulin activity is observed with age. Two studies demonstrated that in vivo zinc supplementation could increase thymulin activity in serum from elderly subjects.[15,48]

Thymulin is present in the plasma in two forms, a zinc-bound active one and a zinc-free, inactive form. In vitro, thymulin activity increased after addition of zinc to the plasma of elderly donors,

Table 1. Main effects of zinc supplementation in the elderly

Parameter	Effect of Zinc	Ref.
DTH	Increased number of positive reactions	15, 44-46
Thymulin	Increased plasma thymulin activity due to formation of the zinc-bound (active) form	15, 48
Vaccination	Increased antibody production in response to vaccination	44
T-cells	Increased number	43, 44

indicating that the reduced activity was due to a lack of zinc, but not reduced peptide levels as a consequence of thymic involution.[48,49]

Response to Vaccination

With increasing age, the response to vaccines decreases, e.g., in the case of tetanus[50] or influenza[51] vaccination. An early study investigated the IgG antibody response after tetanus vaccination and found a significant improvement in 11 subjects aged over 70 years who had received a daily dose of 100 mg zinc (as zinc sulfate) for one month, compared to a control group that had not been given zinc.[44] In a larger, placebo-controlled study,[52] 140 institutionalized patients >65 years who had received 20 mg zinc daily (as zinc sulfate) together with 100 µg selenium were vaccinated against influenza after 15 to 17 months of supplementation. The number of seroprotected patients after vaccination was significantly higher in the group that had received selenium and zinc.[52]

Conclusions

There are many similarities between immunosenescence and the changes of the immune system occurring as a result of zinc deficiency. Given the significant reduction of zinc levels with age, it is reasonable to conclude a connection between these two events. Accordingly, several studies have been performed, investigating the potential benefit of zinc supplementation on immune function of elderly individuals and several parameters were shown to be corrected by zinc supplementation either in vitro or in vivo (Fig. 2). While most studies come to the conclusion

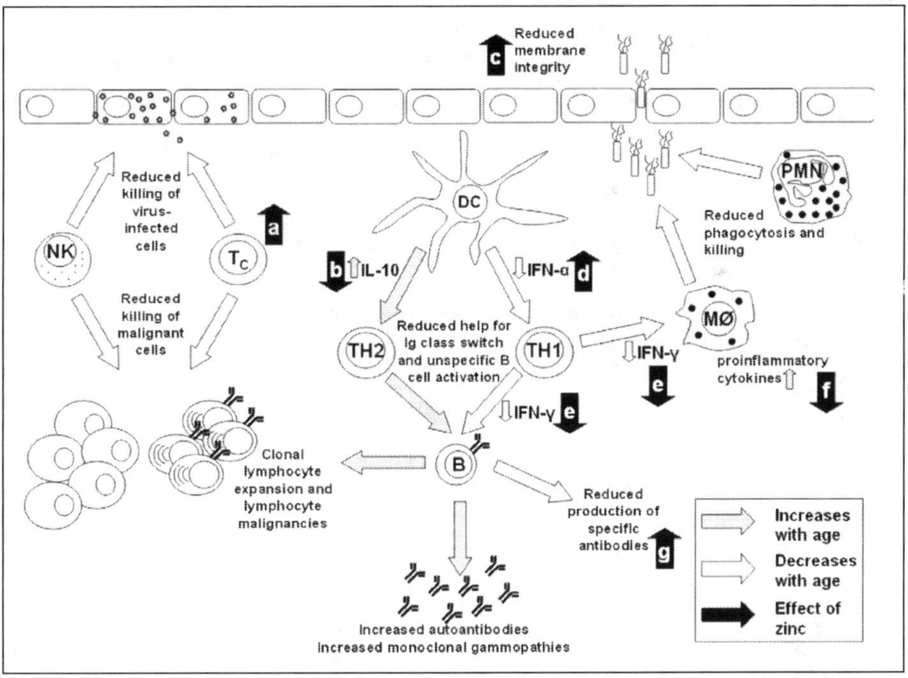

Figure 2. Immunological alterations during immunosenescence and the effect of zinc. Aging leads to many changes in the immune system, including an increase in pro-inflammatory cytokines, decreased protection against neoplasia and viral infections, alteration of the T-helper cell balance leading to changes in B-cell help and antibody production, increased mono-or oligoclonal gammopathy and decreased phagocytosis and killing by the innate immune system, which is further aggravated by increased permeability of physical barriers, like delayed wound healing. Legend continued on following page.

Figure 2 legend continued. Zinc counteracts some of these effects by [a] increasing T-cell number and function and lytic activity of NK-cells,[41] [b] reducing IL-10 production (unpublished observation), [c] promoting wound healing,[41] [d] inducing IFN-γ secretion,[54] [e] restoring IFN-α production,[42] [f] suppressing pro-inflammatory cytokine production[41] and [g] improving antibody protection after vaccination.[44,52] Cell types: Dendritic cells (DC), T helper 1 and 2 cells, (TH1, TH2), monocytes/macrophages (MØ), polymorphnuclear neutrophils (PMN), B-cells (B), cytotoxic T-cells (T_c) and natural killer cells (NK).

that zinc supplementation improves immunity, a few do not find a beneficial effect of zinc. The difference may result from the fact that zinc deficiency is likely to be highest in sick individuals, while supplementation may be ineffective in healthy, zinc-replete individuals. The outcome of a study therefore depends on the population that is investigated. Another point is that high doses of zinc are immunosuppressive[53] and, if taken over a longer period of time, interfere with copper homeostasis,[12] so that the resulting copper deficiency affects immune function.

Even though zinc deficiency is not exclusively found in the elderly, they have significantly reduced zinc levels and a higher incidence of zinc deficiency compared to younger subjects. The adverse effects thereof on the immune system seem to be factors contributing to immunosenescence. While zinc supplementation will not be sufficient to counteract all age-related changes in immune function, moderate zinc supplementation in zinc-deficient individuals can be a cost-effective way to restore immune function with a low risk of toxic side effects.

References

1. Vallee BL, Falchuk KH. The biochemical basis of zinc physiology. Physiol Rev 1993; 73(1):79-118.
2. Wellinghausen N, Kirchner H, Rink L. The immunobiology of zinc. Immunol Today 1997; 18(11):519-521.
3. Rink L, Gabriel P. Extracellular and immunological actions of zinc. Biometals 2001; 14(3-4):367-383.
4. Krabbe KS, Pedersen M, Bruunsgaard H. Inflammatory mediators in the elderly. Exp Gerontol 2004; 39(5):687-699.
5. Brown KH. Effect of infections on plasma zinc concentration and implications for zinc status assessment in low-income countries. Am J Clin Nutr 1998; 68(2):425S-429S.
6. Zoli A, Altomonte L, Caricchio R et al. Serum zinc and copper in active rheumatoid arthritis: correlation with interleukin 1 beta and tumour necrosis factor alpha. Clin Rheumatol 1998; 17(5):378-382.
7. Simkin PA. Oral zinc sulphate in rheumatoid arthritis. Lancet 1976; 2(7985):539-542.
8. Chausmer AB. Zinc, insulin and diabetes. J Am Coll Nutr 1998; 17(2):109-115.
9. Haase H, Maret W. Protein tyrosine phosphatases as targets of the combined insulinomimetic effects of zinc and oxidants. Biometals 2005; 18(4):333-338.
10. Schott-Ohly P, Lgssiar A, Partke HJ et al. Prevention of spontaneous and experimentally induced diabetes in mice with zinc sulfate-enriched drinking water is associated with activation and reduction of NF-kappa B and AP-1 in islets, respectively. Exp Biol Med (Maywood) 2004; 229(11):1177-1185.
11. Briefel RR, Bialostosky K, Kennedy-Stephenson J et al. Zinc intake of the US population: findings from the third National Health and Nutrition Examination Survey, 1988-1994. J Nutr 2000; 130:1367S-1373S.
12. Maret W, Sandstead HH. Zinc requirements and the risks and benefits of zinc supplementation. J Trace Elem Med Biol 2006; 20(1):3-18.
13. Lindeman RD, Clark ML, Colmore JP. Influence of age and sex on plasma and red-cell zinc concentrations. J Gerontol 1971; 26:358-363.
14. Hotz C, Peerson JM, Brown KH. Suggested lower cutoffs of serum zinc concentrations for assessing zinc status: reanalysis of the second National Health and Nutrition Examination Survey data (1976-1980). Am J Clin Nutr 2003; 78:756-764.
15. Prasad AS, Fitzgerald JT, Hess JW et al. Zinc deficiency in elderly patients. Nutrition 1993; 9:218-224.
16. Goode HF, Penn ND, Kelleher J et al. Evidence of cellular zinc depletion in hospitalized but not in healthy elderly subjects. Age Ageing 1991; 20:345-348.
17. Wörwag M, Classen HG, Schumacher E. Prevalence of magnesium and zinc deficiencies in nursing home residents in Germany. Magnes Res 1999; 12:181-189.
18. Pepersack T, Rotsaert P, Benoit F et al. Prevalence of zinc deficiency and its clinical relevance among hospitalized elderly. Arch Gerontol Geriatr 2001; 33:243-253.

19. Girodon F, Blache D, Monget AL et al. Effect of a two-year supplementation with low doses of antioxidant vitamins and/or minerals in elderly subjects on levels of nutrients and antioxidant defense parameters. J Am Coll Nutr 1997; 16:357-365.
20. Bogden JD, Oleske JM, Munves EM et al. Zinc and immunocompetence in the elderly: baseline data on zinc nutriture and immunity in unsupplemented subjects. Am J Clin Nutr 1987; 46:101-109.
21. Kreft B, Fischer A, Kruger S et al. The impaired immune response to diphtheria vaccination in elderly chronic hemodialysis patients is related to zinc deficiency. Biogerontology 2000; 1:61-66.
22. Kaplan J, Hess JW, Prasad AS. Impaired interleukin-2 production in the elderly: association with mild zinc deficiency. J Trace Elem Exp Med 1988; 1:3-8.
23. Ibs KH, Gabriel P, Rink L. Zinc and the immune system of elderly. Adv Cell Aging Gerontol 2003; (13):243-259.
24. Schröder AK, Rink L. Neutrophil immunity of the elderly. Mech Ageing Dev 2003; 124(4):419-425.
25. Allen JI, Perri RT, McClain CJ et al. Alterations in human natural killer cell activity and monocyte cytotoxicity induced by zinc deficiency. J Lab Clin Med 1983; 102(4):577-589.
26. Keen CL, Gershwin ME. Zinc deficiency and immune function. Annu Rev Nutr 1990; 10:415-431.
27. Sansoni P, Cossarizza A, Brianti V et al. Lymphocyte subsets and natural killer cell activity in healthy old people and centenarians. Blood 1993; 82(9):2767-2773.
28. Solana R, Mariani E. NK and NK/T-cells in human senescence. Vaccine 2000; 18(16):1613-1620.
29. Tapazoglou E, Prasad AS, Hill G et al. Decreased natural killer cell activity in patients with zinc deficiency with sickle cell disease. J Lab Clin Med 1985; 105(1):19-22.
30. Fraker PJ, King LE. Reprogramming of the immune system during zinc deficiency. Annu Rev Nutr 2004; 24:277-298.
31. Kury S, Dreno B, Bezieau S et al. Identification of SLC39A4, a gene involved in acrodermatitis enteropathica. Nat Genet 2002; 31(3):239-s40.
32. Beck FW, Prasad AS, Kaplan J et al. Changes in cytokine production and T-cell subpopulations in experimentally induced zinc-deficient humans. Am J Physiol 1997; 272(6):E1002-1007.
33. Liu CG, Zhang L, Jiang Y et al. Modulation of gene expression in precancerous rat esophagus by dietary zinc deficit and replenishment. Cancer Res 2005; 65(17):7790-7799.
34. Doerr TD, Prasad AS, Marks SC et al. Zinc deficiency in head and neck cancer patients. J Am Coll Nutr 1997; 16(5):418-422.
35. Jarrard DJ. Does zinc supplementation increase the risk of prostate cancer? Arch Ophtalmol 2005; 123:102-103.
36. Rink L, Cakman I, Kirchner H. Altered cytokine production in the elderly. Mech Ageing Dev 1998; 102(2-3):199-209.
37. Cakman I, Rohwer J, Schutz RM et al. Dysregulation between TH1 and TH2 T-cell subpopulations in the elderly. Mech Ageing Dev 1996; 87(3):197-209.
38. Prasad AS. Effects of zinc deficiency on Th1 and Th2 cytokine shifts. J Infect Dis 2000; 182(1): S62–S68.
39. Bao B, Prasad AS, Beck FW et al. Zinc modulates mRNA levels of cytokines. Am J Physiol Endocrinol Metab 2003; 285(5):E1095-E1102.
40. Gabriel P, Cakman I, Rink L. Overproduction of monokines by leukocytes after stimulation with lipopolysaccharide in the elderly. Exp Gerontol 2002; 37(2-3):235-247.
41. von Bülow V, Rink L, Haase H. Zinc-mediated inhibition of cyclic nucleotide phosphodiesterase activity and expression suppresses TNF-alpha and IL-1 beta production in monocytes by elevation of guanosine 3', 5'-cyclic monophosphate. J Immunol 2005; 175(7):4697-4705.
42. Cakman I, Kirchner H, Rink L. Zinc supplementation reconstitutes the production of interferon-alpha by leukocytes from elderly persons. J Interferon Cytokine Res 1997; 17(8):469-472.
43. Fortes C, Forastiere F, Agabiti N et al. The effect of zinc and vitamin A supplementation on immune response in an older population. J Am Geriatr Soc 1998; 46:19-26.
44. Duchateau J, Delepesse G, Vrijens R et al. Beneficial effects of oral zinc supplementation on the immune response of old people. Am J Med 1981; 70:1001-1004.
45. Wagner PA, Jernigan JA, Bailey LB et al. Zinc nutriture and cell-mediated immunity in the aged. Int J Vit Nutr Res 1983; 53:94-101.
46. Cossack ZT. T-lymphocyte dysfunction in the elderly associated with zinc deficiency and subnormal nucleoside phosphorylase activity: effect of zinc supplementation. Eur J Cancer Clin Oncol 1989; 25:973-976.
47. Bogden JD, Oleske JM, Lavenhar MA et al. Effects of one year of supplementation with zinc and other micronutrients on cellular immunity in the elderly. J Am Coll Nutr 1990; 9:214-225.
48. Boukaiba N, Flament C, Acher S et al. A physiological amount of zinc supplementation: effects on nutritional, lipid and thymic status in an elderly population. Am J Clin Nutr 1993; 57:566-572.

49. Fabris N, Mocchegiani E, Amadio L et al. Thymic hormone deficiency in normal ageing and Down's syndrome: is there a primary failure of the thymus? Lancet 1984; 1:983-986.
50. Arreaza EE, Gibbons JJ, Siskind GW et al. Lower antibody response to tetanus toxoid associated with higher auto-anti-idiotypic antibody in old compared with young humans. Clin Exp Immunol 1993; 92(1):169-173.
51. Powers DC, Belshe RB. Effect of age on cytotoxic T-lymphocyte memory as well as serum and local antibody responses elicited by inactivated influenza virus vaccine. J Infect Dis 1993; 167(3):584-992.
52. Girodon F, Galan P, Monget AL et al. Impact of trace elements and vitamin supplementation on immunity and infections in institutionalized elderly patients: a randomized controlled trial. MIN. VIT. AOX. geriatric network. Arch Intern Med 1999; 159:748-754.
53. Faber C, Gabriel P, Ibs KH et al. Zinc in pharmacological doses suppresses allogeneic reaction without affecting the antigenic response. Bone Marrow Transplant 2004; 33(12):1241-1246.
54. Driessen C, Hirv K, Rink L et al. Induction of cytokines by zinc ions in human peripheral blood mononuclear cells and separated monocytes. Lymphokine Cytokine Res 1994; 13(1):15-20.

Chapter 12

Zinc-Binding Proteins and Immunosenescence:

Implications as Biological and Genetic Markers

Eugenio Mocchegiani* and Marco Malavolta

Abstract

The ageing process is defined as a decline in performance and fitness with advancing age. The progressive decrease in physiological capacity and the reduced ability to respond to environmental stress leads to increased vulnerability to disease. Consequently, the incidence of many diseases and the mortality rate increases with ageing. Conversely, human centenarians escape many age-related diseases with subsequent healthy ageing and longevity. In order to explain increased mortality with advancing age and at the same time the reasons of an exceptional longevity of centenarians, many theories have been proposed. Among them, "Antagonistic Pleiotropy" seems to fit best with one of the causes of immunosenescence, such as chronic inflammatory status, because some genes related to the inflammatory response, from useful mediators devoted to the neutralization of dangerous/harmful agents early in life and in adulthood, become detrimental in aging where the antigenic load is chronic, resulting in a dangerous self-destructive response. As an example in this context, we discuss here two zinc-binding stress-related proteins, Metallothioneins (MT) and alpha-2 macroglobulin (α2-M). We propose that these alter function from a role in protection against oxidative stress and inflammation in young-adult age, to become harmful agents in ageing, mainly due to their continuous sequestration of zinc allowing low zinc ion bioavailability for a satisfactory inflammatory/immune response. Some MT genetic polymorphisms are related to low zinc status and innate immune dysfunction and with the appearance of cardiovascular diseases and type II diabetes, whereas α-2M polymorphisms are associated with the appearance of Alzheimer's disease and myocardial infarction. Therefore MT and α-2M may be considered as biological and genetic markers of ageing. Their association with zinc transporters (ZnT and Zip family), which are in turn involved in the correct maintenance of intracellular zinc homeostasis, is crucial in order to better understand the genes and the mechanisms involved in longevity and immunosenescence.

Introduction

The ageing process is defined as a decline in performance and fitness with advancing age.[1] Often referred to simply as "ageing", senescence is a nearly universal feature of multicellular organisms and understanding why it occurs is a long-standing problem in biology. However, in humans as well as in all organisms, the aging process seems inexorable. The progressive decrease in physiological capacity and the reduced ability to respond to environmental stress leads to increased susceptibility to disease. Consequently, the incidence of many diseases and the mortality rate increase with ageing.[2] The basis for these dramatic rises in mortality is incompletely understood,

*Corresponding Author: Eugenio Mocchegiani—Immunology Center (Section Nutrition Immunity and Ageing), Research Department, I.N.R.C.A., via Birarelli 8, 60121 Ancona, Italy. Email: e.mocchegiani@inrca.it

Immunosenescence, edited by Graham Pawelec.

but presumably involves changes in the function of many types of cells, which lead to tissue/organ dysfunction and systemic illness. Interestingly, a retrospective study of centenarians demonstrated that they live 90-95% of their lives in good health and with a high level of functional independence.[3] Centenarians do suffer a 30-40 % annual mortality, but this represents a marked compression of morbidity toward the end of life.[2] However, the intrinsic causes of their longevity are not clear. In an effort to adequately explain both the increased mortality with advancing age and at the same time the reasons of the exceptional longevity of centenarians, many theories have been proposed, even if what is supposedly known about the fundamental molecular mechanisms involved in ageing remains controversial and largely unproven.

The major reason for this is the obvious complexity of the problem. Aging changes are manifest from the molecular to the functional levels and some measurable biomarkers are lacking or misleading, especially when these markers refer to downstream physiological cascades that occur during ageing. This is particularly evident when some immune biomarkers of ageing are proposed downstream of the immune response.[4] Therefore, it is relevant to go back upstream in order to find more precise biomarkers of ageing in general, or of immunosenescence in particular. In this context, the genetic background is fundamental because it is the background the downstream homeostatic mechanisms that change with advancing age, including the immune response. Kirkwood proposes that three categories of genes may be involved in regulating longevity and senescence:[5] (a) those that regulate somatic maintenance and repair, (b) negatively pleiotropic genes that enhance early survival but are disadvantageous later in life (antagonistic pleiotropy) and (c) harmful late-acting mutations upon which little evolutionary selection is exerted. The presence of these genes may represent a spectrum from general to species-specific. Genes involved in cell maintenance and repair are likely to be present in all organisms, since such essential processes are similar across species. Late-acting mutations are probably species-specific, because they are likely to be individualistic and random. nonmaintenance pleiotropic genes could be universally found within a population or species. In this last category of pleiotropic genes belong, for instance, many genes involved in inflammatory response, such as pro-inflammatory cytokines, which, from useful mediators devoted to the neutralization of dangerous/harmful agents early in life and in adulthood, become detrimental in aging where the antigenic load is chronic, thus resulting in a dangerous self-destructive response. As such, inflammation becomes a chronic condition that continuously damages the surrounding tissues. This phenomenon, termed "inflamm-aging," is a feature of the old organism that becomes frail and exposed to risk of degenerative age-related diseases.[6] Chronic inflammation indeed is involved in the pathophysiology of Alzheimer's disease, atherosclerosis, diabetes, sarcopenia, cancer, infections and other age-related diseases with relevant inflammatory components[7] and immune dysfunction.[8] In this context, some stress-related proteins may have pleiotropic roles in ageing related to the circulating inflammatory mediators. Here, we review the biological and genetic role played by Metallothioneins (MT) and alpha 2- macroglobulin (α-2M) in ageing. Their role is crucial because both proteins are involved in the homeostasis of a trace element, zinc, that is pivotal for proper functioning of the inflammatory/immune response.[4] The role played by zinc transporters (ZnT and Zip family) is also discussed because their concomitant presence together with MT is fundamental for maintaining optimal intracellular zinc homeostasis.[9] The other area where knowledge in this context is quite advanced, ie., tinc transporters related to the aged brain, is not discussed here, but the reader is referred to a recent review.[10]

Metallothioneins and Ageing

Metallothioneins (MT) are a group of low-molecular-weight metal-binding proteins having high affinity for zinc (kd=1.4×10^{-13} M).[11] MT are present as different isoforms characterized by the length of aminoacid chain: isoform I, II, III e IV mapped to chromosome 16 in man and chromosome 8 in mice, both with complex polymorphisms.[12] The more common isoforms are I and II; the isoform III, also called growth inhibitory factor (GIF), is a brain-specific member of the MT family and isoform IV is restricted in squamous epithelia. MT contain 20 cysteines, all in reduced form and bind seven zinc atoms through mercaptide bonds that have the spectroscopic characteristics of metal thiolate clusters.[13]

MT distribute intracellular zinc because zinc undergoes rapid inter- and intracluster exchange. They act as antioxidants inducing, through the reduced thiol groups, the transfer of zinc to other proteins, such as antioxidant metalloenzymes.[11] Therefore, MT play a protective role during oxidative stress[14] as well as for a prompt immune response against stressor agents and inflammation. Indeed, in this last condition, macrophages produce some cytokines, such as IL-1, IL-6, IFN-α, TNF-α, which provoke a neosynthesis of MT in the liver[15] but, at the same time, an alteration in the zinc status.[16] IL-1 also affects MT mRNA in thymic epithelial cells (TECs) and, at the same time, MT are donors of zinc for thymulin reactivation in TECs.[17] MT act both as a reservoir of zinc during zinc deficiency and as zinc buffering proteins in the presence of excessive amounts of zinc in order to prevent zinc toxicity.[18] Based on these findings, MT are protective agents which also have the task of preventing zinc deficiency during inflammation. The recent discovery that, under inflammatory conditions, MT in the extracellular environment support the beneficial movement of leukocytes to the site of inflammation representing a "danger signal" for the immune cells and modifying the character of the immune response when cells sense cellular stress, is in line with this assumption. However, high levels of MT produced in chronic inflammation may alter normal chemotactic responses that regulate leukocyte trafficking.[19] Taking into account that zinc ions attract leukocytes by inducing and promoting the chemotactic response,[20] high MT production might thus be dangerous for immune responses in the presence of chronic inflammation. That MT can be harmful in immunosenescence is further suggested by i) the existence of high MT and low zinc ion bioavailability in the liver and atrophic thymus from old mice;[21] ii) the presence of high MT in lymphocytes from old and Down's syndrome subjects and low MT in lymphocytes from centenarians[22] and iii) the occurrence of atrophic thymus in stressed MT transgenic mice.[23] Additionally, in the presence of chronic environmental stress, elevated levels of MT can cause dramatic decreases in murine cytotoxic T-lymphocyte (CTL) activity against allogeneic target cells, reduce the proliferative response of CTLL-2 cells to cytokines and decrease the level of major histocompatibility complex (MHC) Class I and CD8 molecules detectable on the surface of lymphocytes.[24] Therefore high MT may also have an immunosuppressive effect worsened by the fact that they are not donors of zinc in ageing but rather they sequester zinc.[22] On the other hand, high MT induce down-regulation of many other biological functions related to zinc, such as metabolism, gene expression and signal transduction.[11]

A major problem that it is still unresolved in ageing is the inability of MT to release zinc. In this context, zinc release from MT under oxidative stress conditions is accompanied by more MT disulfide bond formation.[25] But, an intriguing point is that also NO provokes zinc release by MT, via s-nitrosylation.[26] Despite iNOS increases in ageing, the release of zinc by MT is very limited. One hypothesis might be an imbalance between NO synthases (iNOS and cNOS).[27] However, NO donors and zinc fluorescent probes are useful tools in order to study zinc release from MT and to evaluate intracellular labile zinc in ageing.

Preliminary experiments in our laboratory, using NO donors and the fluorescent probe Zinpyr-1, show a more limited zinc release by MT in PBMC from old subjects then in the young (Malavolta et al unpublished results). Moreover, a zinc-sensitive fluorescent probe, FluoZin-3, was recently used to quantify the amount of labile zinc in PBMC, with the result that the intracellular concentrations of labile zinc in resting cells were estimated to be 0.17 nM in monocytes and 0.35 nM in lymphocytes ($CD4^+$).[28] The combination of these two novel methodological procedures will permit us to study in depth the cause of limited zinc release from MT in ageing and, at the same time, to evaluate intracellular labile zinc. In any event, limited zinc release from MT in ageing provokes a low free zinc ion availability for immune responses. The recent discovery of a novel polymorphism of MT (-209A/G MT2A) may indirectly support this assumption. Indeed, old subjects of the AA genotype display high MT, low zinc ion availability, enhanced IL-6 and impaired innate immune responses with subsequent possible risk for atherosclerosis and diabetes type II.[29] Moreover, old subjects of the GG genotype for another novel polymorphism of MT (+838 C/G MT2A) display low NK cell cytotoxicity and are at risk for cardiovascular diseases (stroke) (Mocchegiani and Giacconi, unpublished results). Therefore, MT may have different roles in immunosenescence, following the concept that several genes/proteins that increase fitness early in life may also have negative effects later in life: i.e., the "Antagonistic Pleiotropy Theory of Ageing".[30] Therefore, MT may be considered as biological and genetic marker of immunosenescence.

Alpha-2 Macroglobulin and Ageing

In recent decades, various biological functions have been described for α-2M. The earliest function to be discovered was its proteinase-binding capacity with an inhibitory action on proteinases themselves.[31] One decade later, its ability to modulate T-cell response was described, thus with a role in the immune system.[32] In particular, α-2M-proteinase complexes, but not native α-2M, decrease T-cell responses to mitogens, antigens and alloantigens.[33] Various hypotheses were offered on the mechanisms involved. One of them is the down-regulation of HLA-DR expression,[34] however not confirmed by others.[33] A more accredited hypothesis is the discovery that the target of α-2M-proteinase complexes was T-cell growth factors (i.e., cytokines). Indeed, it has been shown that the decreased T-cell response is the consequence of IL-2 degradation and inactivation by α-2M-bound proteinases.[35] Other cytokines, such as pro-inflammatory cytokines (IL-1, IL-6 and TNF-α), bind α-2M, but without being degraded.[35] The formation of α-2M-proinflammatory cytokine complexes has been suggested to provide protection from immediate toxic effects of cytokines, protection from extracellular proteolysis and from the loss of cytokines through the kidney and, finally, targeting of cytokines to α-2M-receptor-bearing cells.[35] These phenomena occur as indirect or alternative inhibitory roles of α-2M against increased MMP, which are in turn temporarily produced in response to exogenous signaling by pro-inflammatory cytokines.[36] The binding of α-2M with pro-inflammatory cytokines occurs in order to induce latency of cytokines themselves.[35] As a consequence, minor MMP production occurs and the inflammation is "under control".[37] In addition, the possible over-expression of these cytokines may be also avoided by the α-2M-cytokine complex through α-2MR/LRP.[38] Pro-inflammatory cytokines (IL-1, IL-6 and TNF-α) increase in inflammation[39] and IL-6 regulates α-2M gene expression.[40] Thus, the interrelationship between α-2M and the immune system is evident with a role of protection against toxic effects resulting from overproduction of pro-inflammatory cytokines during inflammation. This role is out not in doubt in transient inflammation, as it may occur in young-adult age, but it may be questionable in chronic inflammation (ageing, cancer and infection) for the following reasons: i) α-2M requires zinc in order to be active[31] and free zinc ions, as reported above, are important for the efficacy of the entire immune system; ii) zinc is also required for the binding of α-2M to cytokines;[41] iii) α-2M plasma levels increase in ageing[42] and in certain type of cancer[43] coupled with damaged central (thymic endocrine activity) and peripheral immune functions (NK cell cytotoxicity) and low zinc availability. In contrast, human centenarians display relatively low plasma levels of α-2M, satisfactory zinc ion availability and thymic endocrine activity, enhanced IL-2 production and NK cell cytotoxicity,[22,4] as well as lower inflammation due to low gp130 expression (subunit of IL-6 receptor) despite high IL-6 circulating levels.[44] Since high α-2M levels are associated with an unfavorable prognosis in cancer,[45] these findings suggest that α-2M may possibly have a protective in young adult-age but acting as a harmful agent in aging and inflammation due to continuous sequestration of zinc. Thus, α-2M may also represent, like MT, a further and interesting biological and genetic marker of immunosenescence following the antagonistic pleiotropy idea. The discovery of one polymorphism for α-2M receptor (LRP-1) associated with myocardial infarction[46] and polymorphisms for α-2M in Alzheimer,[47] which are age-related diseases characterized by altered zinc turnover and immune performances,[48] may support this assumption.

Zinc Transporters and Ageing

The proteins that control zinc influx/efflux and compartmentalization have been almost completely identified and characterized in the last decade. These proteins all have transmembrane domains and are encoded by two solute-linked carrier (SLC) gene families: ZnT (SLC30) and Zip (SLC39). Transporters of the Zip family include at least 15 members in human cells and allow influx of zinc into the cytosol from extracellular fluids or from the lumen of intracellular compartments. Transporters of the ZnT family include 9-10 members and allow efflux of zinc from the cytosol to extracellular fluids and to intracellular compartments. Although many excellent reviews concerning isolation and characterization of Zip and ZnT family members have been published,[49,50] very little attention has been addressed to the relevance of these transporters in aging. In this chapter we therefore attempt to review the few papers dealing with zinc transporters in aging and to extrapolate results that could be of potential interest in aging studies.

Taking into account the current state of the art, it appears that the mechanisms involved in the efflux/influx of zinc through the cell membrane may be similar but differently regulated in different cells and tissues. However, it is surprising to note that despite many papers reporting an impaired intestinal absorption of zinc in the elderly, there is no published report on the expression of zinc transporters in intestinal cells during aging. The involvement of zinc transporters in the age-related malabsorption may be argued for, taking into account that in senescent rats (30 month of age) zinc uptake was found to be significantly lower in the distal part of the intestine and the ileal transport capacity decreased by 50% when compared to adult rats.[51] Conversely, a previous report indicated that intestinal zinc transport was greater in 20 month-old rats than in 12 and 3 month-old rats.[52] Therefore, the zinc transport systems might manifest continuous age-related changes that most probably involve the surface expression of specific importers/exporters of zinc. The zinc transporters that mainly contribute to regulating absorption and distribution of zinc at the intestinal level have been identified. Among them, the main ones seem to be ZnT1 (SLC30A1), ZnT5 or hZnT1 (SLC30A5), Zip4 (SLC39A4) and Zip5 (SLC39A5).[53] Zip4 seems to be main importer of zinc in enterocytes and endoderm cells because it is strongly induced and recruited to the apical surface of these cells during zinc deficiency[54] and it is downregulated during zinc supplementation.[53] Moreover, a genetic defect of this gene is responsible for the human genetic disorder of zinc absorption, acrodermatitis enteropathica.[55] Therefore, it may not be surprising to find downregulation of this zinc transporter during aging. This hypothesis could be complicated by the presence of an inadequate intake of zinc, a common feature in old age, which might result in a compensatory up-regulation of Zip4. Thus it will necessary to carefully study also the dietary intake of zinc, zinc status and the regulation of this transporter after zinc supplementation in order to give a correct picture of the phenomenon in aging.

A key organ in the maintenance of zinc homeostasis is the liver. Unfortunately, also in this case there is no specific study that takes into account hepatic zinc transporters in aging. However, it is possible to foresee a particular aspect of this phenomenon based on recent findings on the regulation of Zip14 in the liver.[56] In fact, in vivo and in vitro experiments demonstrate that Zip14 expression is up-regulated through IL-6 and that this zinc transporter most likely plays a major role in the mechanism responsible for hypozincemia that accompanies the acute-phase response to inflammation and infection. Since chronic inflammation characterized by high levels of IL-6 and an increased incidence of infections are common features of aging, it could be expected that Zip14 expression might be increased in the liver during aging. Further support for this hypothesis derives from previous papers reporting increased levels of MT and zinc in the liver of old mice.[22,23]

The only study that has partially focused attention on zinc transporter in aging is a recent microarray-based examination of the expression of several zinc transporters, before and after zinc supplementation, in blood leukocytes from human subjects, including elderly women.[57] The most relevant changes were observed in the zinc importer Zip1. The expression of Zip1 was reduced upon zinc supplementation as would be expected following the idea that, in the presence of additional zinc, the expression of membrane zinc importers decreases to avoid intracellular accumulation of toxic amount of zinc. Interestingly, greater percentage decreases in Zip1 mRNA levels occurred in elderly women than young, but the baseline values of Zip1 expression were not found to be different between the two age groups. ZnT1 expression was also found not to be different between young and elderly subjects at baseline, neither without or after supplementation. However, the authors noticed an evident upregulation of ZnT1 expression in some subjects, especially in the elderly group. These findings may appear in contrast with the general view of a zinc deficiency in aging, since it could be expected that upregulation of Zip1 and downregulation of ZnT1 would occur in zinc deficiency. However, it has to be taken into account that also MT, another biomarker of zinc status, are increased in aging, thus suggesting that despite the presence of a general zinc deficiency some proteins involved in the regulation of zinc homeostasis are dysregulated. The main factor contributing to explain this situation is the oxidative stress that is generally increased in aging and that might induce a high release of free zinc from MT. These free zinc ions, acting through the pathway MTF-1/MRE, may be the main factors responsible for the increased levels of MT.[58] The unchanged situation regarding Zip1 and ZnT1 expression may arise from the generation of a new homeostatic equilibrium where high levels of MT are necessary to avoid the presence of excessive amounts of intracellular free zinc ions. In this condition, zinc supplementation might contribute to saturate the overexpressed MT in the cells of aged individuals. However, the persistence of oxidative stress in these cells might induce a higher release of free zinc ions than in young, thereby contributing to the greater percentage decreases in Zip1 levels observed in elderly women than in young. Alternatively, it may be hypothesized that these is a dichotomy between the RNA expression and the protein expression of the zinc transporters in aging. In this case, it would have to be assumed that the protein expression of Zip1 is upregulated despite its mRNA remaining stable, thus suggesting that the increased MT does not contribute to reconstitute a new homeostatic equilibrium but, conversely, contributes to the presence of an intracellular free zinc deficiency. Further studies are currently in progress to establish the function of these homeostatic mechanisms in aging (as part of the Zincage Project).[59]

Conclusions and Future Perspectives

From the data herein reported, it emerges that the zinc-binding proteins metallothioneins and alpha-2 macroglobulin may be considered as two potential biological and genetic markers of immunosenescence. In particular, MT polymorphisms are related to the appearance of cardiovascular diseases and type II diabetes, whereas α-2M polymorphisms are associated with the appearance of Alzheimer's dementia and myocardial infarction, which are common pathologies of the elderly. This potential role played by MT and α-2M as biomarkers is largely due to the fact that both proteins are related to intracellular zinc homeostasis that is crucial for good control of the inflammatory/immune response. The inability of MT to release zinc during chronic inflammation, as is common in ageing, also leads to an inefficiency of α-2M in inducing latency to immune inflammatory mediators with subsequent damage to surrounding tissues and continuous worsening of the chronic inflammation. As such, the old organism becomes frail and exposed to the risk of the appearance of degenerative diseases. Conversely, in centenarian subjects with lower inflammatory status, both the zinc-binding proteins are expressed less and this contributes to satisfactory zinc status and good performance of innate and adoptive immune responses. Therefore, MT and α-2M may be considered as biological and genetic markers of ageing following the "Antagonistic Pleiotropy Theory of Ageing", which suggests that a trade-off between early beneficial effects and late negative outcomes can occur at the genetic and molecular level.

Good homeostatic control of MT is strictly linked to the zinc transporters of the ZnT and Zip families, which are involved in zinc efflux and influx, respectively. A preserved regulation of MT and zinc transporters is crucial in maintaining an optimal intracellular zinc concentration with benefit for the inflammatory/immune response. Although there is a paucity of data on the role played by zinc transporters in ageing and in immunosenescence, intriguing and interesting perspectives are foreseeable from studying zinc transporters both at the gene expression level and as polymorphisms in ageing and age-related diseases with a concomitant approach to MT and α-2M polymorphisms. As such, a more extensive picture of the role played by zinc homeostasis, via MT, α-2M and zinc transporters, may be attained in order to better understand the genes and the mechanisms involved in longevity and senescence upstream of immune efficiency. This is one of the specific tasks of the Zincage project (see www.zincage.org).

Acknowledgements

Supported by EC (Project ZINCAGE n. FOOD-CT-2004-506850, Coordinator: Dr. Eugenio Mocchegiani).

References

1. Finch CE. Mechanisms in senescence: some thoughts in April 1990. Exp Gerontol 1992; 27:7-16.
2. Troen BR. The biology of aging. The Mont Sinai J Med 2001; 70:3-22.
3. Hitt R, young-xu Y, silver M et al. Centenarians: the older you get, the healthier you have been. Lancet 1999; 354:652.
4. Mocchegiani E, Costarelli L, Giacconi R et al. Nutrient-gene interaction in ageing and successful ageing. A single nutrient (zinc) and some target genes related to inflammatory/immune response. Mech Ageing Dev 2006; 127:517-525.
5. Kirkwood TB. Human senescence. Bioessays 1996; 18:1009-1016.
6. Franceschi C, Bonafe M, Valensin S et al. Inflamm-aging. An evolutionary perspective on immunosenescence. Ann N Y Acad Sci 2000; 908:244-254.
7. Candore G, Colonna-Romano G, Balistreri CR et al. Biology of longevity: role of the innate immune system. Rejuvenation Res 2006; 9:143-148.
8. Pawelec G, Ouyang Q, Wagner W et al. Pathways to a robust immune response in the elderly. Immunol Allergy Clin North Am 2003; 23:1-13.
9. Eide DJ. Zinc transporters and the cellular trafficking of zinc. Biochim Biophys Acta 2006, [in press].
10. Mocchegiani E, Bertoni-Freddari C, Marcellini F et al. Brain, aging and neurodegeneration: role of zinc ion availability. Prog Neurobiol 2005; 75:367-390.
11. Kagi JH, Schaffer A. Biochemistry of metallothionein. Biochemistry 1988; 27:8509-8515.
12. West AK, Stallings R, Hildebrand CE et al. Human metallothionein genes: structure of the functional locus at 16q13. Genomics 1990; 8:513-518.
13. Maret W, Vallee BL. Thiolate ligands in metallothionein confer redox activity on zinc clusters. Proc Natl Acad Sci 1998; 95:3478-3482.
14. Sato M, Kondoh M. Recent studies on metallothionein: protection against toxicity of heavy metals and oxygen free radicals. Tohoku J Exp Med 2002; 196:9-22.
15. Cui L, Takagi Y, Wasa M et al. Zinc deficiency enhances interleukin-1alpha-induced metallothionein-1 expression in rats. J Nutr 1998; 128:1092-1098.
16. Bui LM, Dressendorfer RH, Keen CL et al. Zinc status and interleukin-1 beta-induced alterations in mineral metabolism in rats. Proc Soc Exp Biol Med 1994; 206:438-444.
17. Coto JA, Hadden EM, Sauro M et al. Interleukin 1 regulates secretion of zinc-thymulin by human thymic epithelial cells and its action on T-lymphocyte proliferation and nuclear protein kinase C. Proc Natl Acad Sci 1992; 89:7752-7756.
18. Kelly EJ, Quaife CJ, Froelick GJ et al. Metallothionein I and II protect against zinc deficiency and zinc toxicity in mice. J Nutr 1996; 126:1782-1790.
19. Yin X, Knecht DA, Lynes MA. Metallothionein mediates leukocyte chemotaxis. BMC Immunol 2005; 6:21-31.
20. Hujanen ES, Seppa ST, Virtanen K. Polymorphonuclear leukocyte chemotaxis induced by zinc, copper and nickel in vitro. Biochim Biophys Acta 1995; 1245:145-152.
21. Mocchegiani E, Giacconi R, Cipriano C et al. Are zinc-bound metallothionein isoforms (I + II and III) involved in impaired thymulin production and thymic involution during ageing? Immun Ageing 2004; 1:5-10.
22. Mocchegiani E, Giacconi R, Cipriano C et al. Metallothioneins (I + II) and thyroid-thymus axis efficiency in old mice: role of corticosterone and zinc supply. Mech Ageing Dev 2002; 123:675-694.
23. Mocchegiani E, Giacconi R, Cipriano C et al. MTmRNA gene expression, via IL-6 and glucocorticoids, as potential genetic marker of immunosenescence: lessons from very old mice and humans. Exp Gerontol 2002; 37:349-357.
24. Youn J, Lynes MA. Metallothionein-induced suppression of cytotoxic T-lymphocyte function: an important immunoregulatory control. Toxicol Sci 1999; 52:199-208.
25. Feng W, Benz FW, Cai J et al. Metallothionein disulfides are present in metallothionein-overexpressing transgenic mouse heart and increase under conditions of oxidative stress. J Biol Chem 2006; 281:681-687.
26. Zangger K, Oz G, Haslinger E et al. Nitric oxide selectively releases metals from the amino-terminal domain of metallothioneins: potential role at inflammatory sites. FASEB J 2001; 15:1303-1305.
27. Mocchegiani E, Muzzioli M, Giacconi R. Zinc and immunoresistance to infection in aging: new biological tools. Trends Pharmacol Sci 2000; 21:205-208.
28. Haase H, Hebel S, Engelhardt G et al. Flow cytometric measurement of labile zinc in peripheral blood mononuclear cells. Anal Biochem 2006; 352:222-230.
29. Giacconi R, Cipriano C, Muti E et al. Novel -209A/G MT2A Polymorphism in Old Patients with Type 2 Diabetes and Atherosclerosis: Relationship with Inflammation (IL-6) and Zinc. Biogerontology 2005; 6:407-413.

30. Williams PD, Day T. Antagonistic pleiotropy, mortality source interactions and the evolutionary theory of senescence. Evolution Int J Org Evolution 2003; 57:1478-1488.
31. Sottrup-Jensen L. Alpha-macroglobulins: structure, shape and mechanism of proteinase complex formation. J Biol Chem 1989; 264:11539-11542.
32. Lysiak JJ, Hussaini IM, Gonias SL. alpha 2-Macroglobulin synthesis by the human monocytic cell line THP-1 is differentiation state-dependent. J Cell Biochem 1997; 67:492-497.
33. Heumann D, Vischer TL. Immunomodulation by alpha 2-macroglobulin and alpha 2-macroglobulin-proteinase complexes: the effect on the human T-lymphocyte response. Eur J Immunol 1988; 18:755-760.
34. Hoffman MR, Pizzo SV, Weinberg JB. Modulation of mouse peritoneal macrophage Ia and human peritoneal macrophage HLA-DR expression by alpha 2-macroglobulin "fast" forms. J Immunol 1987; 139:1885-1890.
35. Borth W. Alpha 2-macroglobulin. A multifunctional binding and targeting protein with possible roles in immunity and autoimmunity. Ann N Y Acad Sci 1994; 737:267-272.
36. Kahari VM, Saarialho-Kere U. Matrix metalloproteinases and their inhibitors in tumour growth and invasion. Ann Med 1999; 31:34-45.
37. James K. Interactions between cytokines and alpha 2-macroglobulin. Immunol Today 1990; 11:163-166.
38. Gonias SL, LaMarre J, Crookston KP et al. Alpha 2-macroglobulin and the alpha 2-macroglobulin receptor/LRP. A growth regulatory axis. Ann NY Acad Sci 1994; 737:273-290.
39. Dinarello CA. Reduction of inflammation by decreasing production of interleukin-1 or by specific receptor antagonism. Int J Tissue React 1992; 14:65-75.
40. Horn F, Wegenka UM, Lutticken C et al. Regulation of alpha 2-macroglobulin gene expression by interleukin-6. Ann N Y Acad Sci 1994; 737:308-323.
41. Gettins PG, Crews BC. Binding of epidermal growth factor to human alpha 2-macroglobulin. Significance for cytokine alpha 2-macroglobulin interactions. Ann NY Acad Sci 1994; 737:383-398.
42. Rink L, Seyfarth M. Characteristics of immunologic test values in the elderly. Z Gerontol Geriatr 1997; 30:220-225.
43. Mocchegiani E, Ciavattini A, Santarelli L et al. Role of zinc and alpha2 macroglobulin on thymic endocrine activity and on peripheral immune efficiency (natural killer activity and interleukin 2) in cervical carcinoma. Br J Cancer 1999; 79:244-250.
44. Moroni F, Di Paolo ML, Rigo A et al. Interrelationship among neutrophil efficiency, inflammation, antioxidant activity and zinc pool in very old age. Biogerontology 2005; 6:271-281.
45. Matoska J, Wahlstrom T, Vaheri A et al. Tumor-associated alpha-2-macroglobulin in human melanomas. Int J Cancer 1988; 41:359-363.
46. Gonzales P, Alvarez R, Reguero JR et al. Variation in the lipoprotein receptor related protein, alpha2-macroglobulin and lipoprotein receptor-associated protein genes in relation to plasma lipid levels and risk of early myocardial infarction. Coron Artery Dis 2002; 13:251-254.
47. Kovacs DM. alpha 2-macroglobulin in late-onset Alzheimer's disease. Exp Gerontol 2000; 35:473-479.
48. Fabris N, Mocchegiani E. Zinc, human diseases and aging. Aging 1995; 7:77-93.
49. Liuzzi JP, Cousins RJ. Mammalian zinc transporters. Annu Rev Nutr 2004; 24:151-172.
50. Kambe T, Yamaguchi-Iwai Y, Sasaki R et al. Overview of mammalian zinc transporters. Cell Mol Life Sci 2004; 61:49-68.
51. Teillet L, Tacnet F, Ripoche P et al. Effect of aging on zinc and histidine transport across rat intestinal brush-border membranes. Mech Ageing Dev 1995; 79:151-167.
52. Mooradian AD, Song MK. The intestinal zinc transport in aged rats. Mech Ageing Dev 1987; 41:189-197.
53. Cragg RA, Phillips SR, Piper JM et al. Homeostatic regulation of zinc transporters in the human small intestine by dietary zinc supplementation. Gut 2005; 54:469-478.
54. Dufner-Beattie J, Kuo YM, Gitschier J et al. The adaptive response to dietary zinc in mice involves the differential cellular localization and zinc regulation of the zinc transporters ZIP4 and ZIP5. J Biol Chem 2004; 279:49082-49090.
55. Wang F, Kim BE, Dufner-Beattie J et al. Acrodermatitis enteropathica mutations affect transport activity, localization and zinc-responsive trafficking of the mouse ZIP4 zinc transporter. Hum Mol Genet 2004; 13:563-571.
56. Liuzzi JP, Lichten LA, Rivera S et al. Interleukin-6 regulates the zinc transporter Zip14 in liver and contributes to the hypozincemia of the acute-phase response. Proc Natl Acad Sci 2005; 102:6843-6848.
57. Andree KB, Kim J, Kirschke CP et al. Investigation of lymphocyte gene expression for use as biomarkers for zinc status in humans. J Nutr 2004; 134:1716-1723.
58. Maret W. Cellular zinc and redox states converge in the metallothionein/thionein pair. J Nutr 2003; 133:1460S-1462S.
59. Description of Zincage project. Available at: www.zincage.org. Accessed, 2006.

Chapter 13

Immunogenetics of Aging

Elissaveta J. Naumova* and Milena I. Ivanova

Abstract

Deterioration of the immune system with aging is associated with an increased susceptibility to infectious diseases, cancer and autoimmune disorders. It has been demonstrated that immunosenescence is associated with chronic, low-grade inflammatory activity. The aging process is very complex and longevity is a multifactorial trait, which is determined by genetic and environmental factors and the interaction of "disease" processes with "intrinsic" ageing processes. It is hypothesized that the level of immune response as well as possibly longevity could be associated with genes regulating immune functions. It is further hypothesized that the diversity of these genes might influence successful aging and longevity by modulating an individual's response to life-threatening disorders. Several studies have focused on the role of genes encoding molecules with immune functions. In this chapter we will review the data on the role of HLA and cytokine gene polymorphisms in human longevity and comment on the future directions in this field.

Aging is a complex process and longevity is a biological phenomenon which shows a large inter-species as well as inter-individual variability that could be determined by the interaction of many factors: genetic background, environment, lifestyle and nutrition. The somatic theory explains aging in terms of accumulation of mutations in the genome of somatic cells leading to cell senescence, cell death or transformation, as well as loss of function. A variety of models in different species demonstrate that mutations in different genes are able to induce a consistent and marked increase of the lifespan. Most of these genetic variations which significantly impact upon longevity address a limited number of pathways highly conserved in evolution. Assuming that *Homo sapiens* is not an exception to this order, many studies over the last few years have been focused on these evolutionarily conserved pathways in order understand the genetics of human aging and longevity. Furthermore, genetic heritability of human lifespan was confirmed by investigations in different populations showing heritability estimates between 0.10 and 0.33 and therefore that genes account for about 25% of longevity determination.

Different approaches have been applied in order to search for genetic determinants of human longevity. One of these approaches, utilized by the Leiden research program on aging,[1] is a population approach that focuses on determinants of age-associated diseases, to explain the majority of disabilities and impaired well-being at old age. Recently, approaches that identify longevity genes in model organisms were combined with approaches in man focusing on genes evolved for somatic maintenance and repair mechanisms aimed to clarify determinants assuring human longevity. Another approach to discover genomic regions associated with longevity is based on methods that allow for an extensive sampling of the genome, without making any a priori assumptions about "candidate" loci. The advantage of such a strategy is that the search for longevity-associated loci is not restricted to the small number of already-known candidate genes, but is potentially extended to the whole genome. Several studies have also focused on polymorphic microsatellite

*Corresponding Author: Elissaveta J. Naumova—Central Laboratory of Clinical Immunology, University Hospital Alexandrovska, 1 Georgy Sofiisky str., 1431 Sofia, Bulgaria. Email: immun@alexandrovska-hospital.bg, mivanova@intech.bg

Immunosenescence, edited by Graham Pawelec. ©2007 Landes Bioscience and Springer Science+Business Media.

loci[2] and indicated an association between increased microsatellite instability and aging. One of the possible causes for this could be age-associated impaired mismatch repair capacity, leading to the accumulation of DNA damage, rsulting in alterations of cellular functions and increased incidence of diseases in elderly. Another investigation was based on screening of Alu sequences, repetitive elements interspersed on all chromosomes and constituting more than 10% of the human genome. Genomic regions enriched in Alu sequences are potentially unstable and mutations may cause the exonization of intronic Alu sequences, with important effects on gene expression and functionality.[3] Interestingly, Alu elements are not randomly scattered throughout the genome. On the contrary, as many as 45% of Alu elements are contained within genes; and they are found to be highly clustered in genes that are involved in metabolism, transport and signaling processes.[4] Thus, Alu sequences may be seen as good markers of highly variable and potentially unstable loci in functionally important genomic regions.

In order to identify genes responsible for human longevity, some recent studies were based on the model of centenarians, as these individuals represent the best example of successful aging. These studies have shown that centenarians largely escape most of the major age-related diseases[5] and that they are characterized by a complex remodeling of immune responses,[6] and particularly by largely conserved or even upregulated innate immunity.[7] Chronic low-grade inflammation appears to be a major component of the most common age-related diseases, such as diabetes, osteoporosis and osteoarthritis, dementia, cardiovascular diseases and cancer. Moreover, there is evidence that glucose utilization is remarkably well-conserved and insulin resistance remarkably absent or is very low in centenarians, suggesting that they are also characterized by a well-conserved IGF1/insulin pathway.[8] Data in animal models and evidence in centenarians suggest that longevity is associated with the capability of cells to cope with a variety of stressors, including oxidative stress.[9]

Using these approaches, different investigations have shown that human longevity might be associated with several functionally different genes such as genes involved in DNA repair, cell proliferation and apoptosis (p 53 p 66);[10,11] Insulin/IGF1 signaling pathway;[12] genes that counteract oxidative stress (e.g., Paraoxonase1);[9] polymorphic genes related to immune responses and inflammation. However, the data on genes involved in the regulation of immune response are still limited and here we focus on the impact of such genetic factors.

Immunity and Aging

The ageing process seems to be directly correlated with immune deterioration, resulting from the combination of different genes and environmental factors.[13] Several studies showed that senescence of the immune system is related to a decrease of cellular and humoral responses and an increased frequency of infectious, autoimmune and malignant diseases.[14,15] The T-cell-mediated immune responses are more strongly affected in comparison with humoral immune responses. T-cell proliferation to antigens and mitogens, as well as cell-mediated cytotoxicity, decrease with age. Senescence negatively influences the membrane structures participating in the early stages of T-cell activation, as well as the transcription factors regulating gene expression. These age-associated changes also affect the distribution of the T-cell subpopulations. A decrease in absolute number of $CD4^+$ and $CD8^+$ subpopulations was found with age.[16] An increase in the relative proportion of CD8 lymphocytes not expressing CD28 is commonly found in the elderly.[17,18] "An immune risk phenotype" (a low number of CD4, a high number of CD8, decreased production of IL-2 and poor proliferative responses to mitogens, less $CD28^+$ and CD57- cells) which is predictive of a shorter remaining lifespan in the very elderly has been described.[19-21] Decreased antibody responses to exogenous antigens result from both suppressed T-cell function and B cell functional activity. Although there is an increased absolute number of NK cells in the peripheral blood, their functional capacity is reduced with age, which contributes to a higher susceptibility to malignant and viral diseases. Nowadays, it is accepted that many aspects of senescence are characterized by inflammatory status and that senescence is associated with a chronic low-level inflammatory activity leading to tissue damage. The established "immune risk phenotypes" are likely to be related to an inability to control the sytemic inflammation. However, the extent to which these changes depend on the genetic background needs to be explored. Family segregation analysis has shown that numbers of CD4 and CD8 cells and the CD4/CD8 ratio, are under genetic control.[22,23] Immunophenotypic analysis in families with long-lived members in the Bulgarian population indicated lower absolute and relative numbers of $CD3^+8^+$, $CD8^+28^+$ and $CD8^+57^+$ cells in elderly people. Although more pronounced in the elderly group, lower numbers of $CD8^+$ T-cells were also found in middle aged and young members of those families compared to age-matched controls. A progressive decline of the $CD8^+28^+$ subset was seen with aging. In addition, the "immune risk phenotype", a marker of increased inflammatory reactivity, was not observed, which is consistent with the healthy immune status of elderly individuals in this study.[24]

All these data imply that aging processes may be associated with alterations in the immune system, suggesting that genetic determinants of senescence also reside in those polymorphisms for the immune system genes that regulate immune responses. Genes encoding molecules involved in the development of protective immunity are highly polymorphic and present significant variation in populations, possibly resulting from an evolutionary adaptation of the organism facing an ever-evolving environment. These genes include HLA genes, genes encoding "unusual" HLA-like molecules (CD1), killer cell immunoglobulin-like receptor genes (KIR), leukocyte Fcγ receptor genes, cytokine and cytokine receptor genes; Toll-like receptor genes and TNF-receptor associated factors. Several studies have focused on the role of HLA and cytokine gene polymorphisms and human longevity. The diversity of these genes might influence successful aging and longevity by modulating an individual's response to life-threatening disorders.

Why MHC?

In the past 25 years, many studies have focused on the possible association between the MHC and human longevity. HLA is the largest highly polymorphic genetic complex, located on chromosome 6p,25 and spanning about 4000 kb of DNA (approximately 1/750 of the human genome). It has been suggested that this region of the human genome has been generated by extensive duplications of genes with redundant functions but subtle differences. Further diversification of the MHC is a result of recombination, gene conversion and point mutations. It is likely that this evolution is pathogen-driven and the polymorphic variants are subjected to natural selection. Therefore, the genes in HLA complex are under strong evolutionary pressure.

Table 1. Gene clusters in the HLA region

Cluster	Loci	Protein Encoding Loci	Pseudo Genes
Gene superclusters			
Histones	66	55	11
HLA class I	26	7	17
tRNA	157	151	6
Butyrophilin	8	8	0
Olfactoryreceptor	34	14	20
Zink finger proteins	36	26	10
Gene Clusters			
Vomeronasal receptor	5	0	5
Tumor necrosis factor	3	3	0
Lymphocyte antigen 6	5	5	0
Heat shock proteins	3	3	0
HLA class II	24	15	9

www.anthonynolan.org/uk.hig

Another main characteristic of the MHC region is the clustering of genes with similar functions. Different clusters and subclusters containing protein-encoding functional loci and pseudogenes have been identified (Table 1). Presently, more that 200 genes with different functions such as antigen processing and presentation, inflammation, leukocyte maturation, immune regulation, as well as stress-inducible proteins and complement components have been discovered in the human HLA (Table 2).[25] These genes could be divided into three regions known as class I, II and III. The HLA class I region, that is the closest to the telomere, contains 6 functional genes—3 highly polymorphic and highly expressed classical HLA class I genes (*HLA-A, -B* and *-C*), encoding the heavy chains of HLA class I molecules and 3 less polymorphic nonclassical genes (*HLA-E, G, F*), encoding the heavy chains of nonclassical HLA molecules; 2 genes MICA and MICB with limited expression on epithelia and fibroblasts and highly conserved in evolution; and, finally, class I noncoding sequences, including pesudogenes. The HLA class II region initially had been divided into three subregions: *DP, DQ* and *DR*, each containing at least two genes—A and B, encoding α and β chains of class II molecules. Later, two additional subregions—*DO, DN* and *DM* were identified in the HLA class II region. *DMA* and *DMB* genes show about 30% sequence homology with classical class II genes and encode a heterodimer involved in peptide loading of classical class II molecules. It is thought that *DNA (DOA)* and *DOB* genes regulate binding of peptides to class II molecules although the exact role of these genes in the immune response remains to be clarified. Additionally, a cluster of 4 genes (*LMP2, LMP7, TAP1, TAP2*), encoding molecules involved in processing those antigens presented by classical HLA class I molecules. The HLA class III subregion is located between class I and class II and contains more than 40 genes, including genes encoding complement components (C4a and C4B, C2), the gene for properdin factor B, genes encoding TNF α and β, three genes encoding proteins of the HSP70 family, as well as other genes with non-immunological functions.

HLA class I and class II molecules are highly polymorphic heterodimeric cell surface glycoproteins. HLA class I molecules consist of an α- chain, encoded by class I genes and a non-MHC-encoded β2 microglobulin. Similarly, class II molecules have 2 polypeptide chains α and β, but both are encoded by class II genes. These molecules present processed antigenic peptides to T-cell receptors. Peptides presented by class I molecules to CD8-positive cytotoxic T-lymphocytes are derived from processed intracellularly synthesized antigens, eg. viral proteins. HLA class II molecules present peptides obtained from proteolytic cleavage of extracellular antigens, eg. bacterial antigens to CD4-positive T-cells. HLA class I and class II molecules play a crucial role in the development of the immune response by restricting and regulating T-cell responses against specific

Table 2. Function of genes in the HLA region

Function	Genes
Antigen processing and presentation	HLA-A,-B,-C,-DMA, -DMB, -DOA, -DOB, -DPA1, -DPB1, -DQA1, -DQB1, -DRA, -DRB1, -DRB3, -DRB4, -DRB5, PRSS16, PSMB8, PSMB9, TAP1, TAP2, TAPBP, UBD
Immunoglobulin superfamily	AGER, BTN1A1, BTN2A1, BTN2A2, BTN2A3, BTN3A1, BTN3A2, BTN3A3, BTNL2, MOG
Inflammation	ABCF1, AIF1, DAXX, IER3, LST1, LTA, LTB, NCR3, TNF
Leukocyte maturation	DDAH2, LY6G5B, LY6G5C, LY6G6D, LY6G6E, LY6G6C
Complement system	BF, C2, C4A, C4B
Non classical HLA class I	HLA-E, HLA-F, HLA-G
Immune regulation	NFKBIL1, RXRB, FKBPL
Stress-inducible proteins	HSPA1A, HSPA1B, HSPA1L, MICA, MICB

www.anthonynolan.org/uk.hig

antigens, because each allele can bind and presents only a limited number of peptides, especially the class I alleles. These have a key role in the phenomenon known as determinant selection. HLA molecules are generally conserved in the domains of the protein responsible for the interactions with conserved parts of the T-cell receptors and their co-receptors, while domains responsible for antigen binding and interactions in variable regions of the T-cell receptor are characterized with extensive polymorphism. Therefore polymorphism of HLA molecules occurs preferentially in their functional domains and has dramatic effects on epitope selection and antigen presentation.

One of the major benefits provided by extensive MHC polymorphism is an increased likelihood that an individual will be heterozygous and consequently will carry two different alleles at each HLA locus. As polymorphisms occur in functionally-important domains, heterozygosity doubles the antigen presenting potential of an individual. Since individuals within the same ethnic group generally express different HLA phenotypes, the overall repertoire of the group is exponentially broadened by the presence of extensive polymorphism increasing the likelihood of the species of surviving a wide variety of pathogens. Thus evolutionary advantage of broadening HLA repertoire might explain why it has been so difficult to pinpoint associations between particular HLA alleles or haplotypes and infectious diseases. Only a few examples of such associations have been reported, such as a reduced risk of developing severe malaria in children caring HLA-B*5301 and DRB1*1302,[26,27] and even in such cases antigen-presenting efficiency does not seem to be the whole explanation. Slightly cleare evidence has accumulated on the role of HLA polymorphism in immune-mediated diseases such as autoimmune diseases.[28] Currently there are data on many diseases showing associations with HLA alleles. For many of them, however, these associations are weak, but for others the HLA associations are so strong that there is no doubt that the disease is a result of direct involvement of one or more genes within the HLA complex. Classical examples of HLA-disease associations include HLA-B27 with ankylosing spondylitis, A29 with Bidshot's retinopathy and the DRB1*1501, DQA1*0102, DQB1*0602 haplotype with narcolepsy. For other autoimmune diseases such as IDDM, multiple sclerosis, rheumatoid arthritis, etc., the evidence for primary association to some peptide-presenting HLA molecules is less strong, but probable. Moreover, it has been demonstrated that different HLA molecules predisposing to or protective for a given disease share common amino acid motifs, mainly in the peptide-binding groove or in the regions recognized by the T-cell receptor.

HLA and Longevity

Due to the central role of HLA molecules in the development of protective immunity and to the significant polymorphism of HLA genes, some studies addressed the possible impact of these genes on human longevity. However, due to methodological problems, the results are still controversial. Nonetheless, most of the data available so far are consistent with a possible role of HLA class II specificities in human longevity. The association studies performed in the Dutch population suggest that longevity is associated with positive selection for the HLA-DR11 allele in women >85 years old.[29] An increased frequency of the DRB1*11 allele was recorded among elderly Mexican Mestizo women.[30] Henon et al[31] also reported a higher frequency of the same allele in familial nonagenarians from the French population. In centenarians from the same population, Ivanova et al found statistically significant increases in the frequency of DR7, DR11 and DR13 specificities.[32] DR7 was increased in elderly men, DR11 was found with higher frequency in women with familial longevity, while DR13 showed no interaction with gender. The established increased frequency of DR7 in centenarians is of particular interest since this allele is also associated with some diseases including viral infections. On the other hand, the increased frequency of DR13 in centenarians is consistent with a protective role of this allele in infectious diseases. In agreement with that notion, the positive association of HLA-DR11 with longevity corresponds to the decreased frequency of this allele in infectious and autoimmune diseases. Furthermore, the results obtained by evaluating HLA-DR frequencies in centenarians from Sardinia, which is a genetically relatively homogeneous population, showed no significant differences between centenarians and controls except for HLA-DRB1*15 that was increased in the former, as also did not confirm findings observed in other Caucasoid populations.[33] Studies in the Bulgarian population did not show significant age-associated differences in HLA allele distribution, although a tendency towards increased frequencies of A*0101, *0301; B*3501, *0801, *1302; DRB1*1101, 1001 alleles was seen in the elderly group. An increased frequency of several HLA class II alleles, DRB1*0101, 1201, *1401, DQB1*0503, DQA1*0101, *05 and decreased frequencies of DRB1*0403 and *1302 were found in Japanese centenarians.[34]

Studies addressing other genes within the HLA complex are very limited and the available data were obtained using less sophisticated serological techniques. In elderly Italians increased frequencies of B7 and Cw6 were found.[35] Association of the HLA-B locus with longevity was also found for Greeks (increased B16 and decreased B15 in elderly),[36] and in elderly Dutch women (decreased B40)[29] (Table 3). These heterogeneous results in different populations suggest that HLA longevity associations are population-specific and are likely affected by population-specific genetic and environmental factors.

The MHC-restriction of T-cell responses against pathogens confers a positive benefit on HLA heterozygosity. Several investigations demonstrate this in longevity, whereas in other studies a similar association was not found, although most reports were limited to heterozygosity at the HLA-A and –B loci. Decreased HLA-DR heterozygosity in centenarians was found by Takata et al.[37] Similar data were established by Ivanova et al.[32] These results support the hypothesis that in the lack of disease-associated recessive alleles, homozygosity could be associated with a longer lifespan.

Comparatively small numbers of investigations have explored possible associations of HLA haplotypes, rather than single alleles, with longevity, despite the fact that they have a markedly

Table 3. HLA alleles associated with longevity

HLA specificity	Effect	Population	Reference
DRB1*11	Increased in elderly women	Dutch	Lagaay et al (1991)
	Increased in elderly women	Mexican mestizo	Soto-Vega E et al (2005)
	Increased in familial nonagenerians	French	Henon W et al (1997)
	Increased in women from sibships, no change in centenarians	French	Ivanova R et al (1998)
DRB1*07	Increased in long-lived men	French	Ivanova R et al (1998)
	Increased in elderly	Greek	Papasteriades C et al (1997)
DRB1*13	Increased in centenarians	French	Ivanova R et al (1998)
DRB1*15	Increased in centenarians	Sardinians	Lio D et al (2003)
DRB1*0101, *1201,*1401 DQB1*0503; DQA1*0101,*05	Increased in centenarians	Japanese	Akisaka M, Suzuki M (1998)
DRB*0403, *1302	Decreased in centenarians	Japanese	Akisaka M, Suzuki M (1998)
B7, Cw7	Increased in elderly	Italians	Ricci G. et al (1998)
B16	Increased in elderly	Greek	Papasteriades C. et al (1997)
B15	Decreased in elderly		
B40	Decreased in elderly women	Dutch	Lagaay et al (1991)

Table 4. HLA haplotypes, associated with longevity

HLA haplotype	Effect	Population	Reference
A1-B8-C7-DR3	Increased in centenarian males	Irish	Rea and Middleton (1994)
	No association	Irish	Ross et al (2003)
A*0101-B*5101-DRB1*1101	Increased in elderly	Bulgarians	Naumova et al (2007)
A*0201-B*1801-DRB1*1601	Increased in elderly	Bulgarians	Naumova et al (2007)

clearer role in the genetic predisposition to HLA-associated diseases. Rea and Middleton[38] showed an increased frequency of the serologically-defined A1-B8-C7-DR3 haplotype in male centenarians. However, this observation was not confirmed by later studies.[39] Further, other studies find a decreased frequency of the HLA-B8-DR3 haplotype in centenarian females. The possible mechanisms responsible for the different associations of this haplotype in centenarians of both sexes are not clarified yet, although likely to be related to the influence of the haplotype on immune reactivity. In Caucasians, the HLA-B8-DR3 haplotype is associated with a higher risk for the development of autoimmune diseases. Individuals carrying this haplotype are predominantly characterized by type 2 immune responses. It could be speculated that the HLA-B8-DR3 haplotype is related to a shift of the immune response from type 1 to type 2, which is one if the hallmarks of the immune alterations in senescence. A study of the Bulgarian population showed significantly increased frequencies of DRB1*11 and DRB*16—related haplotypes: A*0101-B*5101-DRB1*1101 and A*0201-B*1801-DRB1*1601 in elderly individuals compared to young controls. It is interesting to note that DRB1*11 and DRB1*16-related haplotypes were also found to be protective for different autoimmune diseases in this population. This is in agreement with the hypothesis that the lifespan prolongation is associated with a delay of immunodeficiency of normal aging or with possible amelioration of autoimmunity that develops with age.

Why Cytokines?

Cytokines are an integral part of the immune response stimulated by antigen presentation in the context of HLA. Many studies have shown that the pathology of some infectious, autoimmune and malignant diseases is influenced by the profiles of cytokine production in pro-inflammatory (Th1) and anti-inflammatory (Th2) T-cells. Additionally, some authors have shown differences in cytokine levels in elderly and possible associations with age-related diseases. Ferrucci et al,[40] and Harris et al[41] demonstrated that increased IL-6 serum levels could be a marker for functional disability and a predictor of mortality in the elderly. Increased levels of IL-6 are also observed under conditions of stress, one of the characteristics of ageing.[42] Elevated levels of this cytokine associated with development of frailty and susceptibility to diseases in elderly were also observed.[43] This inflammatory marker could be involved in the low-grade inflammation that develops with age.[43-46] Additionally, dysregulation of IL-6 has been proposed to be involved in the pathogenesis of a variety of age-related diseases, such as diabetes and atherosclerosis, which have a substantial inflammatory pathogenesis.[47,48] At the same time, a decreased capacity to produce IFN-γ, IL-2, IL-4 upon stimulation has been reported in the edlerly.[49] Higher levels of TNF-α also correlate with decreased functional status and survival in the elderly.[40,41,43,44] Moreover, dysregulation and, in particular, overproduction, of TNF has been implicated in a variety of human diseases including sepsis, cerebral malaria and autoimmune diseases such as multiple sclerosis, rheumatoid arthritis (RA), systemic lupus erythematosus and Crohn's disease, as well as cancer. Interestingly, in a very large study in the Italian population, IL-1Ra plasma levels were increased with age in both male and female subjects.[50] Because of the pivotal role of anti-inflammatory cytokines TGF-β1 and IL-10 in regulation of immune responses, variability in the levels of these cytokines may greatly affect the low grade inflammation that develops with age. It has been shown that elevated levels of anti-inflammatory cytokines such as IL-10 and TGF-β in serum is associated with increased resistance against septic shock in the elderly.[43] Increased ex vivo capacity of macrophages from the elderly to produce anti-inflammatory IL-10 was also found.

Similarly to other genes coding for molecules with immune functions, cytokine genes are highly polymorphic. However, most of the polymorphic sites identified so far are located in the noncoding regions, containing regulatory sequences, while exon sequences are highly conserved. Three main forms of polymorphisms have been identified in cytokine genes: single nucleotide polymorphisms (SNPs),[51] variable numbers of tandem repeats and micro-satellites[52,53] Although still controversial, polymorphic variants observed for some cytokine genes have been correlated to the level of gene expression. Thus, cytokine gene polymorphism may be responsible for observed inter-individual differences in cytokine production and may be one possible mechanism for perturbation of the Th1/Th2 balance. Some polymorphisms may have direct functional significance by altering directly or indirectly the level of gene expression and/or function; others may only be useful for the determination of genetic linkage to a particular haplotype associated in turn with a given clinical condition.

An increasing number of studies investigated associations between cytokine gene polymorphism, cytokine gene expression in vitro and susceptibility to various immunologically mediated disorders.[53] The polymorphisms in cytokine genes are probably implicated in genetic regulation of inflammatory responsse and resistance or susceptibility to infectious, autoimmune diseases, allergy, cancer and degenerative diseases.[54] Although the data are limited and controversial,[55] it has been hypothesized that longevity could be associated with cytokine gene polymorphisms correlated with different levels of cytokine expression and modulating immune response to several diseases.[44,56,57] Taking into account the integral role of cytokine genes in immune responses, the regulation of cytokine expression levels and their polymorphic nature, investigating the genetic variations of these loci with functional significance could be appropriate as immunogenetic candidate markers for studying mechanisms of successful aging and longevity.

Several studies have reported associations of longevity with gene polymorphisms in different pro- and anti-inflammatory cytokines (Tables 5, 6). Genetic variations correlating with elevated levels of pro-inflammatory cytokines have been negatively associated with ageing.[58] Several studies showed that cytokine polymorphisms related to different level of secretion were associated with longevity. Genetic polymorphisms, associated with high level of IL-10 expression were increased,[59] while polymorphisms possibly related to increased expression of pro-inflammatory cytokines-IFN-γ, TNF-α and IL-6 were decreased in elderly individuals.[60,61] These data confirmed the hypothesis that longevity is related to anti-inflammatory genotype profile. Additionally, the pro-inflammatory cytokine profile was correlated with decreased of human life span in elderly. However, in elderly with different ethnical background, investigations reported contradicting results on associations with cytokine gene polymorphism.[62,63] Additionally, the majority of data were associated with investigation of single polymorphisms in single cytokine genes. The analysis of extended haplotypes which include several polymorphisms in the cytokine gene, as well as haplotypes which consist of SNPs in different cytokine genes will help to determine the precise immunogenetic basis of longevity.

Gene Polymorphism of Proinflammatory Cytokines and Aging

IL-6

The human IL-6 gene is mapped to chromosome 7p21–24. Different studies identified three SNPs (−597G/A, −572G/C и −174G/C) and one AT polymorphism (−373(A)n(T)m) in 5' regulatory region in the gene.[64-66] It was demonstrated that the IL-6 – 174C allele was significantly associated with lower plasma concentrations of this cytokine.[65] Despite a significant number

Table 5. Gene polymorphisms of pro-inflammatory cytokines associated with aging

Cytokine gene polymorphism	Effect	Population	Reference
IL-6 (–174 C/G)	Increased (C-low) in male centenarians	Italian	Bonafe et al (2001)
	Decreased (G/G-high) in octogenarians and nonagenerians	Irish	Rea et al (2003); Ross O et al (2003)
	No association in nonagenarians	Finish	Wang X et al (2001)
	No association	Bulgarian	Naumova et al (2004)
IFN-G (+874 T/A)	Increased (T/T-high) in female centenarians	Italian	Lio D et al (2002)
	No associations	Bulgarian	Naumova et al (2004)
TNF-A (–308 G/A)	Decreased (A-high) in centenarians	Danish	Buunsgaard H et al (2004)
	No association in nonagenarians	Finish	Wang X et al (2001)
	No association in centenarians	Italian	Lio D et al (2003)
	No association in elderly	Bulgarians	Naumova et al (2004)

Table 6. Gene polymorphisms of anti-inflammatory cytokines associated with aging

Cytokine gene polymorphism	Effect	Population	Reference
TGF-B1 (915 C/G)	Decreased C allele and C/G genotype in centenarians	Italian	Carrieri G et al (2004)
TGF-B1 (cdns 10, 25)	No associations in elderly	Bulgarians	Naumova et al (2004)
IL-10 (–1082 A/G)	Increased (G/G-high) in male centenarians	Italians	Lio D et al (2002)
	No association with longevity	Finish	Wang X et al (2001)
	No association with longevity	Irish	Ross O et al (2003)
	No association with longevity	Sardinian	Pes G et al (2004)
IL-10 (–819 C/T)	No association with longevity	Italian	Lio D et al (2002)
IL-10 (–592 C/A)	No association with longevity	Italian	Lio D et al (2002)
IL-10 (–1082G,–819C,–592C)	Increased in elderly	Bulgarians	Naumova et al (2004)
IL-10 (–1082A,–819T,–592A)	Decreased in elderly	Bulgarians	Naumova et al (2004)

of studies on the possible role of IL-6 gene polymorphisms in different diseases, associations still remain to be clarified.[65,67-69] Several studies focused on IL-6 – 174C/G polymorphism and susceptibility to common causes of morbidity and mortality among the elderly, such as type 2 diabetes, cardiovascular diseases and dementia. Data on centenarians and elderly individuals from Italy showed an increased frequency of C alleles in male centenarians and this seemed to be a gender-specific effect on longevity.[70] Correlations with the serum levels showed that men carrying the GG genotype had higher IL-6 serum levels than subjects with CC or CG genotypes. The authors hypothesized that individuals predisposed to produce high level of IL-6 (men carrying GG) have a reduced capacity to reach the extreme limits of human life-span. Additionally, they demonstrated that age-related increases of IL-6 serum levels in women were quite independent of the −174C/G genotype. It has also been shown that the proportion of IL-6 high-producers (GG genotypes) was increased in individuals affected by age-related diseases with an inflammatory pathogenesis, i.e., diabetes, atherosclerosis, osteoporosis and neurodegenerative diseases. Similarly, Rea et al and Ross et al reported a decreased frequency of IL-6 – 174GG carriers in Irish octogenarian and nonagenarian subjects from the BELFAST elderly longitudinal ageing study.[71,72] However, in Finish nonagenarians, analysis of IL6-174, IL1a-889, IL1b-511, IL1Ra VNTR, IL10-1082 and TNFa-308 did not show any associations, alone or in combinations.[73] A similar lack of association of IL-6 gene polymorphism with longevity was observed in the Bulgarian population.[24] These conflicting data could be partly explicable by population-specific factors including genetic background, environmental factors and lifestyle. Most studies analyzed the effect of isolated polymorphisms and this could partly contribute to the contradictory results. Recently, Terry et al[69] demonstrated that haplotypic combinations of promoter polymorphisms in the IL-6 gene are more informative markers associated with the level of gene expression than −174G/C in isolation. Additionally, studies have demonstrated that CC or CG carriers have an increased risk of Alzheimer's disease (AD) and cardiovascular diseases.[74]

IFN-γ

Because of the key role of proinflammatory IFN-γ in immune responses, genetic control of its production has been the focus of several studies of longevity. Among numerous intronic polymorphisms in the IFN-γ gene, a variable length CA repeat sequence and polymorphism in intron 1 at position +874 relative to the transcriptional start site has been implicated to influence the level of gene expression in vitro.[75] +874 T/A SNP is one well-known single-nucleotide polymorphism at the 5' end of the CA repeat region in the first intron of the IFN-γ gene. Specific binding of the nuclear transcription factor-κB to the DNA sequence containing the +874 T allele has been reported; this could have functional consequences for the transcription of the IFN-γ gene and could then influence the rate of expression. Studies in Italian centenarians showed an increased frequency of +874 T/T in females. On the other hand, investigations in elderly individuals from the Bulgarian population did not show significant differences in IFN-γ (+874) allele distribution compared to young controls.[24]

TNF-α

TNF-α is a proinflammatory cytokine also involved in the immune response. The gene for TNF-α is located within the class III region of the major histocompatibility complex, which is a highly polymorphic region and its expression is tightly controlled at the transcriptional and post-transcriptional level. Several biallelic polymorphisms have been described within the TNF-α gene, including six in the promoter region at positions −1031T > C, −863C > A, −857C > T, −376G > A, −308G > A and −238G > A. Moreover, a number of studies have shown that the TNF-α promoter polymorphisms have a significant effect on transcriptional activity. Susceptibility to many diseases is thought to have a genetic basis and the TNF gene is considered a candidate predisposing gene. However, unraveling the importance of genetic variation in the TNF locus to disease susceptibility or severity is complicated by its location within the MHC and the strong linkage disequilibrium with other genes. Several investigations reported associations of MHC haplotypes with different TNF-α phenotypes: DR3 and DR4 haplotypes were correlated with high levels of TNF-α,[76,77] while DR2 haplotypes were associated with low expression.[76,78] These findings implied the existence of a functional polymorphism involved in the regulation of TNF-α production. Single nucleotide polymorphisms at position –308 have been commonly studied with respect to their influence on TNF-α expression. Transfection studies in human B-cell lines showed that the presence of the rare TNF2 allele (A at position –308) results in higher constitutive and inducible levels of TNF expression compared with the common TNF1 allele (G at position –308), confirming the importance of this site for the transcriptional regulation of the TNF gene.[79-82] The functional relevance of this SNP has been confirmed by its involvement in determining susceptibility to immune-inflammatory diseases.[80-87] Although the polymorphism –308 G/A associated with differential gene expression is one of the most widely investigated in different diseases, no correlation of this SNP and longevity was found in centenarians from Finnish or Italian populations.[59,73] Similar results were found in elderly individuals from the Bulgarian population.[24] In Danish population, however, a decreased frequency of the A allele was present in centenarians. Positive association with the process of successful ageing was observed also in male centenarians from the Italian population when the two SNPs –308 G/A from TNFA gene and –1082 G/A SNP from IL10 gene were analyzed simultaneously.[73] It was reported that an anti-inflammatory genotype TNFA GG (low)/ IL10 GG (high) has a protective role in longevity.

TGF-B1

The available data in the literature seem limited to only one investigation of an association between the functional genetic variants +915 G/C in exon 1, –800 G/A and –509 C/T in the promoter of the TGF-β1 gene with longevity. The TGF-β1 gene consists of 7 exons and 6 introns. So far, eight polymorphisms in the promoter (–509 C→T, –800 G→A and –988 C→T) and coding (+896 T→C, +915 G→C, +788 C→T, +652 C→T and +673 T→C) regions and one deletion (713 del C) in inton 4 of the TGF-β1 gene have been found. Polymorphisms +896 (codon 10) and +915 (codon 25), associated with different levels of expression, are the most commonly studied. For polymorphism +915 G/C, the presence of the C allele is generally associated with lower TGF- β synthesis in vitro and in vivo. Associations between the presence of a particular TGF- β1 allele and the level of the product indicates that the G–800 A and C–509 T polymorphisms may also be involved in the modulation of expression of the TGF- β1 gene. The –509 T allele has been reported to be associated with marginally higher transcriptional activity of TGF- β compared to the –509 C allele. The TGF-β –800 G→A polymorphism is in a consensus CREB half-site. The presence of the A allele was suggested to result in a reduced affinity for the CREB family of transcription factors, resulting in a lower level of total TGF-β1 in the circulation. Analysis of these three SNPs +915 G→C, –509 C→T and –800 G→A by the group of Carrieri et al found that only the +915 C allele and GC genotype were present with significantly lower frequency than in controls. Additionally they also found a decreased frequency of extended haplotype G –800/C –509/C 869/C 915 and elevated levela of TGF-β1 in the elderly, but correlation with investigated genotypes in the TGF-B1 gene was not found.[88] Similarly no associations of TGF-B1 codons 10 and 25 genotypes with longevity were observed in Bulgarians.[24] It has been hypothesized that genetically-determined cytokine profiles of TGF-β1 could be involved in mechanisms of successful ageing but more data are needed to confirm this hypothesis.

IL-10

Since one of the main functions of IL-10 may be to limit inflammatory responses,[89] polymorphisms in the regulatory region of this gene could possibly be related to longevity. Stimulation of human blood samples with bacterial lipopolysaccharide showed variation of IL-10 production, suggesting a genetic component of approximately 75%.[90] Inter-person differences in the regulation of IL-10 production may be critical with respect to the final outcome of an inflammatory response.

The IL-10 gene is located on chromosome 1 at q31-32. Several polymorphisms in the human IL-10 5' flanking region and two microsatelites associated with differential IL-10 production have been identified. The most extensively investigated SNPs are in the promoter region at positions –1082, –819 and –592[91-93] correlated with different transcriptional activity. The three dimorphisms appear in three potential haplotypes: GCC, ACC and ATA (G/A at position −1082, C/T at position −819 and C/A at −592 correspondingly) and relate to different levels of gene expression. The ability of individuals to produce high levels of IL-10 is evidently controlled by a G at position -1082, as this variant is found in the highest producers.[91,93-96] Several studies reported a linkage between the sites −819 and −592. The A allele of the −592 SNP was found to be associated with lower stimulated IL-10 release. In the presence of allele −1082A, stimulation of lymphocytes with concanavalin A resulted in lower IL-10 production than in allele −1082A-negative cells.[91,97] The functional relevance of this SNP has been shown by its involvement in determining susceptibility to immune-inflammatory diseases.[95,98-103] The two dimorphisms −819 and −592 exhibit strong linkage disequilibrium.

The IL-10 −1082 A/G polymorphism has been reported to be a male-specific marker for longevity,[94] while no differences were found regarding the −819 and −592 polymorphisms. The −1082GG genotype, associated with high IL-10 production, was argued to confer an anti-inflammatory status.[94] Studies in Bulgarians demonstrated significant differences for two IL-10 haplotypes: one of them (−1082A, −819T, −592A), possibly associated with low level of gene expression, was decreased in the elderly, while the other (−1082G, −819C, −592C), maybe associated with high cytokine gene expression, was significantly more frequent among healthy elderly compared to young controls. This effect was more pronounced in GCC homozygous individuals as indicated by the analysis of IL-10 genotypes. However, studies in two other populations—Irish nonagenarians[39] and Finish nonagenarians[73]-did not show any association with longevity. A possible explanation for the negative results in the Irish and Finnish studies could be the younger age of the old subjects investigated in comparison with the Italian study. Note that the IL-10 −1082 GG genotype is much less frequent in patients affected by Alzheimer's disease.[59]

Conclusions

Although several studies mentioned above have focused on the role of HLA and cytokine genes for successful aging, available data do not allow at present to clarify their actual role. A broad family-based analysis on the role of genes with immune functions on longevity in different ethnic groups would be very informative and would allow clarification of the impact of these genetic markers in successful aging. One of these studies in families with longevity members (at least two generations, with octogenarians or nonagenarians) from the Bulgarian population showed increased frequencies of haplotypes, found to be protective for different autoimmune diseases. Furthermore alleles and haplotypes positively associated with autoimmunity were not observed in families with long-lived members. Prevalence of genotypes associated with anti-inflammatory cytokine gene profiles was also observed in these families. Based on the inheritance in families, extended immunogenetic profiles associated with longevity can be identified (Fig. 1).

An international collaboration was established within the 14th International HLA and Immunogenetics Workshop to clarify further the contribution of HLA, cytokine genes and other MHC-encoded loci to successful aging and an increased capacity to reach extreme limits of life-span. The main objectives of the component "Immunogenetics of aging" were to analyze two data sets from different ethnic groups: 1) families with long-lived members (octogenarians and nonagenarians) and unrelated elderly individuals compared to ethnically-matched young controls; 2) to investigate the effect of classical HLA class I and class II loci and cytokine polymorphisms in regulatory and/or coding regions, with a possible impact on the level of gene expression of pro- and anti-inflammatory cytokines in these datasets. These approaches are based on linkage and association analyses to identify extended immunogenetic profiles that could be relevant to better understanding the mechanisms of successful aging and longevity. Preliminary analyses carried out in different populations included in this component suggested that longevity is associated with positive selection for HLA haplotypes shown to be protective for diseases and an increased frequency of anti-inflammatory cytokine gene profiles. A consensus was reached on recommendations for future progress: It was concluded that additional samples/data from different ethnic groups should be collected and high-resolution HLA typing should be performed whenever possible in order to clarify further population-specific HLA associations in longevity. Another potential opportunity

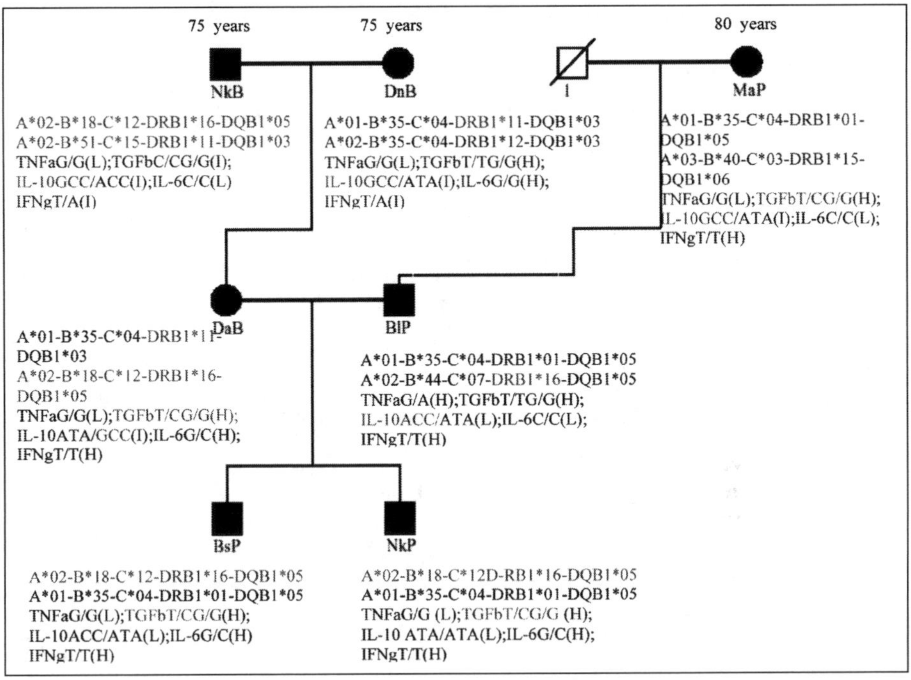

Figure 1. Extended immunogenetic profiles in Bulgarian families with long-lived members.

is to perform extensive characterization of the extended HLA region in selected samples using micro satellites and SNP markers. It is also critical to incorporate in the future studies additional, possibly functionally relevant, polymorphisms in cytokine genes and to extend studies to other relevant immune molecules such as KIR, TLR4, CD14, CCR5 and MMP3.

References

1. Westendorp R. Leiden research program on aging. Exp Gerontol 2002; 37:609-14.
2. Neri S, Gardini A, Facchini A et al. Mismatch repair system and aging: microsatellite instability in peripheral blood cells from differently aged participants. J Gerontol A Biol Sci Med Sci 2005; 60:285-92.
3. Lev-Maor G, Sorek R, Shomron N et al. The birth of an alternatively spliced exon: 3' splice-site selection in Alu exons. Science 2003; 300:1288-91.
4. Grover D, Mukerji M, Bhatnagar P et al. Alu repeat analysis in the complete human genome: trends and variations with respect to genomic composition. Bioinformatics 2004; 20:813-7.
5. Bonafe M, Valensin S, Gianni W et al. The unexpected contribution of immunosenescence to the leveling off of cancer incidence and mortality in the oldest old. Crit Rev Oncol Hematol 2001; 39:227-33.
6. Franceschi C, Bonafe M, Valensin S et al. Inflamm-aging. An evolutionary perspective on immunosenescence. Ann NY Acad Sci 2000; 908:244-54.
7. Franceschi C, Bonafe M, Valensin S. Human immunosenescence: the prevailing of innate immunity, the failing of clonotypic immunity and the filling of immunological space. Vaccine 2000; 18:1717-20.
8. Paolisso G, Barbieri M, Rizzo M et al. Low insulin resistance and preserved beta-cell function contribute to human longevity but are not associated with TH-INS genes. Exp Gerontol 2001 ; 37:149-56.
9. Monti D, Salvioli S, Capri M et al. Decreased susceptibility to oxidative stress-induced apoptosis of peripheral blood mononuclear cells from healthy elderly and centenarians. Mech Ageing Dev 2000; 121:239-50.
10. van Heemst D, Mooijaart SP, Beekman M et al. Variation in the human TP53 gene affects old age survival and cancer mortality. Exp Gerontol 2005; 40:11-5.
11. Donehower LA. p 53: guardian AND suppressor of longevity? Exp Gerontol. 2005; 40:7-9.

12. van Heemst D, Beekman M, Mooijaart S et al. Reduced insulin/IGF-1 signalling and human longevity. Aging Cell 2005; 4 :79-85.
13. De Benedictis G. Genes and longevity. Aging (Milano) 1996; 8:367-9.
14. Franceschi C, Valensin S, Lescai F et al. Neuroinflammation and the genetics of Alzheimer's disease: the search for a pro-inflammatory phenotype. Aging (Milano) 2001; 13:163-70.
15. Pawelec G, Koch S, Griesemann H et al. Immunosenescence, suppression and tumour progression. Cancer Immunol Immunother 2006; 55:981-6.
16. Doria G, Frasca D. Genes, immunity and senescence: looking for a link. Immunol Rev 1997; 160:159-70
17. Fagnoni FF, Vescovini R, Mazzola M et al. Expansion of cytotoxic CD8+ CD28- T-cells in healthy ageing people, including centenarians. Immunology 1996; 88:501-7.
18. Nociari MM, Telford W, Russo C. Postthymic development of CD28-CD8$^+$ T-cell subset: age-associated expansion and shift from memory to naive phenotype. J Immunol 1999; 162:3327-35.
19. Ferguson FG, Wikby A, Maxson P et al. Immune parameters in a longitudinal study of a very old population of Swedish people: a comparison between survivors and nonsurvivors. J Gerontol A Biol Sci Med Sci 1995; 50:B378-82.
20. Wikby A, Maxson P, Olsson J et al. Changes in CD8 and CD4 lymphocyte subsets, T-cell proliferation responses and nonsurvival in the very old: the Swedish longitudinal OCTO-immune study. Mech Ageing Dev 1998; 102:187-98.
21. Olsson J, Wikby A, Johansson B et al. Age-related change in peripheral blood T-lymphocyte subpopulations and cytomegalovirus infection in the very old: the Swedish longitudinal OCTO immune study. Mech Ageing Dev 2000; 121:187-201.
22. Amadori A, Zamarchi R, De Silvestro G et al. Genetic control of the CD4/CD8 T-cell ratio in humans. Nat Med 1995; 1:1279-83.
23. Clementi M, Forabosco P, Amadori A et al. CD4 and CD8 T lymphocyte inheritance. Evidence for major autosomal recessive genes. Hum Genet 1999; 105:337-42.
24. Naumova E, Mihaylova A, Ivanova M et al. Immunological markers contributing to sucsessful aging in Bulgarians. Exp Gerontol 2004; 39:637-44.
25. Klein J, Sato A. The HLA system. N Engl J Med 2000; 343:702-709; 782–786
26. Hill A, Elvin J, Willis A et al. Molecular analysis of the association of HLA-B53 and resistance to severe malaria. Nature 1992; 360:434-9
27. Hill A. The immunogenetics of human infectious diseases. Ann Rev Immunol 1998; 16: 593–617.
28. Thorsby E. Invited Anniversary Review: HLA associated diseases. Hum Immunol 1997; 53:11-11.
29. Lagaay AM, D'Amaro J, Ligthart GJ et al. Longevity and heredity in humans. Association with the human leucocyte antigen phenotype. Ann N Y Acad Sci 1991; 621:78-89.
30. Soto-Vega E, Richaud-Patin Y, Llorente L. Human leukocyte antigen class I, class II and tumor necrosis factor-alpha polymorphisms in a healthy elder Mexican Mestizo population. Immun Ageing 2005; 32:13.
31. Henon N, Schachter F. Genetics of longevity. C R Seances Soc Biol Fil 1997; 191:553-62.
32. Ivanova R, Henon N, Lepage V et al. HLA-DR alleles display sex-dependent effects on survival and discriminate between individual and familial longevity. Hum Mol Genet 1998; 7:187-94.
33. Lio D, Pes GM, Carru C et al. Association between the HLA-DR alleles and longevity: a study in Sardinian population. Exp Gerontol 2003; 38:313-7.
34. Akisaka M, Suzuki M. Okinawa Longevity Study. Molecular genetic analysis of HLA genes in the very old] Nippon Ronen Igakkai Zasshi 1998; 35:294-8.
35. Ricci G, Colombo C, Ghiazza B et al. Association between longevity and allelic forms of human leukocyte antigens (HLA): population study of aged Italian human subjects. Arch Immunol Ther Exp (Warsz) 1998; 46:31-4
36. Papasteriades C, Boki K, Pappa H et al. HLA phenotypes in healthy aged subjects. Gerontology 1997; 43:176-81
37. Takata H, Suzuki M, Ishii T et al. Influence of major histocompatibility complex region genes on human longevity among Okinawan-Japanese centenarians and nonagenarians. Lancet 1987; 2:824-6.
38. Rea IM, Middleton D. Is the phenotypic combination A1B8Cw7DR3 a marker for male longevity? J Am Geriatr Soc 1994; 42:978-83.
39. Ross O, Curran M, Rea I. HLA haplotypes and TNF polymorphism do not associate with longevity in Irish. Mechanisms of Aging and Development 2003; 124:563-7.
40. Ferrucci L, Harris TB, Guralnik JM et al. Serum IL-6 level and the development of disability in older persons. J Am Geriatr Soc 1999; 47:639-46.
41. Harris TB, Ferrucci L, Tracy RP et al. Associations of elevated interleukin-6 and C-reactive protein levels with mortality in the elderly. Am J Med 1999; 106:506-12.

42. Heinrich PC, Behrmann I, Haan S et al. Principles of interleukin (IL)-6-type cytokine signalling and its regulation. Biochem J 2003; 374:1-20.
43. Forsey RJ, Thompson J, Ernerudh J et al. Plasma cytokine profiles in elderly humans. Mech Ageing Dev 2003; 124:487-93.
44. Ershler WB, Keller ET. Age-associated increased interleukin-6 gene expression, late-life diseases and frailty. Annu Rev Med 2000; 51:245-70.
45. Bruunsgaard H, Skinhoj P, Qvist J et al. Elderly humans show prolonged in vivo inflammatory activity during pneumococcal infections. J Infect Dis 1999; 180:551-4.
46. Cohen HJ, Pieper CF, Harris T et al. The association of plasma IL-6 levels with functional disability in community-dwelling elderly. J Gerontol A Biol Sci Med Sci 1997; 52:M201-8.
47. Chamorro A. Role of inflammation in stroke and atherothrombosis. Cerebrovasc Dis 2004; 17 Suppl 3:1-5.
48. Dandona P, Aljada A, Bandyopadhyay A. Inflammation: the link between insulin resistance, obesity and diabetes. Trends Immunol 2004; 25:4-7.
49. Franceschi C, Bonafe M, Valensin S et al. Inflamm-aging. An evolutionary perspective on immunosenescence. Ann N Y Acad Sci 2000; 908:244-54.
50. Cavallone L, Bonafe M, Olivieri F et al. The role of IL-1 gene cluster in longevity: a study in Italian population. Mech Ageing Dev 2003; 124:533-8.
51. Kruglyak L. Prospects for whole-genome linkage disequilibrium mapping of common disease genes. Nat Genet 1999; 22:139-44.
52 Weber JL, May PE. Abundant class of human DNA polymorphisms which can be typed using the polymerase chain reaction. Am J Hum Genet 1989; 44:388-96.
53. Bidwell J, Keen L, Gallagher G et al. Cytokine gene polymorphism in human disease: on-line databases, supplement 1. Genes Immun 2001; 2:61-70.
54. Bidwell J, Keen L, Gallagher G et al. Human cytokine gene nucleotide sequence alignments: supplement 1. Eur J Immunogenet 1999; 26:135-223.
55. Caruso C, Candore G, Colonna Romano G et al. HLA, aging and longevity: a critical reappraisal. Hum Immunol 2000; 61:942-9.
56. Volpato S, Gyralink J, Ferrucci L et al. Cardiovascular disease, interleukin-6 and risk of mortality in older women: the women's health and aging study. Circulation 2001; 103:947-953.
57. Bruunsgaard H, Pedersen M, Pedersen BK. Aging and proinflammatory cytokines. Curr Opin Hematol 2001; 8:131-6.
58. Bhojak TJ, DeKosky ST, Ganguli M et al. Genetic polymorphisms in the cathespin D and interleukin-6 genes and the risk of Alzheimer's disease. Neurosci Lett 2000; 288:21-4.
59. Lio D, Scola L, Crivello A et al. Inflammation, genetics and longevity: further studies on the protective effects in men of IL-10 -1082 promoter SNP and its interaction with TNF-alpha -308 promoter SNP. J Med Genet 2003; 40:296-9.
60. Lio D, Scola L, Crivello A et al. Allele frequencies of +874T—>A single nucleotide polymorphism at the first intron of interferon-gamma gene in a group of Italian centenarians. Exp Gerontol 2002; 37:315-9.
61. Bruunsgaard H, Benfield TL andersen-Ranberg K et al. The tumor necrosis factor alpha -308G>A polymorphism is associated with dementia in the oldest old. J Am Geriatr Soc 2004 A; 52:1361-6.
62. Bonafe M, Olivieri F, Cavallone L et al. A gender—dependent genetic predisposition to produce high levels of IL-6 is detrimental for longevity. Eur J Immunol 2001; 31:2357-61.
63. Lio D, Scola L, Crivello A et al. Gender-specific association between -1082 IL-10 promoter polymorphism and longevity. Genes Immun 2002; 3:30-3.
64. Georges JL, Loukaci V, Poirier O et al. Interleukin-6 gene polymorphisms and susceptibility to myocardial infarction: the ECTIM study. Etude Cas-Temoin de l'Infarctus du Myocarde. J Mol Med 2001; 79:300-5.
65. Terry CF, Loukaci V, Green FR. Cooperative influence of genetic polymorphisms on interleukin 6 transcriptional regulation. J Biol Chem 2000; 275:18138-44.
66. Fishman D, Faulds GJeffery R et al. The effect of novel polymorphisms in the interleukin-6 (IL-6) gene on IL-6 transcription and plasma IL-6 levels and an association with systemic-onset juvenile chronic arthritis. J Clin Invest 1998; 102:1369-76.
67. Nauck M, Winkelmann BR, Hoffmann M et al. The interleukin-6 G(-174)C promoter polymorphism in the LURIC cohort: no association with plasma interleukin-6, coronary artery disease and myocardial infarction. J Mol Med 2002; 80:507-13.
68. Humphries SE, Luong LA, Ogg M et al. The interleukin-6 -174 G/C promoter polymorphism is associated with risk of coronary heart disease and systolic blood pressure in healthy men.Eur Heart J 2001; 22:2243-52.

69. Rauramaa R, Vaisanen SB, Luong LA et al. Stromelysin-1 and interleukin-6 gene promoter polymorphisms are determinants of asymptomatic carotid artery atherosclerosis. Arterioscler Thromb Vasc Biol 2000; 20:2657-62.
70. Bonafe M, Olivieri F, Cavallone L et al. A gender—dependent genetic predisposition to produce high levels of IL-6 is detrimental for longevity. Eur J Immunol 2001; 31:2357-61.
71. Rea IM, Ross OA, Armstrong M et al. Interleukin-6-gene C/G 174 polymorphism in nonagenarian and octogenarian subjects in the BELFAST study. Reciprocal effects on IL-6, soluble IL-6 receptor and for IL-10 in serum and monocyte supernatants. Mech Ageing Dev 2003; 124:555-61.
72. Ross OA, Curran MD, Meenagh A et al. Study of age-association with cytokine gene polymorphisms in an aged Irish population. Mech Ageing Dev 2003; 124:199-206.
73. Wang XY, Hurme M, Jylha M et al. Lack of association between human longevity and polymorphisms of IL-1 cluster, IL-6, IL-10 and TNF-alpha genes in Finnish nonagenarians. Mech Ageing Dev 2001; 123:29-38.
74. Zhang Y, Hayes A, Pritchard A et al. Interleukin-6 promoter polymorphism: risk and pathology of Alzheimer's disease. Neurosci Lett 2004; 362:99-102.
75. Pravica V, Asderakis A, Perrey C et al. In vitro production of IFN-gamma correlates with CA repeat polymorphism in the human IFN-gamma gene. Eur J Immunogenet 1999; 26:1-3.
76. Jacob CO, Fronek Z, Lewis G et al. Heritable major histocompatibility complex class II-associated differences in production of tumor necrosis factor alpha: relevance to genetic predisposition to systemic lupus erythematosus. Proc Natl Acad Sci USA 1990; 87:1233-7.
77. Abraham LJ, French MA, Dawkins RL. Polymorphic MHC ancestral haplotypes affect the activity of tumour necrosis factor-alpha. Clin Exp Immunol 1993; 92:14-8.
78. Bendtzen K, Morling N, Fomsgaard A et al. Association between HLA-DR2 and production of tumour necrosis factor alpha and interleukin 1 by mononuclear cells activated by lipopolysaccharide. Scand J Immunol 1988; 28:599-606.
79. Wilson AG, Symons JA, McDowell T et al. Effects of a polymorphism in the human tumor necrosis factor alpha promoter on transcriptional activation. Proc Natl Acad Sci USA 1997; 94:3195-9.
80. Makhatadze NJ. Tumor necrosis factor locus: genetic organisation and biological implications. Hum Immunol 1998; 59(9):571-9.
81. Lio D, Candore G, Colombo A et al. A genetically determined high setting of TNF-alpha influences immunologic parameters of HLA-B8,DR3 positive subjects: implications for autoimmunity. Hum Immunol 2001; 62:705-13.
82. Hajeer AH, Hutchinson IV. Influence of TNF alpha gene polymorphisms on TNF alpha production and disease. Hum Immunol 2001; 62:1191-9.
83. Witte JS, Palmer LJ, O'Connor R et al. Relation between tumour necrosis factor polymorphism TNFalpha-308 and risk of asthma. Eur J Hum Genet 2002; 10:82-5.
84. Sakao S, Tatsumi K, Igari H et al. Association of tumor necrosis factor-alpha gene promoter polymorphism with low attenuation areas on high-resolution CT in patients with COPD. Chest 2002; 122:416-20.
85. Dalziel B, Gosby AK, Richman R et al. Association of the TNF-alpha -308 G/A promoter polymorphism with insulin resistance in obesity. Obes Res 2002; 10:401-7.
86. O'Keefe GE, Hybki DL, Munford RS. The G—>A single nucleotide polymorphism at the -308 position in the tumor necrosis factor-alpha promoter increases the risk for severe sepsis after trauma. J Trauma 2002; 52:817-25.
87. Heijmans BT, Westendorp RG, Droog S et al. Association of the tumour necrosis factor alpha -308G/A polymorphism with the risk of diabetes in an elderly population-based cohort. Genes Immun 2002; 3:225-8.
88. Carrieri G, Marzi E, Olivieri F et al. The G/C915 polymorphism of transforming growth factor beta1 is associated with human longevity: a study in Italian centenarians. Aging Cell 2004; 3:443-8.
89. Moore KW, de Waal Malefyt R, Coffman RL et al. Interleukin-10 and the interleukin-10 receptor. Annu Rev Immunol 2001; 19:683-765.
90. Westendorp RG, Langermans JA, Huizinga T et al. Genetic influence on cytokine production in meningococcal disease. Lancet 1997; 349:1912-3.
91. Turner DM, Williams DM, Sankaran D et al. An investigation of polymorphism in the interleukin-10 gene promoter. Eur J Immunogenet 1997; 24:1-8.
92. D'Alfonso S, Rampi M, Rolando V et al. New polymorphisms in the IL-10 promoter region. Genes Immun 2000; 1:231-3.
93. Kube D, Rieth H, Eskdale J et al. Structural characterisation of the distal 5' flanking region of the human interleukin-10 gene. Genes Immun 2001; 2:181-90.
94. Lio D, Scola L, Crivello A et al. Gender-specific association between -1082 IL-10 promoter polymorphism and longevity. Genes Immun 2002; 3:30-3.

95. Crawley E, Kay R, Sillibourne J et al. Polymorphic haplotypes of the interleukin-10 5' flanking region determine variable interleukin-10 transcription and are associated with particular phenotypes of juvenile rheumatoid arthritis. Arthritis Rheum 1999; 42:1101-8.
96. Eskdale J, Gallagher G, Verweij C et al. Interleukin 10 secretion in relation to human IL-10 locus haplotypes. Proc Natl Acad Sci USA 1998; 95:9465-70.
97. Hutchinson IV, Pravica V, Perrey C et al. Cytokine gene polymorphisms and relevance to forms of rejection. Transplant Proc 1999; 31:734-6.
98. Wu MS, Huang SP, Chang YT et al. Tumor necrosis factor-alpha and interleukin-10 promoter polymorphisms in Epstein-Barr virus-associated gastric carcinoma. J Infect Dis 2002; 185:106-9.
99. Howell WM, Turner SJ, Bateman A et al. IL-10 promoter polymorphisms influence tumour development in cutaneous malignant melanoma. Genes Immun 2001; 2:25-31.
100. Hajeer AH, Lazarus M, Turner D et al. IL-10 gene promoter polymorphisms in rheumatoid arthritis. Scand J Rheumatol 1998; 27:142-5.
101. Tagore A, Gonsalkorale WM, Pravica V et al. Interleukin-10 (IL-10) genotypes in inflammatory bowel disease. Tissue Antigens 1999; 54:386-90.
102. Girndt M, Kaul H, Sester U et al. Anti- inflammatory interleukin-10 genotype protects dialysis patients from cardiovascular events. Kidney Int 2002; 62:949-55.
103. Shoskes DA, Albakri Q, Thomas K et al. Cytokine polymorphisms in men with chronic prostatitis/chronic pelvic pain syndrome: association with diagnosis and treatment response. J Urol 2002;168:331-5.
104. Naumova E, Pawelec G, Ivanova H et al. Fourteenth annual international HLA and immunogenetics workshop: Report on the immunogenetics of aging. Tissue Antigens 2007; 69:304-310.

Chapter 14

The Genetics of Innate Immunity and Inflammation in Ageing, Age-Related Diseases and Longevity

Calogero Caruso,* Carmela Rita Balistreri, Antonino Crivello, Giusi Irma Forte, Maria Paola Grimaldi, Florinda Listì, Letizia Scola, Sonya Vasto and Giuseppina Candore

Abstract

Inflammation is a key component of age-related diseases such as atherosclerosis and Alzheimer's disease (AD) and genes coding for inflammatory or anti-inflammatory molecules are, therefore, good candidates for influencing the risk of developing these pathologies. Findings discussed in this chapter suggest that different alleles of genes coding for pro- or anti-inflammatory genes may affect individual life-span expectancy by influencing the type and intensity of immune-inflammatory responses against environmental stressors involved in the development of age-related diseases. Our immune system has evolved to control pathogens and so pro-inflammatory responses are likely to be evolutionarily programmed to resist fatal infections in earlier life. However, this may have a deleterious effect on cardiovascular and other inflammatory diseases in later life, such that cardiovascular diseases are a late consequence of an evolutionary beneficial pro-inflammatory response programmed to resist infections in earlier life. Genetic polymorphisms responsible for a low inflammatory response might result in an increased chance of a long life-span in an environment with a reduced antigen (i.e., pathogens) load, such as a modern day healthy environment and may also permit a lower grade survivable inflammatory response to atherogenesis and atherosclerosis-related disease. Here, we review the available data in the literature on inflammatory gene polymorphisms in successful and unsuccessful ageing.

Introduction

The innate immune system is an evolutionally conserved host defence mechanism against pathogens. Innate immune responses are initiated by pattern recognition receptors (PRRs), which recognize specific structures of micro-organisms, defined as pathogen-associated molecular patterns (PAMPs). To date, three families of PRRs, usually defined as "the trinity of pathogen sensors": Toll-like receptors (TLRs), NOD-like receptors and RIG-like receptors, have been identified.[1,2] This constitutive expression of a limited set of PRRs by many cell types of the innate immune system does not require clonal expansion of specific cell populations. These germ cell-encoded proteins recognize microbial pathogens or ligands from damaged tissues based on shared molecular structures and induce host responses that localize the spread of infection and enhance systemic

*Corresponding Author: Calogero Caruso—Gruppo di Studio sull'Immunosenescenza, Dipartimento di Biopatologia e Metodologie Biomediche, Università di Palermo, Corso Tukory 211, 90134 Palermo, Italy. Email: marcoc@unipa.it

Immunosenescence, edited by Graham Pawelec. ©2007 Landes Bioscience and Springer Science+Business Media.

resistance to infection. Therefore, the expression of a limited number of highly active genes during the activation of innate immunity is able to induce rapid efficient defensive immune responses.[3] Several cell types contribute to innate immunity and the mononuclear phagocyte lineage plays a pivotal role in innate immunity.[4] Their receptor engagement induces transmembrane signals that activate NF-κB and mitogen dependent protein kinase pathways,[4] hence inducing the expression of a wide range of genes encoding proteins, such as pro-inflammatory cytokines and tissue inflammation.[5] Therefore, the capacity of each individual organism to regulate the activation of innate immunity and local inflammatory responses is crucial for initiating defensive action against pathogens, limiting tissue damage and enhancing fast recovery and tissue healing.

It is quite clear that inflammation is an important and necessary part of the normal host responses to pathogens. Our immune system has evolved to control pathogens, so pro-inflammatory responses are likely to be evolutionarily programmed to resist fatal infections. On the other hand, the overproduction of inflammatory molecules due to the life-long pathogen load is responsible for the pro-inflammatory status of ageing that is related to all-cause mortality risk in older persons. Systemic inflammation interacting with the genetic constitution of the organism may cause defined organ-specific illnesses as atherosclerosis and Alzheimer's disease.[6,7] It is now widely accepted that inflammation plays a major role in the development and progression of atherosclerosis, complications of which contribute to a large extent to morbidity and mortality especially in older people. The initial injury that occurs in atherosclerosis is damage to the endothelial cells lining the blood vessels. Some of the factors leading to the injury include increased levels of oxidized low density lipoproteins (ox-LDL) found in dyslipidemia, free radicals formed by cigarette smoking, possible infectious agents and the shearing stress placed on endothelial cells due to hypertension.[8] However, our understanding of the pathogenetic mechanisms underlying atherosclerosis and its complications remains incomplete, since more than half of patients with atherosclerosis do not show classical risk factors, such as hypercholesterolemia, hypertension, smoking history, diabetes, obesity or sedentary life style.[9,10] Recent advances in basic science have established a fundamental role for innate immunity and inflammation in mediating all stages of this disease from initiation through progression and, ultimately, to the thrombotic complications. Low-grade chronic inflammation, as indicated by levels of the inflammatory marker reactive C protein (CRP) and pro-inflammatory cytokines, prospectively defines risk of atherosclerotic complications, thus adding to prognostic information provided by traditional risk factors. In fact, levels of CRP or Interleukin (IL)-6 have been suggested to be significant predictive risk factors for future development of cardiovascular events.[10,11] In atherosclerosis, the initiating event is the accumulation of lipids in the vessel wall, which subsequently become modified by oxidation, glycation and aggregation. Their subsequent association with proteoglycans or incorporation into immune complexes initiates and triggers an inflammatory process. Ox-LDL are taken up by macrophages through scavenger receptors. Further, other monocytes, attracted from the blood, differentiate into macrophages, take up modified ox-LDL and form lipid-laden foam cells, which initiate atherosclerotic plaque development. Later on, inflammatory mediators increase, other immune cells are attracted and smooth muscle cells become activated and involved. More advanced stages of plaque development are characterized by increased deposition of extra cellular lipid cores, fibrous material and often necrosis. Subsequently, these macrophages are progressively activated, leading to the production of a wide range of cytokines and growth factors. Myocardial infarction (MI) may occur as a result of erosion or uneven thinning and rupture of the fibrous cap, often at the shoulders of the lesion where macrophages enter, accumulate and are activated and where apoptosis may occur.[12,13] Inflammation is a key component of atherosclerosis and genes coding for inflammatory or anti-inflammatory cytokines are, therefore, good candidates for risk factors in atherosclerosis. Because genetic traits contribute significantly to the risk of coronary heart disease,[14] a number of studies have now addressed the hypothesis that allelic variations in genes regulating innate immunity may increase the risk of disease.[15,16] Differences in the genetic regulation of inflammatory processes might partially explain why some people, but not others, develop the disease and why some develop a greater inflammatory response than others (Fig. 1 shows the inflammatory factors believed to play a central role in atherosclerosis).

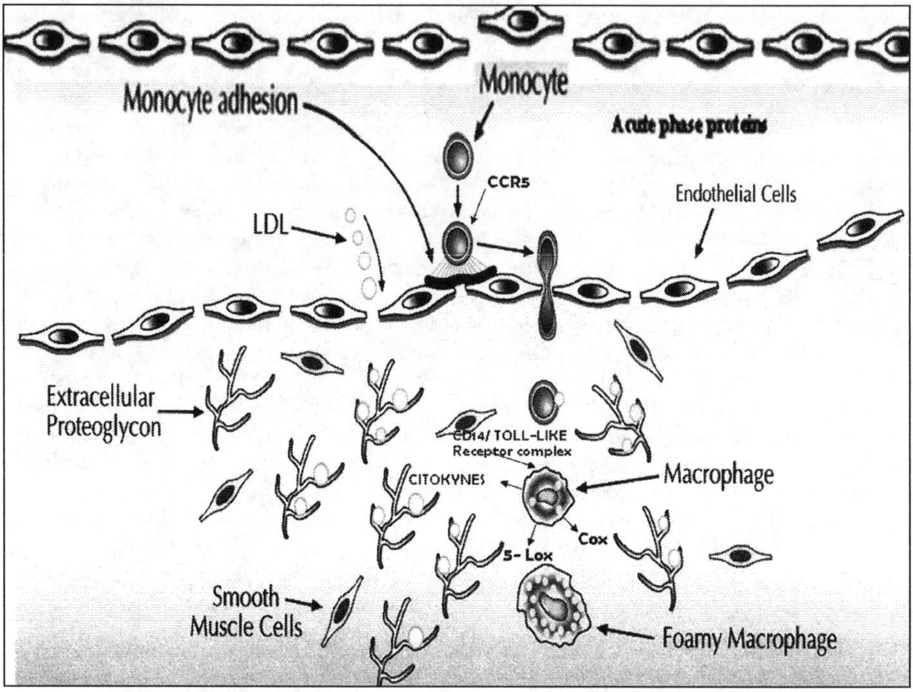

Figure 1. Representation of the section of a blood vessel and schematic representation of inflammatory molecules involved in pathogenesis of atherosclerosis and plaque formation.

Besides atherosclerosis, another significant example in which immunogenetics and inflammation are involved is Alzheimer's disease (AD), a progressive neurodegenerative disorder characterized by extensive cerebral tissue loss mainly surrounding the medial temporal lobe, near the hippocampus.[17] The pathological hallmarks found in patients affected by AD are extracellular senile plaques and neurofibrillary tangles. Senile plaques are precipitations of circular β amyloid peptide (Aβ) of about 100-200 μ in diameter, encircled by neuritic like-tangles, degenerated structure and glial cells. Jointly with astrocytes and microglial activated cells, which are the neuronal immune system cells, there are neurofibrillary tangles, composed of a hyperphoshorylated form of microtubular protein tau. Aβ is around 40-42 amino acids in length (Aβ40, Aβ42). Aβ42 can more easily form aggregates than Aβ40. The APP molecule is the Aβ peptide precursor; it is a trans-membrane glycoprotein ubiquitously expressed (it is also present on platelets), produced by the endoplasmatic reticulum and mainly involved in neuronal and dendritic growth and synapse formation.[18] The mechanisms by which the peptide Aβ can induce neuron toxicity involves the production of inflammatory molecules including free radicals which oxidize membrane phospholipids. Senile plaques are enriched by inducing inflammatory cytokine as IL-1, IL-6, complement proteins, lysosomal enzymes and others factors inducing inflammation. Aβ can stimulate the production and secretion of IL-1, IL-6 and IL-8 from microglial cells, switched on by the inflammatory response. Usually, glial cells and neurons produce basal levels of cytokines but when they are stimulated, they are induced to produce high levels of pro-inflammatory cytokines. Additionally, in the nervous tissue, cytokines induce APP production and subsequently Aβ formation which will increase cytokine production by glial and neurons cells (a vicious circle).[19,20] The APP modification begins from 2 to 3 decades ahead of disease onset. During this period the microglia and astrocyte cells are active. The microglia activation in AD can be due to mimetic properties between Aβ and the CD14 receptor and may be enhanced by chronic long-life pathogen load. The binding of Aβ to the receptor induces transcription mediated by TLRs. After activation, the microglia cells modify their morphology and became tissue macrophages producing inflammatory cytokine, complement factors, acute phase protein, different enzymes and eicosanoids.[20,21] Also in AD a number of studies, as discussed below, have now addressed the hypothesis that allelic variations in genes of innate immunity may increase the risk of disease[20] (Fig. 2 outlines the role of inflammation in AD).

Taking into account the role of inflammation in age-related diseases, it is not surprising that successful ageing (i.e., longevity) has been shown to be associated with a better control of inflammation.[7,20] Accordingly, centenarians are characterized by marked delay or escape from age-associated diseases that, on average, cause mortality at earlier ages. Moreover, centenarian offspring, that have an increased likelihood of surviving up to 100 years, show a reduced prevalence of age-associated diseases, such as cardiovascular diseases (CVD) and less prevalence of cardiovascular risk factors.[22,23] So, genes involved in CVD, the leading worldwide cause of morbidity and death, should play an opposite role in human longevity. Because inflammation is a key component of atherosclerosis and AD, genes coding for inflammatory or anti-inflammatory molecules are, therefore, good candidates for the risk of developing these pathologies. Findings discussed in this chapter suggest that different alleles of genes coding for pro- or anti-inflammatory genes may affect individual life-span expectancy by influencing the type and intensity of the immune-inflammatory responses against environmental stressors involved in the development of age-related diseases.

CD14 and Toll-Like Receptor 4 (TLR4)

The CD14/TLR4 receptor complex (Fig. 3) interacts with several ligands, including lipopolysaccharide (LPS, endotoxin) of Gram-negative bacteria, but it is also activated by endogenous ligands, such as ox-LDL, Aβ peptide and those produced in response to tissue injury.[24-29] Its activation induces transmembrane signals that activate NF-κB and mitogen-dependent protein kinase pathways, determining the expression of a wide range of genes encoding proteins such as cytokines, with regulatory functions in leukocyte activation and tissue inflammation.[2] So, this

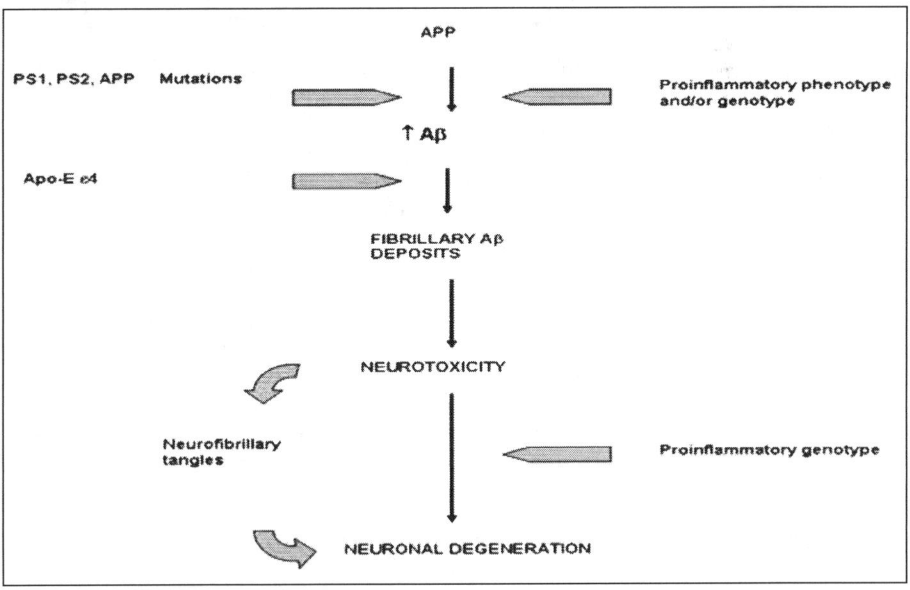

Figure 2. Alzheimer's disease: amyloid deposition is one the main pathogenetic mechanism. Accumulation of AB peptide may be caused by (1) gene mutations (PS1, PS2 and APP human mutations in familial Alzheimer's disease) (2) genotype (and/or phenotype) favouring unbalanced inflammatory responses (pro-inflammatory genotype/anti-inflammatory genotype).

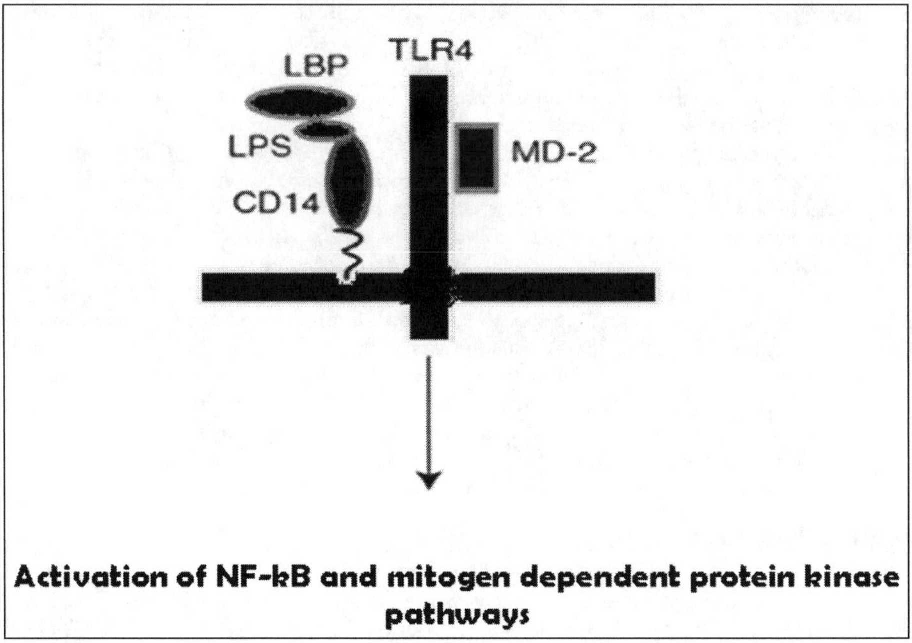

Figure 3. Activation CD14/TLR4 receptor complex TLR4 by LPS (or other agents) induces transmembrane signals that activate NF-kB and mitogen dependent protein kinase pathways, determining the expression of a wide number of genes encoding proteins, such as cytokines, with regulatory functions upon leukocyte activation and tissue inflammation. MD2 referred to accessory MD(Myeloid Differentiation)-2 protein, a small secreted glycoprotein that binds with cytokine-like affinities to both the hydrophobic portion of LPS and to the extracellular domain of TLR4; LBP referred to Lipopolysaccharide-binding protein a 50-kDa polypeptide mainly synthesized in hepatocytes and is released as a 58- to 60-kDa glycoprotein into the bloodstream after glycosylation.

receptor complex plays a key role in both innate and clonotypic immunity to Gram-negative bacteria and to other agents and it seems to be the hub of inflammatory pathophysiology of age-related diseases like atherosclerosis and AD.[30,31] CD14/TLR4 receptor complex activity and function may be modulated by genetic polymorphisms (for the most part, single nucleotide polymorphisms, SNPs). An SNP has been identified in the human TLR4 gene, an A-G base transition at position +896 base pairs from the transcriptional start site, resulting in an aspartic acid to glycine exchange at position 299 in the amino-acid sequence (referred to as Asp299Gly or +896A/G). This SNP causes hyporesponsiveness to LPS as well as an increased risk and susceptibility to gram-negative infections both in humans and experimental animals.[31,32] As already mentioned, there is evidence for a role of TLR4 in initiation and the progression of atherosclerosis.[31,33] The association between TLR4 and atherosclerosis is consistent with findings showing that TLR4 mRNA and protein are more abundant in atherosclerosis lesions than in unaffected vessels. Among cellular components presented in atherosclerotic plaques are several TLR4-expressing cells, including macrophages, endothelial cells, smooth muscle cells, T-cells and dendritic cells.[31] TLR4 in atherosclerotic lesions seems to be activated by ox-LDL and other endogenous ligands that are expressed during arterial injury and also by pathogen ligands, determining the production of inflammatory mediators.[34] However, discrepant data have been published on the pathogenic role of microbial stimuli or endogenous mediators triggering TLR4 receptor in vessel lesions.[34,35]

To date, there is a large body of genetic data implicating the involvement of Asp299Gly SNP in atherosclerosis development. Ultrasound analysis of carotid arteries in a large Italian population showed that the Asp299Gly was found less frequently in people with progressive lesions representing carotid atherosclerosis, compared with a control group.[36] This result was confirmed by two studies that found a protective effect of the TLR4 variants on acute coronary events.[37,38] However, other recent studies investigating a potential association of this SNP with CVD, as MI, did not yield significant associations.[30]

Concerning the association between SNPs of CD14 gene and atherosclerosis, data in the literature are again inconsistent. An SNP in position −260 of the CD14 promoter, C(−260)T, has been proposed to be associated with decreased affinity of Sp protein binding and enhanced transcriptional activity.[39] A higher density of mCD14 and higher serum levels of sCD14 were shown in the homozygous carriers of the T allele of this polymorphism and this allele has also been reported to be associated with a higher risk of MI.[40,41] It has also been recently reported that peripheral blood mononuclear cells from T/T homozygotes release a large amount of tumour necrosis factor (TNF)-α when challenged with LPS, more than C/T or C/C genotypes.[42] Moreover, in the same study, it was reported that coronary plaques of patients carrying the T/T genotype have a tendency to rupture because of the presence of activation-prone monocytes/macrophages, even when there is only a small amount of coronary atheroma.[42] This finding supports the role of CD14 promoter polymorphism in atheromatous plaque vulnerability. In contrast, other studies have not corroborated these findings and found no association between C(−260)T CD14 promoter and risk of MI[43,44] and susceptibility to atherosclerosis.[45] Finally, a recent study, performed in a Central Italy elderly population with atherosclerotic carotid stenosis, showed an increased TT homozygote frequency of C−260T polymorphism, providing insight into the pathogenetic role of this polymorphism in atherosclerosis.[46]

Several data have also recently suggested involvement of the CD14/TLR4 receptor complex in neurodegeneration in AD.[47,48] Some studies have suggested that activation of microglial cells may be induced through the binding of Aβ peptides.[21,29,49] Several membrane proteins expressed on microglial cells seem to be implicated in Aβ peptide binding. It has been demonstrated that the CD14/TLR4 receptor complex binds highly hydrophobic aggregates of Aβ peptides, because of a structural mimicry between highly hydrophobic fibrillar Aβ and PAMPs, suggesting the production of neurotoxic substances.[21,29,49] CD14 expression was increased on microglia in a transgenic murine model of AD and microglia derived from CD14-deficient mice demonstrated reduction in activation by Aβ peptide, suggesting that CD14 is necessary for Aβ-induced microglia activation.[25]

Another, not mutually exclusive, alternative explanation for the key role of microglial activation may be related to the role of this receptor complex as an LPS receptor. In fact, some studies have linked infections to other relevant age-related inflammatory diseases, such as atherosclerosis and AD.[20,21] It is therefore plausible that functional variations in the CD14 and TLR4 genes might influence susceptibility to sporadic AD. This might be the case for the allelic variants of the TLR4 gene, such as the Asp299Gly polymorphism, associated, as described above, with attenuated receptor signalling and a blunted inflammatory response. An association between this polymorphism and AD has been reported by Minoretti et al in an Italian sample.[50] Our own preliminary results of a recent study have confirmed that the Asp299Gly polymorphism of TLR4 gene is associated with AD.[49]

Regarding the CD14 receptor, Combarros et al hypothesized that subjects with the CD14 (−260) T/T genotype exhibit a greater degree of microglial activation in the brain and consequently be at increased risk of AD. In that study, however, there was no association between the CD14 (−260) polymorphism and AD, neither through an independent effect nor through interaction with APOE or cytokine polymorphisms.[51]

Evidence is also accumulating on associations of SNPs of the TLR4 gene and longevity. We have recently demonstrated that the anti-inflammtory allele, +896G, is overrepresented in male Sicilian centenarians and underrepresented in men affected by AMI.[37] Thus, our results suggest a role for the innate immune defence system and particularly TLR4 in CVD and our comparison with the oldest old may help elucidate the role of genetics in age-associated diseases characterized by a multifactorial aetiology. Accordingly, TLR4 polymorphisms, which attenuate receptor signalling, enhance the risk of infections, but decrease that of atherogenesis, presumably by limiting inflammatory responses.[32,36] Hence, the mutation might result in an increased chance of longevity in a modern environment with reduced pathogen load and improved control of severe infections by antibiotics. However, a recent study has excluded a noteworthy influence of Asp299Gly SNP upon human longevity or myocardial infarction in German men.[52] The causes of the discrepancies seem unclear, but the inclusion criteria, the studied populations and the measured endpoint substantially differed between these studies.

Studies on associations between CD14 and aging in rodents examined gene expression of major receptors for LPS (CD14 and TLR4) in the tissues of LPS-treated young and aged animals. The expression of CD14 in rat cardiac tissues was more increased in aged animals after LPS treatment, suggesting that innate immune response would be augmented with aging.[53] It also been demonstrated that even physiological aging of the brain is associated with a regulation of innate immune receptors expression in that TLR4 and CD14 expression was up-regulated in mouse brain in correlation with age.[54]

On the whole, future investigations should be focused on understanding the exact role of the genetic variants of CD14 genes in successful and unsuccessful ageing, whereas the role of TLR4 SNP already seems to clearer.

IL-1 Cluster

The "prototypic" inflammatory cytokine IL-1 is a primary mediator of systemic inflammatory responses, such as hypophagia, slow-wave sleep, sickness behaviour and neuroendocrine changes. It is a family of at least three closely related proteins that are the products of separate genes. The IL-1 gene cluster is located on the long arm of chromosome 2 (2q13) and contains the genes for IL-1α, IL-1β and IL-1 receptor antagonist (RA) within a region of 430 Kb. A number of biallelic and multiallelic markers in the region surrounding these IL-1 genes have been identified and some seem to be functional[55] (Fig. 4 summarizes the activation events following the binding of the cytokines to membrane specific receptors).

IL-1 promotes the interaction of endothelial cells with circulating leucocytes, induces the activation and proliferation of monocytes/macrophages and stimulates smooth muscle cell mitogenesis and the synthesis of plasminogen activator inhibitor 1. For these reasons, it is believed to play a key part in atherogenesis and thrombosis. In an English study, allele frequencies of IL 1-α (−889), IL-ß (−511 and +3593) and the IL-1RA intron 2 VNTR were measured in 827 healthy blood donors, in 232 patients with angiographically unobstructed coronary arteries and in 674 patients with single vessel or multivessel disease. IL-1 gene variants were found not to be significantly associated with the presence or extent of disease, whereas homozygosity for IL-1Ra*2 was significantly associated with single vessel disease. To explain the association with single but not with multivessel disease, the authors suggest that the IL-1ra VNTR may act as a "disease modifying" rather than a causative polymorphism.[56] Two subsequent Italian studies found no significant association between IL-1RA*2 and infarction: the first compared 115 patients with CVD (74 with first myocardial infarction) with 80 matched controls and the second compared 148 infarct survivors with 153 matched controls. In the latter studies, the frequencies of IL-RA*2 were 0.25 in survivors versus 0.25 in controls and those of IL-1RA*1 were 0.72 versus 0.73, respectively. Thus, the association between the IL-1 system and atherosclerosis is probably complex and may vary with clinical phenotype and extent of disease.[57,58]

Other investigations showed that gene polymorphisms of IL-1α and IL-1β are associated with increased risk of AD and influence the age at onset of the disease.[59,60] In fact, the polymorphism of IL-1α is associated with an early presentation of AD and that of IL-1β with a late age at onset of AD.[59] Another independent investigation confirmed the association of IL-1β gene polymorphism with AD in a group of patients with neuropathological diagnosis of the disease.[61]

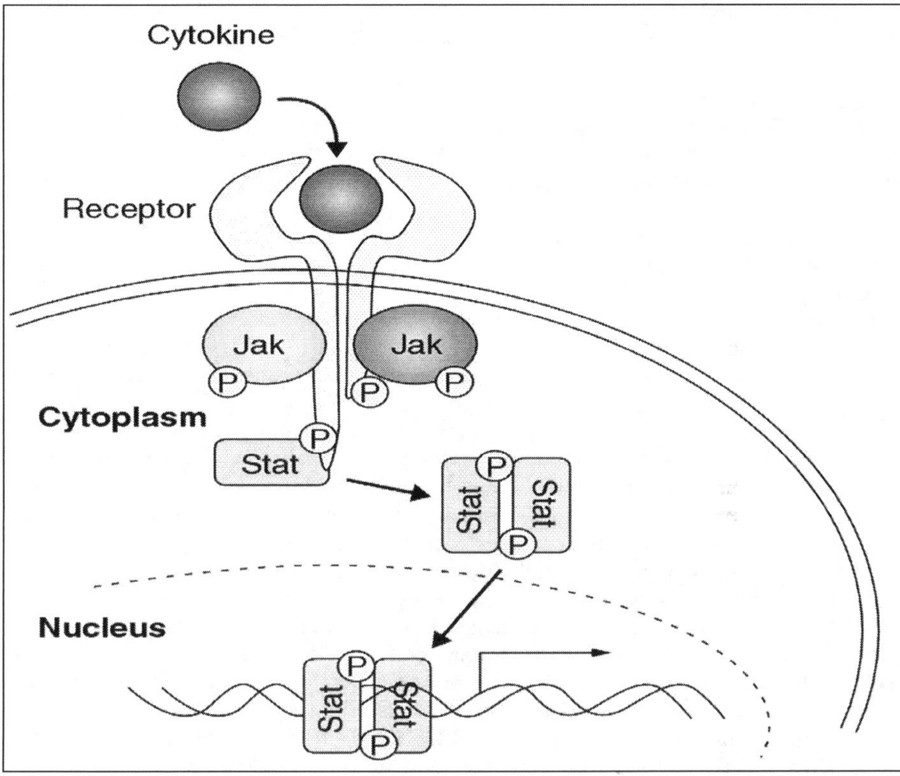

Figure 4. The binding of the cytokines to membrane specific receptors leads to intracellular activation of cytoplasmic tyrosine kinases that constitutively associate with the cytokine receptor, Janus Kinases (JAK) and as well as members of the Signal Transducer and Activator of Transcription (STAT) family. STATs are recruited via their SH2 domains to phosphotyrosine motifs of activated receptors and subsequently become tyrosine phosphorylated by Janus kinases. Phosphorylated STAT proteins dimerize and translocate to the nucleus where they act as transcriptional activators of specific target genes.

The association of allele polymorphism of IL-1α with increased risk of AD and an early age at onset of the disease has also been confirmed.[62] Accordingly, two recent meta-analysis showed a significant association with early AD onset and IL-1α T/T genotype.[63,64] In addition, IL-1β −511 T/T , another functional SNP extensively studied, seems also to be associated with AD.[49] (our unpublished meta-analysis).

In spite of apparent evidence for associations of the IL-1 gene cluster with age-related diseases, the two studies which have related this to ageing have not shown decreased gene frequency with increasing age. In a Finnish study of 250 nonagenarians there was no statistically significant difference in the allele frequencies for IL-1α, IL-1β or IL-1RA gene polymorphisms between nonagenarians compared to younger subjects from a wide range of age groups.[65] In addition there was no difference between male and female subjects in either age group. In a very large Italian study of 1131 subjects, including 300 subjects over 65 years and 134 centenarians, there was also no difference in IL-1α, IL-1β or IL-1RA allele frequency between the aged and younger subject groups nor between males and females at any age group. Serum IL-1RA increased with age in both male and female subjects but there was no age-related change in IL-1β.[66]

Together, these studies suggest that the pleiotropic antagonism among functional effects does not allow to delineate the role of the IL-1 gene cluster in successful ageing.

IL-6

IL-6 is a pleiotropic cytokine capable of regulating proliferation, differentiation and activity of a variety of cell types and plays a major role in bone remodeling, neuro-endocrine homeostasis, hemopoiesis and immune system regulation. In particular, IL-6 plays a pivotal role in acute phase response and in the balancing of the pro-inflammatory/anti-inflammatory pathways.[67] Serum IL-6 has been proposed as a reliable marker for functional decline, as a predictor of morbidity and mortality in old age and it has been associated with functional disability, cognitive decline and stroke in older people.[68-70]

Since elevated IL-6 plasma levels have been implicated in the pathogenesis of CVD, the functional polymorphism –174 was analysed in several groups of patients affected by MI. Some groups have showed an influence for this IL-6 SNP as predictor of death or plaque instability in old patients affected by CVD; in particular a recent study reports that the IL-6 –174 GG genotype is a strong predictor of cardiovascular death after one year of follow-up in old male patients affected by acute coronary syndrome, such as MI and unstable angina.[71] Accordingly, an under-representation of homozygosity for the C allele at –174 locus has been found in a group of Swedish men having a MI under age 40. Moreover, in middle-aged men, homozygosity for the G allele at –174 C/G is associated with increased artery intima-media thickness in the carotid bifurcation, a predilection site for atherosclerosis. On the other hand, in another study, although the –174C allele was not associated with incident events, associations of the genotype with inflammation and infarcts, combined with the plasma IL-6 results, suggest that IL-6 may chronically predispose an individual to develop atherosclerosis.[55] However, the association of –174 C/G with cardiovascular diseases is far from being clear, as shown by a number of reports which demonstrate that an extreme inter-study variation affects the association of –174 C/G locus with cardiovascular risk or with other important parameters associated with it, such as arterial responsiveness, CRP serum levels among others. Indeed the study design (cohort-based, of patients follow-up, population based study, healthy volunteers), ethnicity, age, gender, life-style (alcohol consumption and smoking habits) were all found to be potent confounding factors for these investigations.[55]

Microglia, astroglia, neurons and endothelial cells seem capable of synthesizing IL-6 and elevated levels can cause significant CNS damage and behavioural impairment, suggesting an association of IL-6 with AD.[72,73] However contrasting results have been reported and our recent unpublished meta-analysis clearly demonstrates no association between IL-6–174 SNP and AD.[49]

There has been considerable interest in the 174 IL-6 C/G SNP in longevity because it could be argued that if IL-6 associates with functional decline and age-related disease, then there may be attrition of IL-6GG homozygotes, who produce higher levels of IL-6 in serum, in older survivors in a population. Three studies have looked at the IL-6 174 C/G SNP with respect to ageing. In the earliest of these, Bonafe et al noted a marked reduction in GG homozygotes in male though not female Italian centenarians compared to elderly (60-80 years) and long-lived (80-99 years) subjects.[74] In contrast, Wang et al in a study of Finnish nonagenarians, detected no significant change in IL-6 frequencies between nonagenarians and 400 healthy blood donors, aged 18-60 years, though a reduction of 2% was noted in GG frequency in comparison with their widely age-selected younger control group.[65] In a recent study in 200 Irish nonagenarians, Rea et al reported a nonsignificant trend for a reduced frequency of the GG polymorphism of IL-6 in the octo/nonagenarians from the BELFAST study group (56%) compared to local population younger subjects (61%). This trend appeared more marked in elderly males compared to females but both groups showed an almost 10% decrease in frequency in the oldest subjects.[70] This trend for a reduction in GG homozygosity in elderly males in three countries across Europe is intriguing since it appears to confirm in different study populations and with a different study design, the earlier findings obtained in the Italian population and would seem to be an important and consistent finding which deserves further focused attention and priority study.[74] However, as in the case of the association between the polymorphism and cardiovascular diseases, important effects due to ethnicity, life-style are expected to affect the role of the −174 C/G polymorphism on longevity across the populations examined.[70]

TNF

The TNF cluster genes, which map within the Major Histocompatibility Complex encode three inflammation-related proteins, TNF-α, TNF-β and lymphotoxin-β. Several polymorphic areas are documented within the TNF gene cluster. Notably, some two-allele polymorphisms for the TNF promoter region and several microsatellite polymorphic sites have been described. Polymorphisms in the TNF promoter region have been observed to result in differences in the rate of gene transcription and in the rate of protein production. The TNF-α −308 polymorphism, substituting G/A and designated the TNF2 allele, influences TNF-α production in vitro.[75,76] TNF genes are thought to influence the strength, effectiveness and duration of local and systemic inflammatory reactions as well as repair and recovery from infectious and toxic agents and as such they are prime candidates to be involved in age-related disease processes and ageing itself. Major effects on the cardiovascular system include increased expression of adhesion molecules and human leukocyte antigen proteins, release of endothelial cytokines and nitric oxide, enhanced vascular permeability, negative inotropism, reduced lipoprotein lipase activity, increased hepatic fatty acid synthesis, involvement in obesity-related insulin resistance and prothrombotic effects.[55] Nevertheless, none of the studies performed in patients affected by CVD found any significant associations between −308A and either frequency of healed myocardial infarction and coronary thrombosis or number and severity of coronary stenoses or other phenotypic symptoms.[55] In the central nervous system, TNF-α is produced by activated microglia and astrocytes. TNF-α plays a role as a potent pro-inflammatory, cytotoxic polypeptide also in the brain. However, the pathophysiologic actions of TNF-α in AD are controversial. The highly conflicting data regarding TNF-α polymorphism and AD risk preclude any clear conclusion on the possible role of TNF-α in this disease. At the moment it may be suggested that this gene can act as disease modifier that may influence the age at onset of the disease.[75,77]

The TNF-308A/G polymorphism has been assessed in 3 studies in European nonagenarian and centenarian subjects to assess association with ageing. These studies including 350 Irish and Finnish nonagenarians and 172 Italian centenarians appear to demonstrate no major shift in the genotype frequency of TNF-α −308 polymorphism as a function of ageing.[70]

IL-10

Interleukin-10, a cytokine with anti-inflammatory and B cell stimulating activity, is produced by activated T-cells, B cells, monocytes/macrophages and dendritic cells. IL-10 is thought to block the ability of monocytes/macrophages and dendritic cells to act as antigen presenting cells by down regulation of MHC products and costimulatory molecules via suppression of the MAPK cascade. Thus, the principal function of IL-10 appears to be to limit and ultimately terminate the inflammatory signal.[55,70]

The IL-10 gene is located on chromosome 1 at 1q31-32 and is highly polymorphic. Stimulation of human blood samples with LPS showed large interindividual variations of IL-10 production, suggesting a genetic component of approximately 75%. Interindividual differences in the regulation of IL-10 production may be critical regarding the final outcome of an inflammatory response, i.e., within physiological limits or pathological. In fact, the presence of multiple SNPs in the human IL-10 5' flanking region has been demonstrated and some of these (i.e., −592, −819, −1082) combine with microsatellite alleles to form haplotypes associated with differential IL-10 production.[55,70,78,79]

From the point of view of age-related diseases, case-control studies across Europe do not support any role for IL-10 in either atherosclerotic-related disease or MI.[80] However more recent data from Italian samples of patients affected by MI suggests a role for IL-10 polymorphism linked to low cytokine production in the occurrence of the disease.[81]

Expression of IL-10 is elevated during the course of most major diseases in the brain and its actions consist of promoting survival of neurons and glial cells, expressing cell survival signals and limiting inflammation in the brain. The presence of the −1082A allele and in particular of the −1082A/−819T/−592A haplotype, associated with a low production of the cytokine has been suggested to be an additive and independent genetic risk factor for AD.[82,83] A significant increase of the percentage of −1082A carriers among AD patients is mainly due to an increased frequency of the IL-10 promoter ATA haplotype, which also was associated with low production of IL-10.[83] These data suggest that the presence of low responder IL-10 alleles might be suggested to be an additive and independent genetic risk factor for AD[49] (and unpublished meta-analysis).

The homozygote −1082GG genotype of IL-10 G 1082A/G polymorphism has been reported to be a male-specific marker for longevity.[84,85] The −1082GG genotype, associated with high IL-10 production, was argued to confer an anti-inflammatory status which it was postulated enhanced the possibility of extreme longevity. However, this result was not replicated in an Irish study of 100 healthy nonagenarians[70] where comparable frequencies in aged and younger men have been described, nor in the Finnish nonagenarian study [65] for all nonagenarian subjects or males alone.

The suggestion that enhanced male life expectancy is associated with IL-10-1082GG homozygote status is puzzling in view of the findings suggesting the importance of a good inflammatory response in the control of infections. It might be argued that IL-10-1082GG homozygous males who are lucky enough not to contact serious bacterial infection earlier in life may have an increased chance of long life survival (trade-off). However the same appears not to be true for female life expectancy.[70]

IL-18

IL18 has recently been resequenced in its entirety, enabling the tagging-single-nucleotide polymorphism (tSNP) methodology to be adopted. This approach has yielded interesting results, with genetic variation being shown to affect protein levels and risk for ageing associated diseases. The role of common variation in the IL-18 gene on serum concentrations and muscle-skeletal functioning in old age was evaluated in 1671 participants from two studies: the InCHIANTI study and wave 6 of the Iowa-Established Populations for Epidemiological Study of the Elderly (EPESE). The frequency of the C allele of the IL-18 polymorphism rs5744256, associated with a reduced serum concentration of IL-18, was found to be increased in subjects with shorter walk time performances both in the InCHIANTI ($n = 662$, $p = 0.016$) and Iowa-EPESE ($n = 995$, $p = 0 .026$) studies.[86] These results support further investigation to assess the role of IL-18 polymorphisms in successful and unsuccessful ageing.

Interferon (IFN)-γ

IFN-γ, an important regulator of the development and function of the immune system, plays a key role in defence against intracellular pathogens. The IFN-γ gene is located on 12q14 and its polymorphisms, including the transcription region, might affect host resistance to infectious disease. The IFN-γ microsatellite polymorphism within intron 1 has one common allele (allele2 with 12CA repeats), which is associated with higher expression levels of IFN-γ [87] and is reported to be in absolute correlation with the IFN-γ +874T allele.[87,88]

Concerning age-related diseases, no study has been published on the association between the age-related inflammatory disease atherosclerosis and IFN-γ polymorphisms and a recent report suggests no association between IFN-γ polymorphisms and AD.[89] In centenarians, Lio and colleagues[88] first reported that possession of the +874A allele conferred an overall anti-inflammatory status promoting longevity, particularly in centenarian females. This finding could not be replicated by Ross et al[90] in nearly 200 Irish nonagenarians, where similar frequencies for the CA 12 allele repeat in control and aged subjects were found. The small decrease in the CA 12 repeat in aged Irish nonagenarians vs. control females was not significant, but does demonstrate a similar trend to the findings of Lio and colleagues[88] in their Italian centenarian female group. These findings suggest that bigger repeat studies need to be considered to see if these findings can be replicated, particularly where sex differences need to be taken into account.[70]

Transforming Growth Factor (TGF)-β1

TGF-β1 has marked inhibitory effects on the immune system but also serves as a costimulatory factor in the development of T-cells with down-regulatory activities.[91] TGF-β1 seems to have an important role in human ageing. In fact, TGF-β1 gene over-expression has been observed in human fibroblasts that display a senescent-like phenotype following exposure to oxidative stress.[92] Polymorphisms in the human TGF-β1 gene influencing the cytokine production have been identified (G800A, C509T in the promoter region, Leu10Pro, Arg25Pro in exon 1 and Thr263Ile in exon 5).[93] In spite of this functional relevance, data on associations with age-related pathologies are inconsistent and our and other research groups have shown that these polymorphisms are not associated with cardiovascular disease susceptibility,[94-96] although they have been linked to AD.[97] On the other hand, results obtained in Italian centenarians showed significant associations of TGF-β1 polymorphisms with longevity.[98] Two polymorphisms, G/A −800 and C/T −509, located in the 5′region and the two miss-sense polymorphisms, T/C 869(Leu > Pro)10 and G/C 915 (Arg > Pro)25, respectively, located in the signal peptide, were analysed in 419 subjects from Northern and Central Italy, including 172 centenarians and 247 younger controls. Significant differences were found at the +915 site as far as the C allele and GC genotype were concerned, both of them being lower in centenarians than in young controls. In addition, the G −800/C −509/C 869/C 915 haplotype combination was notably lower in centenarians than in younger individuals. These results suggest that, at least in this population, the variability of the TGF-β1 gene is associated to longevity.

Chemokine-CC-Motif-Receptor 5 (CCR5)

Cellular development and migration of cells to different tissues are important elements of immunity. Ultimately, these processes determine how effectively the host is able to respond to infection or injury. Efficient migration of cells requires extensive cellular communication between various components of the immune system. Molecules involved in the directional migration of leukocytes include selectins, integrins and chemokines. The latter currently include approximately 50 cytokines and 20 receptors.[99] CCR5 (Chemokine-CC-motif-receptor 5 provided by HUGO Gene Nomenclature Committee) is involved in the migration, toward an increasing concentration of β chemokines, of monocytes, NK cells and Th1 cells.[99] Thus, this chemotatic response results in recruitment of leukocytes to sites of inflammation.[99] CCR5 came into prominence a decade ago when it was identified as the major coreceptor for macrophages (M) and dual (T-cell and macrophage)-tropic immunodeficiency viruses.[99] Evidence that CCR5 plays a role in HIV infection came with the demonstration that all three ligands for the CCR5 receptor, CCL3, CCL4 and CCL5, were potent inhibitors of viral entry into cells.[100] Notably, a fraction of the Caucasian population is resistant to infection by M-tropic HIV. A variant of CCR5 gene (number accession of GenBank: NM-00579), a nonfunctional allele resulting from a 32-bp deletion in exon 4 (CCR5Δ32), determines a loss of expression of functional CCR5, protecting from infection with HIV-1 transmitted through sexual contact.[99,100] Additionally, it has been suggested that CCR5Δ32 deletion is involved in the pathogenesis of different inflammatory diseases. More interesting is the involvement of CCR5 in development of atherosclerosis and risk of its complications (as CVD).[99] Concerning the role of CCR5 in atherosclerosis, the accumulation of macrophages and T lymphocytes in vessel walls is a hallmark of atherogenesis. Some recent studies have evaluated the association of CCR5Δ32 deletion with CVD, since this allele variant, conferring natural deficiency in CCR5 molecules, might protect individuals from cardiovascular diseases (as acute MI and severe coronary heart diseases) as consequence of an attenuated inflammatory response that would determine a slower progression of atherosclerotic lesion among CCR5Δ32 carriers. However, conflicting data have been published.[99] We have recently analysed the distribution of CCR5Δ32 in Sicilian MI patients and centenarians. The frequencies of pro-inflammatory alleles of

CCR5 (CCR5Δ32-) were significantly higher in AMI patients and lower in centenarians whereas age-related controls displayed intermediate values[6] (and unpublished data).

Concerning the role of CCR5 in inflammatory AD neurodegeneration, very few studies have examined the expression of chemokines and chemokine receptors in AD brains. However, immunohistochemical analysis of tissue from human brains with AD have revealed the expression of CCR3, CCR5, CXCR2 and CXCR3 together with an increased presence of their ligands.[101] Hence, some studies have evaluated whether the anti-inflammatory allele CCR5Δ32 is a component of the genetic protective background protective versus AD neurodegeneration. The data clearly demonstrate that CCR5 is not a protective factor for AD.[99,102] On the whole, therefore, CCR5Δ32 deletion seems to be a protective factor for CD (hence favouring longevity) but not for AD.

Cyclooxygenase (Cox), Lipoxygenase (Lox)

COX and LOX are key enzymes in the conversion of arachidonic acid to prostaglandins (PGs) and leukotrienes (LTs) and are implicated in a wide variety of inflammatory disorders (Fig. 5). COX-1 and COX-2 play a key role in pathophysiological processes of inflammatory diseases and are the main target for nonsteroidal anti-inflammatory drugs. COX-1 is constitutively expressed whereas COX-2, the inducible isoform, has been shown to be expressed at low levels in most tissues, but can be stimulated by LPS, growth factors and pro-inflammatory cytokines, being implicated in inflammatory processes, including atherosclerosis, rheumatoid diseases and carcinogenesis.[103,104]

PGs have potent actions on vascular smooth muscular cells by controlling contractility as well as cholesterol metabolism (favouring the formation of foam cells) and proliferation. In particular, antiproliferative and antimigration effects may imply the evolution of the plaque toward a more vulnerable one, depleted of smooth cells and enriched in macrophages. Moreover, PGE2 is actively involved in metalloproteinase 2 and 9 production which are tightly linked to atherosclerosis and plaque instability.[103,105,106]

Genetic polymorphisms have been described in the COX-2 gene that likely regulate its expression, prostanoid biosynthesis and functionality, but their functional relevance and pathophysiological role remain to be elucidated.[107,108] Recently, Papafili et al have identified a new variant in the COX-2 promoter, a guanine to cytosine substitution at position –765 G/C, located within a putative binding site for the transcription factor Sp1. They have shown that patients carrying the –765C allele had a reduction of approximately 30% in in vitro promoter activity and this was associated with lower plasma levels of inflammatory markers such as CRP.[108] In several studies, mutation –765GC in the promoter region of the COX-2 gene, resulting in a significantly lower promoter activity, was found to be associated with reduced risk of MI and stroke.[108-111]

The COX-2 enzyme has been detected in different cell types of the CNS, but its expression seems to be primarily neuronal. Reports on astrocyte expression are conflicting. Moreover, in the AD patient brain its expression correlates with amyloid plaque density and neurofibrillary tangles. In recent studies, mutation –765GC in the promoter region of the COX-2 gene, resulting in a significantly lower promoter activity, was found to be associated with reduced risk of AD.[112,113]

An alternative pathway of arachidonic acid generates, through the action of 5-LOX, LCTs that are implicated in a wide variety of inflammatory disorders. LOX can be induced by pro-inflammatory cytokines and its expression in endothelial cells is relatively low in the basal condition, but increased by pro-inflammatory cytokines. Recent studies have outlined crucial new roles for leukotrienes in the development of atherosclerotic lesions. In particular, LTB4 is a chemoactractant for monocytes and activates gene expression in inflammatory cells with a positive loop.[114,115] Two polymorphisms, –1708 G/A and –1761 G/A, of this enzyme[116] that could either modify 5-LOX gene transcription or modify the putative protein derived from translation of 5-LOX mRNA have been described. A few genetic studies have identified variants of the 5-LOX gene promoter as risk factors in atherosclerosis and MI. These studies have shown that 5-LOX polymorphisms, involved in a decreased expression of 5-LOX, are less represented in patients affected by MI and severe atherosclerosis.[117,118] 5-LOX also has been described in neurons and some glial cells throughout the cerebrum, basal ganglia and hippocampus. Compared with controls, a significant increase

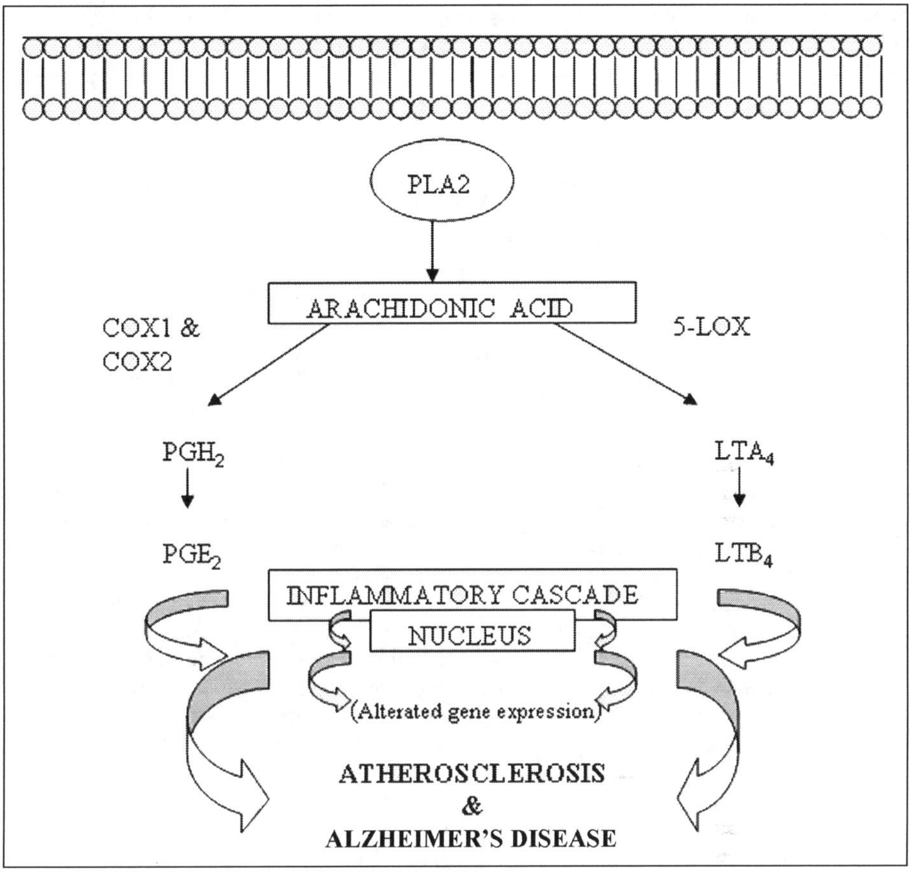

Figure 5. The eicosanoids synthesis and their putative role in chronic inflammation. 5-Lipoxygenase (5-LOX) catalyzes the oxygenation of arachidonic acid (AA) to 5(*S*)-hydroper oxy-6-*trans*-8,11,14- *cis*-eicosatetraenoic acid (5-HPETE) and further dehydration to the allylic epoxide leukotriene A4. LTA4 is further converted by LTA4 hydrolase to the dihydroxyacid LTB4 and by LTC4 synthases to the glutathione conjugateLTC4. LTs are inflammatory mediators causing phagocyte chemotaxis (LTB4) and increased vascular permeability (LTC4 and the other cys-LTs). Cyclooxygenases (COXs, also known as prostaglandin H_2 synthases or PGH_2s) are the rate-limiting enzymes in the conversion of arachidonic acid into prostaglandins (PGs). The precise reaction catalyzed by COXs is the conversion of arachidonic acid into PGH_2, a metabolite that then becomes the substrate of cell-specific prostaglandin and thromboxane synthases that generate PGs and thromboxane A_2, respectively. The oxygenated intermediate PGH2 is in turn metabolized by cell-specific synthases and isomerases into PGD2, PGE2, PGF2a, PGI2 and TXA2.

of LTs was observed in cerebrospinal fluid from AD and mild cognitive impairment. The activation of this enzyme occurs early in the course of AD, before the onset of overt dementia, thereby implicating lipid peroxidation in the pathogenesis of AD.[119,120] Consistent with this notion, recent studies have shown that polymorphisms involved in a decreased expression of 5-LOX are less represented in AD patients. It has been proposed that overexpressed 5-LOX could significantly increase the brain's vulnerability to neurodegeneration.[120]

We have tested the hypothesis that anti-inflammatory variants of these genes confer genetic resistance to AD or MI and conversely favour longevity. For this, we analyzed cohorts of AD and

MI patients, age-matched controls and centenarians. The pro-inflammatory alleles of COX-2 and 5-LOX were overrepresented both in AMI and AD patients and under-represented in centenarians whereas age-matched controls displayed intermediate values[6,49] (and unpublished observations).

Conclusions

Common gene variants with mildly deleterious effects are present in the human population and are expected to make up a significant part of the genetic contribution to the variation in life span. Such gene variants may have reached high frequencies in present populations if they increase the risk of disease only late in life and thus escape the force of natural selection. Studying loci implicated in multiple age-related diseases may prove successful in elucidating the genetic contribution to mortality. Gene variants mediating inflammation may be relevant in this respect since inflammation appears to be a common pathway in the development of age-related diseases such as atherosclerosis and dementia.[6,20,49] Persons who attain extreme old age might simply lack the genetic risk factors for late-onset diseases that are associated with mortality in the general population. A very attractive alternative hypothesis is that extreme longevity might be mediated by a limited number of major protective gene effects. So, extreme longevity might be determined by other genetic and environmental factors than mortality in the general population.

In any case, human longevity appears to be inextricably linked with optimal functioning of the immune system. Accordingly, a higher proportion of centenarians had well-preserved immune function compared to less elderly cohorts. In people over 80 years old, infection causes the majority of deaths and further implicates defective immune function in life-span determination.[121] However, the question to be asked is whether people live longer because of "good" immune function, or they possess good immune function because other factors have enabled them to survive longer. Thus, to better understand the role of the so-called immunological risk phenotype, predictive of remaining life span expectancy, we have to search for immunogenetic markers of longevity considering the opposite genetic background associated to the major ageing-related diseases. In other words, whether the immune system plays a key role in the attainment of successful ageing, then polymorphisms for the immune system genes that regulate immune responses might be critical to achieve a healthy ageing. Reciprocally, the lack of ability to control systemic inflammation may be a good marker of the immunological risk phenotype.[6,49,55]

Inflammatory gene polymorphisms warrant consideration as factors explaining variation in the human immune and inflammatory responses and as candidate susceptibility genes for related pathological states. In several genes, polymorphisms (mostly SNPs or microsatellites) located within the critical promoter or other regulatory regions, affect gene transcription resulting in inter-individual variation in levels of mediator production. The polymorphic nature of these genes may confer flexibility on the immune response with certain alleles promoting differential production of mediators that may influence the outcome of viral and bacterial infections or increase susceptibility and resistance to autoimmune disorders.[55,70]

Some recent evidence has linked inflammatory gene polymorphisms both with successful and unsuccessful ageing. Because genetic traits contribute significantly to the global risk of these diseases, a number of studies have now addressed the hypothesis that variations in the genetics of the inflammatory system may increase their risk. Differences in the genetic regulation of inflammatory processes might explain why some people but not others develop these diseases and why some develop a greater inflammatory response than others.[7]

In this chapter, we have reviewed the available data in the literature on inflammatory gene polymorphisms in successful and unsuccessful ageing and show that the immunogenetics of ageing and longevity is both complex and intriguing. Indeed, the final balance among chronic disease and life span expectancy is strongly affected by life-style and environmental factors and by a network of epistatic and pleiotropic gene effects. The genetics of ageing and longevity is highly unusual and most probably represents a post reproduction genetic scenario, where the force of selection progressively fades in the later decades of life. These emerging studies looking at cytokine polymorphisms and haplotypes clearly indicate that alleles involved in regulation of immune response and inflammation might affect life expectancy and healthy ageing. The findings while intriguing lack adequate statistical power and need to be repeated in large pan-European studies.

Our data prompt consideration of the role that antagonistic pleiotropy plays in disease and in longevity.[122] Our immune system has evolved to control pathogens and so pro-inflammatory responses are likely to be evolutionarily programmed to resist fatal infections, yet the genetic background promoting pro-inflammatory responses may have opposite effects in cardiovascular diseases and in longevity, such that cardiovascular diseases are a late consequence of strong pro-inflammatory responses programmed to resist infections in earlier life. Genetic polymorphisms responsible for a low inflammatory response might result in an increased chance of long life-span in an environment with a reduced antigen (i.e., pathogens) load, such as a modern day healthy environment and may also permit a lower grade survivable inflammatory response to atherogenesis and atherosclerosis-related disease.[7]

References

1. Medzhitov R, Janeway C Jr. Innate immunity. N Engl J Med 2000; 343:338-344.
2. Uematsu S, Akira S. Toll-like receptors and innate immunity. J Mol Med 2006; 84:712-725.
3. Janeway C Jr, Medzhitov R. Viral interference with IL-1 and toll signaling. Annu Rev Immunol. 2002; 20:197-216. Krieger M. The other side of scavenger receptors: pattern recognition for host defense. Curr Opin Lipidol 1997; 8:275-280.
4. Muzio M, Natoli G, Saccani S et al. The human toll signaling pathway: divergence of nuclear factor kappaB and JNK/SAPK activation upstream of tumor necrosis factor receptor-associated factor 6 (TRAF6). J Exp Med 1998; 187:2097-2101.
5. Guha M, Mackman N. LPS induction of gene expression in human monocytes. Cell Signal 2001; 13:85-94.
6. Vasto S, Candore G, Balistreri CR et al. Inflammatory networks in ageing, age-related diseases and longevity. Mech Ageing Dev 2007; 128:83-91.
7. Caruso C, Candore G, Colonna-Romano G et al. Inflammation and life-span. Science 2005; 307:208-209.
8. Ross R. Atherosclerosis. An inflammatory disease. N Engl J Med 1999; 340:115-126.
9. Braunwald E. Shattuck lecture—cardiovascular medicine at the turn of the millennium: triumphs, concerns and opportunities. N Engl J Med 1997; 337:1360-1369.
10. Libby P. Inflammation in atherosclerosis. Nature 2002; 420:868-874.
11. Ridker PM, Hennekens CH, Buring JE et al. C-reactive protein and other markers of inflammation in the prediction of cardiovascular disease in women. N Engl J Med 2000; 342:836-843.
12. Ross R. Atherosclerosis is an inflammatory disease. Am Heart J 1999; 138:419-420.
13. Lusis A J. Atherosclerosis. Nature 2000; 407:233-241.
14. Nabel EG. Cardiovascular disease. N Engl J Med 2003; 349:60-72.
15. Andreotti F, Porto I, Crea F et al. Inflammatory gene polymorphisms and ischaemic heart disease: review of population association studies. Heart 2002; 87:107-112.
16. Auer J, Weber T, Berent R et al. Genetic polymorphisms in cytokine and adhesion molecule genes in coronary artery disease. Am J Pharmacogenomics 2003; 3:317-328.
17. Selkoe DJ. Alzheimer's disease: genes, proteins and therapy. Physiol Rev 2001; 81:741-766.
18. Auld DS, Kornecook TJ, Bastianetto S et al. Alzheimer's disease and the basal forebrain cholinergic system: relations to beta-amyloid peptides, cognition and treatment strategies. Prog Neurobiol 2002; 3:209-245.
19. Feghali CA, Wright TM. Cytokines in acute and chronic inflammation. Front Biosci 1997; 2:12-26.
20. Licastro F, Candore G, Lio D et al. Innate immunity and inflammation in ageing: a key for understanding age-related diseases. Immun Ageing 2005; 18:2-8.
21. Finch CE, Morgan TE. Systemic inflammation, infection, ApoE alleles and Alzheimer disease: a position paper. Curr Alzheimer Res 2007; 4:185-1899.
22. Perls TT, Wilmoth J, Levenson R et al. Life-long sustained mortality advantage of siblings of centenarians. Proc Natl Acad Sci USA 2002; 99:8442-8447.
23. Perls T, Terry D. Genetics of exceptional longevity. Exp Gerontol 2003; 38:725-730.
24. Medzhitov R, Preston-Hurlburt P, Janeway CA JR. A human homologue of the Drosophila Toll protein signals activation of adaptive immunity. Nature 1997; 388:394-397.
25. Fassbender K, Walter S, Kuhl S et al. The LPS receptor (CD14) links innate immunity with Alzheimer's disease. FASEB J 2004; 18:203-215.
26. Goyert SM, Ferrero E, Rettig WJ et al. The CD14 monocyte differentiation antigen maps to a region encoding growth factors and receptors. Science 1988; 239:497-500.
27. Simmons DL, Tan S, Tenen DG et al. Monocyte antigen CD14 is a phospholipid anchored membrane protein. Blood 1989; 73:284-289.

28. Bazil V, Horejsi V, Baudys M et al. Biochemical characterization of a soluble form of the 53 kDa monocyte surface antigen. Eur J Immunol 1986; 16:1583-1589.
29. Tsan MF, Gao B. Endogenous ligands of Toll-like receptors. J Leukoc Biol 2004; 76:514-519.
30. Schroder NW, Schumann RR. Single nucleotide polymorphisms of Toll-like receptors and susceptibility to infectious disease. Lancet Infect Dis 2005; 5:156-164.
31. Cook DN, Pisetsky DS, Schwartz DA. Toll-like receptors in the pathogenesis of human disease. Nat Immunol 2004; 5:975-979.
32. Arbour NC, Lorenz, E, Schutte BC et al. TLR4 mutations are associated with endotoxin hyporesponsiveness in humans. Nat Genet 2000; 25:187-191.
33. Pasterkamp G, Van Keulen JK, De Kleijn DP. Role of Toll-like receptor 4 in the initiation and progression of atherosclerotic disease. Eur J Clin Invest 2004; 34:328-334.
34. Michelsen KS, Doherty TM, Shah PK et al. TLR signaling: an emerging bridge from innate immunity to atherogenesis. J Immunol 2004; 173:5901-5907.
35. Michelsen KS, Doherty TM, Shah PK et al. Role of Toll-like receptors in atherosclerosis. Circ Res 2004; 95:e96-97.
36. Kiechl S, Lorenz E, Reindl M et al. Toll-like receptor 4 polymorphisms and atherogenesis. N Engl J Med 2002; 347:185-189.
37. Balistreri CR, Candore G, Colonna-Romano G et al. Role of Toll-like receptor in acute myocardial infarction and longevity. JAMA 2004; 292:2339-2340.
38. Ameziane N, Beillat T, Verpillat P et al. Association of the Toll-like receptor 4 gene Asp299Gly polymorphism with acute coronary events. Arterioscler Thromb Vasc Biol 2003; 23:e61-64.
39. LeVan TD, Bloom JW, Bailey TJ et al. A common single nucleotide polymorphism in the CD 14 promoter decreases the affinity of Sp protein binding and enhances transcriptional activity. J Immunol 2001; 167:5838-5844.
40. Hubacek JA, Pit'ha J, Skodova 'Z et al. C(-260)—>T polymorphism in the promoter of the CD 14 monocyte receptor gene as a risk factor for myocardial infarction. Circulation 1999; 99(25):3218-3220.
41. Unkelbach K, Gardemann A, Kostrzewa M et al. A new promoter polymorphism in the gene of lipopolysaccharide receptor CD14 is associated with expired myocardial infarction in patients with low atherosclerotic risk profile. Arterioscler Thromb Vasc Biol 1999; 19:932-938.
42. Kondo T, Ohno M, Shimokata K et al. CD14 promoter polymorphism is associated with acute myocardial infarction resulting from insignificant coronary artery stenosis. Heart 2003; 89:931-932.
43. Zee RY, Lindpaintner K, Struk B et al. A prospective evaluation of the CD14 C(-260)T gene polymorphism and the risk of myocardial infarction. Atherosclerosis 2001; 154:699-702.
44. Longobardo MT, Cefalu AB, Pezzino F et al. The C(−260)NT gene polymorphism in the promoter of the CD14 monocyte receptor gene is not associated with acute myocardial infarction. Clin Exp Med 2003; 3:161-165.
45. Elghannam H, Tavackoli S, Ferlic L et al. A prospective study of genetic markers of susceptibility to infection and inflammation and the severity, progression and regression of coronary atherosclerosis and its response to therapy. J Mol Med 2000; 78:562-568.
46. Giacconi R, Caruso C, Lio D et al. CD14 C (−260)T polymorphism, atherosclerosis, elderly: role of cytokines and metallothioneins. Int J Cardiol 2007; 120:45-51.
47. Bsibsi M, Ravid R, Gveric D et al. Broad expression of Toll-like receptors in the human central nervous system. J Neuropathol Exp Neurol 2002; 61:1013-1021.
48. Lucas SM, Rothwell NJ, Gibson RM. The role of inflammation in CNS injury and disease. Br J Pharmacol 2006; 147 (Suppl 1):S232-240.
49. Candore G, Balistreri CR, Grimaldi MP et al. Polymorphisms of pro-inflammatory genes and Alzheimer's disease risk: a pharmacogenomic approach. Mech Ageing Dev 2007; 128:67-75.
50. Minoretti P, Gazzaruso C, Vito CD et al. Effect of the functional toll-like receptor 4 Asp299Gly polymorphism on susceptibility to late-onset Alzheimer's disease. Neurosci Lett 2006; 391:147-149.
51. Combarros O, Infante J, Rodrìguez E et al. CD14 receptor polymorphism and Alzheimer's disease risk. Neuroscience Letters 2005; 380:193-196.
52. Nebel A, Flachsbart F, Schafer A et al. Role of the toll-like receptor 4 polymorphism Asp299Gly in longevity and myocardial infarction in German men. Mech Ageing Dev 2007; 128:409-411.
53. Rosas GO, Zieman SJ, Donabedian M et al. Augmented age-associated innate immune responses contribute to negative inotropic and lusitropic effects of lipopolysaccharide and interferongamma. J Mol Cell Cardiol 2001; 33:1849-1859.
54. Letiembre M, Hao W, Liu Y et al. Innate immune receptor expression in normal brain aging. Neuroscience 2007; 46:248-254.
55. Candore G, Vasto S, Colonna-Romano G et al. Atherosclerosis. In: Vandenbroeck K, ed Cytokine gene polymorphisms in multifactorial conditions. CRC Press (USA) 2006:363-378.

56. Francis SE, Camp NJ, Dewberry RM et al. Interleukin-1 receptor antagonist gene polymorphism and coronary artery disease. Circulation 1999; 99(7):861-866.
57. Manzoli A, Andreotti F, Varlotta C et al. Allelic polymorphism of the interleukin-1 receptor antagonist gene in patients with acute or stable presentation of ischemic heart disease. Cardiologia 1999; 44(9):825-830.
58. Iacoviello L, Di Castelnuovo A, Gattone M et al. Polymorphisms of the interleukin-1beta gene affect the risk of myocardial infarction and ischemic stroke at young age and the response of mononuclear cells to stimulation in vitro. Arterioscler Thromb Vasc Biol 2005; 25(1):222-227.
59. Grimaldi LM, Casadei VM, Ferri C et al. Association of early-onset Alzheimer's disease with an interleukin-1alpha gene polymorphism. Ann Neurol 2000; 47(3):361-365.
60. Licastro F, Veglia F, Chiappelli M et al. A polymorphism of the interleukin-1 beta gene at position +3953 influences progression and neuro-pathological hallmarks of Alzheimer's disease. Neurobiol Aging 2004; 25(8):1017-1022.
61. Nicoll JA, Mrak RE, Graham DI et al. Association of interleukin-1 gene polymorphisms with Alzheimer's disease. Ann Neurol 2000; 47:365-368.
62. Rebeck GW. Confirmation of the genetic association of interleukin-1A with early onset sporadic Alzheimer's disease. Neurosci Lett 2000; 293(1):75-77.
63. Rainero I, Bo M, Ferrero M et al. Association between the interleukin-1alpha gene and Alzheimer's disease, a meta-analysis. Neurobiol Aging 2004; 25:1293-1298.
64. Combarros O, Llorca J, Sanchez-Guerra M et al. Age-dependent association between interleukin-1A (-889) genetic polymorphism and sporadic Alzheimer's disease. A meta-analysis. J Neurol 2003; 250:987-989.
65. Wang XY, Hurme M, Jylha M et al. Lack of association between human longevity and polymorphisms of IL-1 cluster, IL-6, IL-10 and TNF-alpha in Finnish nonagenarians. Mech Ageing Dev 2001; 123:29-38.
66. Cavallone L, Bonafè M, Olivieri F et al. The role of IL-1 gene cluster in longevity: a study in Italian population. Mech Ageing Dev 2003; 124(4):533-538.
67. Ershler WB, Keller ET. Age-associated increased interleukin-6 gene expression, late-life diseases and frailty. Annu Rev Med 2000; 51:245.
68. Ershler WB, Interleukin-6: a cytokine for gerontologists. J Am Geriatr Soc 1993; 41: 176-181.
69. Volpato S, Guralnik JM, Ferrucci L et al. Cardiovascular disease, interleukin-6 and risk of mortality in older women the women's health and aging study. Circulation 2001; 103:947-953.
70. Rea IM, Candore G, Cavallone L et al. Longevity. In: Vandenbroeck K, ed Cytokine gene polymorphisms in multifactorial conditions. USA; CRC Press, 2006:379-394.
71. Antonicelli R., Olivieri F, Bonafe` M et al. The interleukin-6 –174GC promoter polymorphism is associated with an higher risk of death after an acute coronary syndrome in male elderly patients. Int J Cardiol 2005; 103:266-271.
72. Licastro F, Grimaldi LM, Bonafe M et al. Interleukin-6 gene alleles affect the risk of Alzheimer's disease and levels of the cytokine in blood and brain. Neurobiol Aging 2003; 24:921-926.
73. Capurso C, Solfrizzi V, D'Introno A et al. Interleukin 6-174 G/C promoter gene polymorphism and sporadic Alzheimer's disease, geographic allele and genotype variations in Europe. Exp Gerontol 2004; 39:1567-1573.
74. Bonafe M, Olivieri F, Cavallone L et al. A gender-dependent genetic predisposition to produce high levels of IL-6 is detrimental for longevity. Eur J Immunol 2001; 31:2357-2361.
75. Candore G, Balistreri CR, Colonna-Romano G et al. Major histocompatibility complex and sporadic Alzheimer's disease, a critical reappraisal. Exp Gerontol 2004; 39:645-652.
76. Candore G, Lio D, Colonna Romano G et al. Pathogenesis of autoimmune diseases associated with 8.1 ancestral haplotype, effect of multiple gene interactions. Autoimmun Rev 2002; 1:29-35.
77. Lio D, Annoni G, Licastro F et al. Tumor necrosis factor α-308A/G polymorphism is associated with age at onset of Alzheimer disease. Mech Ageing Dev 2006; 127(6):567-571.
78. Turner D, Grant SCD, Yonan N et al. Cytokine gene polymorphism and heart transplant rejection. Transplantation 1997; 64:776-779.
79. Kube D, Rieth H, Eskdale J et al. Structural characterisation of the distal 5□ flanking region of the human interleukin-10 gene. Genes Immunity 2001; 2:181-190.
80. Donger C, Georges JL, Nicaud V et al. New polymorphisms in the interleukin-10 gene—relationships to myocardial infarction. Eur J Clin Invest 2001; 31:9-14.
81. Lio D, Candore G, Crivello A et al. Opposite effects of interleukin 10 common gene polymorphisms in cardiovascular diseases and in successful ageing: genetic background of male centenarians is protective against coronary heart disease. J Med Genet 2004; 41:790-794.
82. Arosio B, Trabattoni D, Galimberti L et al. Interleukin-10 and interleukin-6 gene polymorphisms as risk factors for Alzheimer's disease. Neurobiol Aging 2004; 25:1009-1015.
83. Lio D, Licastro F, Scola L et al. Interleukin-10 promoter polymorphism in sporadic Alzheimer's disease. Genes Immun 2003; 4:234-238.

84. Lio D, Scola L, Crivello A et al. Inflammation, genetics and longevity: further studies on the protective effects in men of IL-10 −1082 promoter SNP and its interaction with TNF-alpha −308 promoter SNP. J Med Genet 2003; 40:296-299.
85. Lio D, Scola L, Crivello A et al. Gender-specific association between −1082 IL-10 promoter polymorphism and longevity. Genes Immun, 2002; 3:30-33.
86. Frayling TM, Rafiq S, Murray A et al. An interleukin-18 polymorphism is associated with reduced serum concentrations and better physical functioning in older people. J Gerontol A Biol Sci Med Sci 2007; 62(1):73-78.
87. Pravica V, Perrey C, Stevens A et al. A single Nucleotide polymorphism in the first intron of the human IFN-γ gene: Absolute correlation with a polymorphic CA microsatellite marker of high IFN-γ gene. Human Immunol 2000; 61:863-866.
88. Lio D, Marino V, Serauto A et al. Genotype frequencies of the +874T—>A single nucleotide polymorphism in the first intron of the interferon-gamma gene in a sample of Sicilian patients affected by tuberculosis. Eur J Immunogenet 2002; 29:371-374.
89. Scola L, Licastro F, Chiappelli M et al. Allele frequencies of +874T —>A single nucleotide polymorphism at the first intron of IFN-gamma gene in Alzheimer's disease patients. Aging Clin Exp Res 2003; 15(4):292-295.
90. Ross OA, Curran MD, Meenagh A et al. Study of age-association with cytokine gene polymorphisms in an aged Irish population, Mech. Ageing Dev 2003; 124:199-206.
91. Ohtsuka K, Gray JD, Stimmler MM et al. Decreased production of TGF-beta by lymphocytes from patients with systemic lupus erythematosus. J Immunol 1998; 160(5):2539-2545.
92. Frippiat C, Dewelle J, Remacle J et al. Signal transduction in H2O2-induced senescence-like phenotype in human diploid fibroblasts, Free Radic Biol Med 2002; 33:1334-1346.
93. Gewaltig J, Mangasser-Stephan K, Gartung C et al. Association of polymorphisms of the transforming growth factor-beta1 gene with the rate of progression of HCV-induced liver fibrosis. Clin Chim Acta 2002; 316:83-94.
94. Syrris P, Carter RD, Metcalfe JC et al. Transforming growth factor-beta1 gene polymorphisms and coronary artery disease. Clin Sci (Lond) 1998; 95:659-667.
95. Wang XL, Sim As, Wilcken DE. A common polymorphism of the transforming growth factor-beta1 gene and coronary artery disease. Clin Sci (Lond) 1998; 95:745-746.
96. Crivello A, Giacalone A, Scola L et al. Frequency of polymorphisms of signal peptide of TGF-β1 and −1082G/A SNP at the promoter region of IL-10 gene in patients with carotid stenosis. Ann NY Acad Sci. 2006; 1067:288-293.
97. Luedecking EK, Dekosky ST, Mehdi H et al. Analysis of genetic polymorphisms in the transforming growth factor-beta1 gene and the risk of Alzheimer's disease. Hum Genet 2000; 106:565-569.
98. Carrieri G, Marzi E, Olivieri F et al. The G/C915 polymorphism of transforming growth factor beta1 is associated with human longevity: a study in Italian centenarians. Aging Cell 2004; 3:443-448.
99. Balistreri CR, Caruso C, Grimaldi MP et al. CCR5 receptor: biologic and genetic implications in age-related diseases. Ann N Y Acad Sci 2007; 1100:162-172.
100. Samson M, Libert F, Doranz BJ et al. Resistance to HIV-1 infection in caucasian individuals bearing mutant alleles of the CCR-5 chemokine receptor gene. Nature 1996; 382:722-725.
101. Xia MQ, Qin SX, Wu LJ et al. Immunohistochemical study of the beta-chemokine receptors CCR3 and CCR5 and their ligands in normal and Alzheimer's disease brains. Am J Pathol 1998; 153:31-37.
102. Balistreri CR, Grimaldi MP, Vasto S et al. Association between the polymorphism of CCR5 and Alzheimer's disease: results of a study performed on male and female patients from Northern Italy. Ann N Y Acad Sci 2006; 1089:454-461.
103. Smith WL, Langenbach R. Why there are two cyclooxygenase isozymes? J Clin Invest 2001; 107:1491-1495.
104. Morita I. Distinct functions of COX-1 and COX-2. Prostaglandins other lipid mediat 2002; 68-69:165-175.
105. Cipollone F, Fazia ML. COX-2 and atherosclerosis. J Cardiovasc Pharmacol. 2006; 47 Suppl 1:S26-36.
106. Dubois RN, Abramson SB, Crofford L et al. Cyclooxygenase in biology and disease. FASEB J 1998; 12:1063-1073.
107. Fritsche E, Baek SJ, King LM et al. Functional characterization of cyclooxygenase-2 polymorphisms. J Pharmacol Exp Ther 2001; 299:468-476.
108. Papafili MR, Hill DJ, Brull RJ et al. Common promoter variant in cyclooxygenase-2 represses gene expression: evidence of role in acute-phase inflammatory response, Arterioscler. Thromb Vasc Biol 2002; 22:1631-1636.
109. Cipollone F, Toniato E, Martinotti S et al. Identification of New Elements of Plaque Stability (INES) Study Group, A polymorphism in the cyclooxygenase 2 gene as an inherited protective factor against myocardial infarction and stroke. JAMA 2004; 291:2221-2228.

110. Orbe J, Beloqui O, Rodriguez JA Protective effect of the G-765C Cox-2 polymorphism on subclinical atherosclerosis and inflammatory markers in asymptomatic subjects with cardiovascular risk factors. Clinica Chimica Acta 2006; 368:138-143.
111. Coalizzo D, Fofi L, Tiscia G et al. The COX-2 G/C –765 polymorphism may modulate the occurrence of cerebrovascular ischemia. Blood Coagul Fibrinolysis 2006; 17(2):93-96.
112. Abdullah L, Ait-Ghezala G, Crawford F et al. The cyclooxygenase 2—765C promoter allele is a protective factor for Alzheimer's disease. Neurosci Lett 2006; 395(3):240-243.
113. Ma SL, Tang NL, Zhang YP et al. Association of prostaglandin-endoperoxide synthase 2 (PTGS2) polymorphisms and Alzheimer's disease in Chinese. Neurobiol Aging 2007 [Epub ahead of print].
114. Jala VR, Haribabu B. Leukotrienes and atherosclerosis: new roles for old mediators. Trends Immunol 2004; 25:315-322.
115. Lotzer K, Funk CD, Habenicht AJ. The 5-lipoxygenase pathway in arterial wall biology and atherosclerosis. Biochim Biophys Acta 2005; 1736:30-37.
116. In KH, Asano K, Beier D et al. Naturally occurring mutations in the human 5-lipoxygenase gene promoter that modify transcription factor binding and reporter gene transcription. J Clin Invest 1997; 99(5):1130-1137.
117. Dwyer, JH, Allayee H, Dwyer KM et al. Arachidonate 5-lipoxygenase promoter genotype, dietary arachidonic acid and atherosclerosis. N Engl J Med 2004; 350:29-37.
118. Helgadottir A, Manolescu G, Thorleifsson S et al. The gene encoding 5-lipoxygenase activating protein confers risk of myocardial infarction and stroke. Nat Genet 2004; 36:233-239.
119. Yao Y, Clark CM, Trojanowski JQ et al. Elevation of 12/15 lipoxygenase products in AD and mild cognitive impairment. Ann Neurol 2005; 58:623-626.
120. Manev H, Manev, R. 5-Lipoxygenase (ALOX5) and FLAP (ALOX5AP) gene polymorphisms as factors in vascular pathology and Alzheimer's disease. Med Hypotheses 2006; 66:501-503.
121. Candore G, Colonna-Romano G, Balistreri CR et al. Biology of longevity: role of the innate immune system. Rejuvenation Res 2006; 9:143-148.
122. Nesse RM, Williams GC. Evolution and the origins of disease. Sci Am 1998;279:86-93.

Chapter 15

SELDI Proteomics Approach to Identify Proteins Associated with T-Cell Clone Senescence

Dawn J. Mazzatti,* Robin Longdin, Graham Pawelec, Jonathan R. Powell and Rosalyn J. Forsey

Summary

The immune system undergoes many complex changes as a result of the aging process. Elderly humans have altered cellular redox levels and deregulated immune responses, both key events underlying the progression of chronic degenerative diseases of aging, such as atherosclerosis and Alzheimer's disease. T-cells are one of the major cell types affected by aging. As such, identifying bio-markers of T-cell aging and senescence would aid in identifying and developing novel intervention strategies.

Proteomics has emerged recently as a rapidly expanding and innovative field, investigating protein expression, interactions, localisation and function at a global level. In this context, we used the Ciphergen ProteinChip® PCS4000 surface enhanced laser desorption/ionisation (SELDI) system, a combination of affinity chromatography and mass spectrometry, to study protein profile changes that occur during in vitro T-cell aging and immunosenescence. This technology offers faster, higher throughput analysis of protein expression than the more traditional 2-dimensional-gel electrophoresis method, allowing the screening of larger sample numbers for potential bio-markers. Furthermore, on-chip affinity chromatography reduces sample complexity and permits targeting of protein groups.

Ciphergen H50 chips (reversed-phase chromatography) and Q10 chips (anion-exchange chromatography) were used to target hydrophobic and negatively charged proteins, respectively, in T-cell lysates. Biomarker analysis using the CiphergenExpress™ software identified differential expression of a variety of peaks associated with in vitro T-cell aging. A consistent pattern of differential protein expression was observed between both early and late passage T-cell clones grown in vitro and from T-cell clones derived from young and old donors. The corresponding proteins were identified by a combination of SELDI-TOF-MS, peptide mass fingerprinting MALDI-TOF-MS and Nanospray-IonTrap-MS/MS. Various molecules were demonstrated to be differentially expressed in aging and senescence, with good correlation between SELDI and MALDI data. This suggests that SELDI is a valuable tool in elucidating proteomic differences between cell populations, identifying potential biomarkers in large sample populations that can then be investigated further using more sensitive MS/MS methods for identification.

*Corresponding Author: Dawn Mazzatti—Unilever Corporate Research, Colworth Park, Sharnbrook, Bedfordshire, MK44 1LQ, U.K. Email: dawn.mazzatti@unilever.com

Immunosenescence, edited by Graham Pawelec. ©2007 Landes Bioscience and Springer Science+Business Media.

Introduction

T-cell immune dysfunction—or immunosenescence—is an important feature of the overall immune deregulation observed in elderly individuals, a process that contributes significantly to increased morbidity and mortality in this population.[1-3] Not only do proportions of T-cell subsets change with age but T-cell functions are also dysregulated in elderly humans. Elderly T-cells show poor proliferative capacity, have deregulated signalling pathways, altered adhesion/activation molecule expression and production of many cytokines. The controlled induction of cytokines during an immune response is beneficial. However, the presence of chronic low level inflammation is probably detrimental.[2,4-6] Higher circulating levels of IL-6 predict onset of disability,[7] frailty and mortality[8] and have been implicated in the pathogenesis of patho-physiologically unrelated diseases that are common in old age, including Alzheimer's disease and cardiovascular disease.[9,10]

The concept of an immune risk profile (IRP) provides a bio-marker framework with which to describe and measure age-related immune dysfunction. It emerged from Swedish Longitudinal studies[11,12] whereby the first data were derived using a cluster approach, showing that high CD8, low CD4 numbers, in combination with decreased T-cell proliferative capacity was predictive of 2 year mortality.[11] Subsequent analysis during this longitudinal study has shown that a surrogate IRP can be defined using only a CD4:CD8 ratio of <1.[12] Huppert et al[13] have recently confirmed these findings, also demonstrating that an inverted CD4:CD8 ratio can predict survival in elderly individuals. Individuals with the IRP also exhibit a pro-inflammatory phenotype, providing evidence of age-related alterations in both innate and adaptive immune systems.

In humans, CMV contributes markedly to the persistent clonal expansions of CD8 T-cells commonly seen in the elderly.[14,15] In fact, as demonstrated by the IRP, one of the major characteristic features of immunosenescence is the predominance of clonal expansions of a limited repertoire of $CD8^+$/CD28-negative cells.[14,16] The majority of CD28-negative T-cells can produce pro-inflammatory cytokines[17] but in the elderly these cells are further compromised in the production of all cytokines.[18] These immunosenescent cells therefore fill up the 'immunological space' resulting in a general immuno-supression through lack of provision of secondary activating signals—adhesion/activation and soluble mediators.

At the cellular level, T-cell immunosenescence in aging is accompanied by the accumulation of cells with decreased membrane fluidity and calcium influx[1] and signal transduction defects, particularly in the tyrosine protein kinase $p56^{lck}$ and mitogen-activated protein kinase (MPK) and MEK pathways.[19,20] In addition, increased levels of inhibitory signalling molecules including MAP phosphatase have been reported in aging.[20] Taken together, these data suggest that multiple pathways may be deregulated in the aging immune system, thereby contributing to loss of function. Understanding the mechanistic drivers that lead to the accumulation of dysfunctional cells may therefore also bring practical benefits in immune intervention in the elderly to reconstitute appropriate immune responses. To this aim, in vitro T-cell models of clonal expansion provide useful tools, enabling the production of enough sample material to investigate the complex mechanistic drivers of aging using 'omic' technologies. Using this in vitro system in the discovery of protein bio-markers of aging and immunosenescence would aid in understanding immune dysfunction in the elderly and would allow development of novel intervention strategies to restore function.

Previous investigations of protein bio-markers have been dependent on high-performance liquid chromatography, two-dimensional (2-D) gel electrophoresis, or other chromatographic approaches. 2-D electrophoresis has previously offered an improved protein separation compared with 1-D electrophoresis. However, 2-D electrophoretic analysis of proteins is very time-consuming and is limited by problems in reproducibility. Furthermore, the sensitivity of this method limits the analysis to proteins larger than 8-10kDa. Therefore, improved methods of protein separation and identification were needed to overcome the problems associated with 2-D electrophoretic separations.

The current study investigated the applicability of surface-enhanced laser desorption/ionisation time-of-flight mass spectrometry (SELDI-ToF-MS) ProteinChip Array technology (Ciphergen Biosystems, Inc., Fremont, CA) for protein profiling of aging and senescent T-cells. This method uses chromatographic surfaces to retain proteins based on their physicochemical characteristics, followed by ToF-MS using a ProteinChip Reader (PBSIIc Series 4000; Ciphergen Biosystems, Inc.).

In this system, proteins are separated on different array surfaces including cation/anion-exchangers, hydrophobic and metal-ion affinity surfaces. Proteins that bind and are retained on the surfaces are analysed by mass spectrometry (MS). This technique offers enhanced sensitivity which is ideal for the analysis of small sample volumes and allows screening of low-molecular weight proteins.

Surface Enhanced Laser Desorption/Ionisation (SELDI) was first described by Hutchens and Yip in 1993.[21] Since then, the technology has been steadily developing into a powerful analytical approach. SELDI is distinct in the field of laser desorption/ionisation (LDI). SELDI, LDI and MALDI techniques all rely on the energy inherent in a focused laser beam to promote the creation of gaseous ions from solid-state matter.[22] MALDI (matrix-assisted LDI), as the name suggests, utilises a crystalline matrix material surrounding the sample as an energy absorber, which then transfers thermal energy to the sample. This energy exchange takes place on a sample probe surface, which in the case of MALDI and LDI has a purely passive role, merely presenting the sample to the mass spectrometer.[22] SELDI is also a matrix-assisted technique, but is distinctive as a result of the sample-presenting surface playing an active role in sample processing (surface enhanced LDI). With the inherent complexity of biological materials such as blood, sera and cell lysates, mass spectrometric analysis of these materials almost universally require one or more upstream purification methods.[23] The sample-presenting surface doubles as an extraction device prior to the addition of energy-absorbing matrix and therefore selectively purifies crude samples depending on the chosen chemical or biological activity of the surface.

The SELDI technology has been commercialised by Ciphergen Biosystems, Inc (Fremont, California). The ProteinChip® system uses metal 'chips' (arrays) each accommodating up to eight samples, providing the surface to which samples are selectively bound and presenting the sample to a Ciphergen mass spectrometer. ProteinChip® surfaces range from anionic exchange, cationic exchange, hydrophobic and normal-phase. Immobilised metal affinity capture (IMAC) arrays allow the researcher to attach metal ions of choice prior to the sample. In addition, arrays are available for the attachment of proprietary antibodies for specific protein purification.

Perhaps the most widely used and arguably best proteomic application of the SELDI technology is in the field of high-throughput protein-profiling of samples in order to identify differential protein expression. Such an approach can lead to the discovery of single biomarkers or biomarker patterns indicative of disease states or pharmacodynamic effects, for example. One such study using SELDI led to the discovery of a biomarker panel for pancreatic adenocarcinoma. Koopmann et al[24] analysed serum samples using two types of chromatographic array surfaces and analysed spectra with the Ciphergen Expression Difference Mapping™ software.

In the current study, we demonstrated the use of SELDI-TOF-MS as a method to profile proteins differentially expressed in T-cell clone dysfunction and senescence following chronic antigenic stress. The goal of this investigation was to find protein or peptide bio-markers of immunosenescence by comparing the protein profiles from T-lymphocytes isolated from young and old donors grown to senescence in culture. Using SELDI-ToF-MS analysis, we identified several candidate protein/peptide peaks that were differentially expressed in late-passage T-cell clones. Proteins which were differentially expressed in immunosenescence were subsequently identified by Nano-LC IonTrap MS/MS and MASCOT analysis.

Materials and Methods

Cloning and Propagation of T-Cell Clones

$CD4^+$ human T-cell clones from activated peripheral blood lymphocytes of healthy octogenarian donors were obtained by limiting dilution in the presence of IL-2 as previously described.[25] Five representative clones were selected from each of three different young and old donors at time points earlier in antigen-stimulated in vitro culture and at a time approaching senescence. These clones were not markedly different in any way from the majority of such clones that we have obtained from numerous different donors and express markers of effector memory cells ($CD45RA^+$, $CD45RB^+$, $CD45RO^{lo}$, $CD28^{lo}$, $CD95^+$, $CCR7^{lo}$) and carry markers of activated T-cells (CD80, CD86, PD-L1, MHC class II). For SELDI analysis, two million cells from each clone were pelleted and resuspended in lysis buffer (H50: 8M urea, 2% CHAPS; Q10: 50 mM Tris-HCl, 5 mM EDTA PH6.0, 2 mM PMSF, 1% Triton X-100) at early and late passages. For protein identification, T-cell clones derived from several young and old donors were pooled to provide adequate sample material.

SELDI-MS Protein Profiling of T-Cell Clones Derived from Young and Old Donors

Sample Preparation

A robotic automation station was utilised for automatic handling of chips including all binding and washing steps. The ProteinChip Array BioProcessor (Ciphergen Biosystems, Inc.) was equipped with 12 ProteinChip arrays, A-H format. All ProteinChip arrays were pretreated according to manufacturer protocols. Binding buffers were 10% acetonitrile/0.1% trifluoroacetic acid (H50) and 50 mM Tris, pH 8.0 (Q10). Protein lysates from 400,000 cells (10 ul in 100 ul binding buffer) were applied to reversed-phase hydrophobic surface (H50), strong anion exchange (Q10) and cation-exchange (CM10) surfaces. The arrays were incubated and washed according to manufacturer protocols. After the wells were dry, 1ul saturated sinapinic acid (in 50% acetonitrile/0.5% trifluoroacetic acid) was manually added to each spot. Spots were allowed to air dry and each spot was analysed in a ProteinChip Reader. Each sample was bound to each array surface in triplicate.

Data Acquisition and Processing

ProteinChip arrays were analysed on a ProteinChip Reader using the ProteinChip Software version 4.0 (Ciphergen Biosystems, Inc.). Initial arrays were read at three laser intensities before optimisation at 5000nJ. The protocol averaged 10 laser shots per pixel with a focus mass of 18,000 Da, a matrix attenuation of 2500 Da and a range of 0-50,000 Da. The raw data were transferred to CiphergenExpress™ software for analysis. The baseline was subtracted using a setting of 8 times the expected peak width. Ciphergen protein standard (All-IN 1 Protein Standard II, Ciphergen) was analysed on an NP20 (normal phase) ProteinChip (Ciphergen) using the same analysis protocol. The following peaks were identified in the resulting spectrum and used to create a three-parameter weighted internal calibration using the CiphergenExpress™ software (version 3.0.5.013): hirudin BKHV (7033.61 Da), bovine cytochrome c (12230.92 Da) equine cardiac myoglobin (16951.51 Da) and bovine RBC carbonic anhydrase (29023.66 Da). This internal calibration was copied and applied to the spectra as an external calibration. Data were normalised by total ion current (TIC) to an external normalisation coefficient of 0.2. The low mass cut-off was 2500 Da m/z. Normalised peaks were used exclusively in data analysis. Expression difference mapping (EDM) was utilised with automatic peak detection using the settings of five times the signal-to-noise (S/N) ratio for the first pass and two times the S/N ratio for the second pass. Peaks were detected between 2500 Da and 30,000 Da and a list of peak clusters created for each experimental sample.

Statistical Analysis

Experimental samples were grouped by donor age and passage time for a total of four groups that would allow four comparisons between (1) young donor early passage versus young donor late passage, (2) old donor early passage versus old donor late passage, (3) young donor early passage versus old donor early passage and (4) young donor late passage versus old donor late passage. An additional comparison of (1) all early versus all late passage samples regardless of donor age and (2) all young versus all old donors regardless of passage time, was made. Early-passage and late-passage T-cell clones derived from both young and old donors were grouped for analysis, for a total of 15 samples in each of the four groups, in triplicate. From the list of peak clusters created, p-values were calculated using CiphergenExpress software (version 3.0.5.013). Peaks were manually scored for quality and peaks with signal:noise ratios (S/N) <5.0 in more than 50% of samples were removed from analysis. Wilcoxon signed-rank tests for the matched pairs were performed and $p < 0.05$ was considered statistically significant.

Identification of Differentially Expressed Protein/Peptide Peaks

Prefractionation of the Lysates by RP-HPLC

To reduce the complexity of the protein constituents within the respective lysates and to mimic the conditions of H50 Ciphergen ProteinChip, reversed-phase (RP)-HPLC fractionation on a Shimadzu Micro HPLC system was performed. 20 μl of each lysate was loaded onto a 150 × 2 mm column filled with 5 μm Reprosil C8 material (Grom, Hailfingen, Germany). A binary gradient was used, with system A consisting of 0.1% (v/v) TFA/water and system B consisting of 20% (v/v) System A/acetonitrile (ACN). At a flow rate of 350 μl/min, the gradient was run for 20 min on System A (removal of the high salt content, especially urea) and then from 20-100 min from 0-80% System B. The eluting analytes were detected at 214 nm and 1 minute fractions were collected (350 μl). Before analysis by MALDI-TOF mass spectrometry and tryptic digestion, samples were dried by lyophilisation.

MALDI-TOF Mass Spectrometry of the HPLC Fractions

Lyophilised HPLC fractions were dissolved in 50 μl HPLC grade water. 0.5 μl of each fraction were mixed with 0.5 μl saturated sinapinic acid matrix (in 50% (v/v) ACN/water with 0.1% (v/v) TFA) and crystallised on a gold MALDI target (Bruker Daltonik, Bremen, Germany). MALDI-TOF analysis was performed on a REFLEX IV™ system (Bruker Daltonik, Bremen, Germany) in the linear mode with external calibration with Protein Calibration Standard I (Bruker Daltonik, Bremen, Germany). 5 μl aliquots of the samples were used for confirmation of protein masses and identification of HPLC fractions containing peaks of interest by SELDI (Ciphergen) analyses.

Tryptic Digest from HPLC Fractions and MALDI PMF Analysis.

The dissolved HPLC fractions (44.5 μl each) were adjusted to a concentration of 50 mM NH_4HCO_3 and a pH value of 7.5. 100 ng of sequencing grade trypsin (Roche Diagnostics, Mannheim, Germany) was added and the proteins digested for 4 hours at 37°C. After digestion, the samples were lyophilised and resolved in 10 μl 1% (v/v) TFA/water. 0.5 μl of each fraction were mixed with 0.5 μl 2,5-DHB matrix (10 mg/ml in 50% (v/v) ACN/water with 0.1% (v/v) TFA) and crystallised on a gold MALDI target (Bruker Daltonik, Bremen, Germany). MALDI-ReTOF analysis was performed on a REFLEX IV™ system (Bruker Daltonik, Bremen, Germany) in the reflector mode with external calibration with Peptide Calibration Standard I (Bruker Daltonik, Bremen, Germany). MALDI peptide mass fingerprints (PMFs) were analysed by Mascot database searching (http://www.matrixscience.com/).

Protein Identification by Nanospray-IonTrap-MS/MS

For Nanospray analysis, the samples were lyophilised and dissolved in 2 μl Nanospray solution (50% (v/v) methanol/water with 0.1% (v/v) formic acid). The dissolved proteins were transferred to a Nanospray needle and analysed in the MS/MS mode on an Esquire 3000 plus™ iontrap mass spectrometer (Bruker Daltonik, Bremen, Germany). Fragment ion spectra were analysed by Mascot database searching (www.matrixscience.com) and by manual interpretation of y- and b-ion series.

Prefractionation of the Lysates by SDS-PAGE and Tryptic Digest

To reduce the complexity of the protein constituents within the lysates, SDS-PAGE was performed. 40 μg lysate was separated on a NuPAGE™ 4-12 % Bis-Tris gradient PAGE gel (Invitrogen, Karlsruhe, Germany) with MES running buffer according to the instructions of the manufacturer. Proteins were stained using the SimplyBlue™ Safe Coomassie system (Invitrogen, Karlsruhe, Germany). The entire region below approximately 25 kDa was excised into five consecutive bands, diced with a scalpel and subjected to in-gel tryptic digestion. All incubation steps were performed on a thermo shaker and all chemicals were from Sigma (Taufkirchen, Germany) if not otherwise stated. Briefly, diced gel pieces were washed two times in 50 μl 50 mM NH4HCO3/30% ACN for 10 min at RT and then shrunk in 30 μl ACN for 5 min at RT. ACN was removed by a pipette and subsequent drying in a SpeedVac for 10 min. Reduction of the proteins was performed with 5 µl 10 mM DTT in 50 mM NH4HCO3/30% ACN at 56°C for 40 min. Alkylation was carried out by replacement of the DTT solution by 5 µl 40 mM iodoacetamide in 50 mM NH4HCO3/30% ACN for 30 min at RT in the dark. The gel pieces were washed twice in 50 µl 50 mM NH4HCO3/30% ACN for 10 min. After removal of the buffer, 100 µl ACN was added and the samples dried for 10 min in a SpeedVac. The digest was performed by the addition of 250 ng sequencing grade trypsin (Roche Diagnostics, Mannheim, Germany) in 50 µl 50 mM NH4HCO3 and incubation at 37°C overnight. After centrifugation (10 min 14,000 rpm) the supernatant was collected. The gel pieces were extracted 2 times with 50 µl 50% (v/v) ACN with 5% (v/v) formic acid and the extraction solutions added to the supernatant. After SpeedVac drying the samples were analysed by MALDI and ESI mass spectrometry.

Protein Identification by Nano-LC-IonTrap-MS/MS

Tryptic peptides were separated on an Agilent 1100 Series HPLC system (Agilent, Waldbronn, Germany) by Nano-RP-HPLC with a 100 mm × 75 µm Grom-SIL 120 ODS-3 CP analytical column (Grom, Hailfingen, Germany). Gradient elution was performed at a flow rate of 200 nl/min. The effluent of the analytical column was directly sprayed into an Esquire 3000 plus™ Iontrap MS (Bruker Daltonik, Bremen, Germany) using an on-line nanoES source equipped with an 8 µm PicoTip emitter (New Objective, Woburn, MA, USA). MS/MS data were analysed using Mascot (www.matrixscience.com) for probability-based peptide identification by database matching. Peptide and thus protein identities were considered significant when a MOWSE score greater than 35 was achieved. Sequential y- and b-ion series were assigned manually.

Results

In order to determine which two arrays to use in the main study investigating protein profiles of T-cell clones approaching senescence, we first investigated several potential lysis conditions, binding buffers and ProteinChip array surfaces. For this preliminary investigation, three T-cell clones were lysed with two different lysis buffers: one tris-based and one urea-based. These six sub-samples were then analysed on the CM10, Q10 and H50 protein arrays. The binding buffers used were selected to promote binding (CM10: 50 mM NaAcetate pH4, Q10: 50 mM Tris-HCl pH8, H50: Ciphergen H50 buffer 10% acetonitrile/0.1% trifluoroacetic acid).

Urea lysis buffer (8M urea, 2% CHAPS) was identified as optimal for use with H50 arrays and Tris buffer (50 mM Tris-HCl, 5 mM EDTA PH6.0, 2 mM PMSF, 1% Triton X-100) was identified as optimal with Q10 arrays due to the increased number of peaks with signal:noise ratios greater than 3.0 (data not shown). Of the three array types investigated, CM10 bound the fewest proteins. It was therefore decided to use both the Q10 and H50 arrays in the investigation of protein bio-markers of T-cell clone senescence.

Having identified the optimal arrays and lysis buffers to use, different Q10 buffers were investigated to establish whether any further optimisation was possible. These were: 50 mM Tris-HCl pH8.0 (as used originally) and 10 mM sodium phosphate pH8.0 and 20 mM HEPES pH8.0. After determining the number of peaks generated using each buffer condition, 50 mM Tris-HCl pH8 was identified as the optimal binding buffer because the greatest numbers of peaks were present under these conditions.

In order to determine which proteins are differentially expressed as T-cells clones age and approach senescence, early and late passage T-cell clones from young and old donors were analyzed on Q10 and H50 ProteinChips using optimised experimental conditions as detailed above. Samples were grouped based on passage time and donor age and expression difference mapping (EDM) was performed using CiphergenExpress software. Peaks were assigned in each sample and those which significantly differed ($p < 0.05$) between (1) early and late passage T-cell clones and (2) between young and old donors, were determined.

In order to determine the inter-chip variability, all protein peaks identified in T-cell clone samples were analysed for variability between replicate SELDI analyses. The samples were prepared in triplicate and analysed on separate ProteinChip surfaces. The resulting protein peaks were ranked

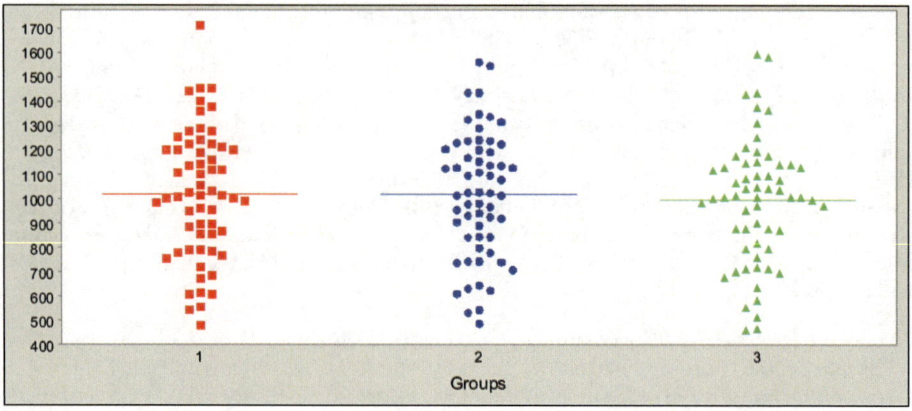

Figure 1. Analysis of variability between triplicates. Protein lysates were prepared and subjected to SELDI-MS analysis in three independent experiments. Mean intensity was determined for each protein/peptide peak. The intensity of one representative protein/peptide peak is shown (on the y-axis) for each of the three independent SELDI-MS analyses (on the x-axis). The mean was determined for each replicate and is indicated by the horizontal bar. Each data point represents protein lysate obtained from one individual T-cell clone.

based on the variability between replicates and the peak with mean variability is shown in Figure 1. Significant variability between replicates was observed in less than 5% of peaks analysed.

Representative spectra were obtained from TCC derived from a young donor at early and late passage (Fig. 2A, 2B respectively) and from an elderly donor at early and late passage (Fig. 2C, 2D respectively). The spectra demonstrate visually apparent differences between young and old donors and at both early and late passage. In order to allow the direct visual comparison of multiple young and old samples, T-cell clone lysates from young and old donors were pooled and overlaid to compare protein expression profiles. Figure 3 depicts the differential protein expression pattern between young and old donors. Several peaks between 16,750 and 18,250 Da differ between young and old donors suggesting there may be differentially expressed proteins associated with aging within this range.

In order to determine which proteins are differentially expressed in young and old donors or early and late passage T-cell clones, samples were grouped based on donor age and passage time and expression difference mapping (EDM) was performed using CiphergenExpress™ software. Peaks were assigned and those which significantly differed ($p < 0.05$) between T-cell clones isolated from young and old donors using H50 and Q10 ProteinChips are shown in Tables 1 and 2, respectively. In addition, protein peaks which were differentially expressed between early and late passage T-cell clones applied to H50 (Table 3) and Q10 (Table 4) ProteinChips are shown ($p < 0.05$). Peak mass and p-value associated with each peak identified are shown.

EDM averages the intensities of each peak for each sample in the group, in order to determine whether the mean intensity differs between two groups. As such, in this approach there is no consideration of how many samples within a group exhibit the trend and it is therefore influenced strongly by outlying samples. In order to directly visualise areas of the proteome that differed between early and late passage T-cell clones, hierarchical clustering was performed on data collected using H50 and Q10 ProteinChips and visualised by heat map, as depicted in Figure 4 and Figure 5, respectively. A heat map is a graphical representation of the entire array in grid form, with columns representing samples and rows representing the m/z of protein/peptide peaks. The intersection of a peak and sample is coloured according to its expression value compared to the mean of the entire sample set. In this analysis red indicates high expression compared to the mean while green indicates low expression compared to the mean intensity value. Black colouring indicates no change from the mean. In addition the intensity of the colour indicates the extent of over- or under-expression compared to the mean.

Table 1. Protein/peptide peaks identified as differentially expressed between young and old donors using H50 ProteinChips

m/z	p-value
7928	0.00001
16042	0.00001
15701	0.00001
15835	0.00001
15273	0.00001
24376	0.00001
17184	0.00001
16899	0.00003
2783	0.0002
16970	0.0002
8364	0.0005
13751	0.002
28326	0.002

Table 2. Protein/peptide peaks identified as differentially expressed between early and late passage TCC using H50 ProteinChips

m/z	p-value
13751	0.001
7763	0.001
4384	0.001
8438	0.001
4263	0.002
4769	0.003
20165	0.004
8364	0.006
13855	0.007
11739	0.01

Table 3. Protein/peptide peaks identified as differentially expressed between young and old donors using Q10 ProteinChips

m/z	p-value
15064	0.00005
14872	0.00005
13850	0.0039
19891	0.0075
4010	0.0136
8540	0.0443

Table 4. Protein/peptide peaks identified as differentially expressed between early and late passage TCC using Q10 ProteinChips

m/z	p-value
13853	0.00010
14875	0.0066
10512	0.0238
19891	0.0294
11948	0.0488

Heat map visualisation of lysates from early and late passage T-cell clones purified on H50 ProteinChips identified a region between 14.5 kDa and 16.5 kDa which was highly conserved between individual samples and significantly differed between the two groups. Furthermore, in general, many proteins/peptides appear to be up-regulated in T-cell clone senescence as compared to early passage cells, as evidenced by a predominance of red-colouring in late passage samples compared to extensive green colouring in early passage samples (Fig. 5).

In order to determine the identities of the differentially expressed peaks in early and late passage T-cell clones obtained using H50 reversed-phase chromatography ProteinChips and SELDI analysis, it was first necessary to enrich protein lysates under the same reversed-phase conditions that were used in the H50 ProteinChip experiments. For this purpose, reversed-phase HPLC was performed to isolate hydrophobic target proteins for use in MALDI mass spectrometric analysis. Because protein masses obtained using SELDI are not as accurate as MALDI-TOF, an initial analysis was made to compare protein mass obtained from SELDI analysis and MALDI. HPLC-purified samples were run in parallel on SELDI and MALDI mass spectrometers, allowing comparison of mass deduced by each technique. The m/z of protein peaks initially identified using

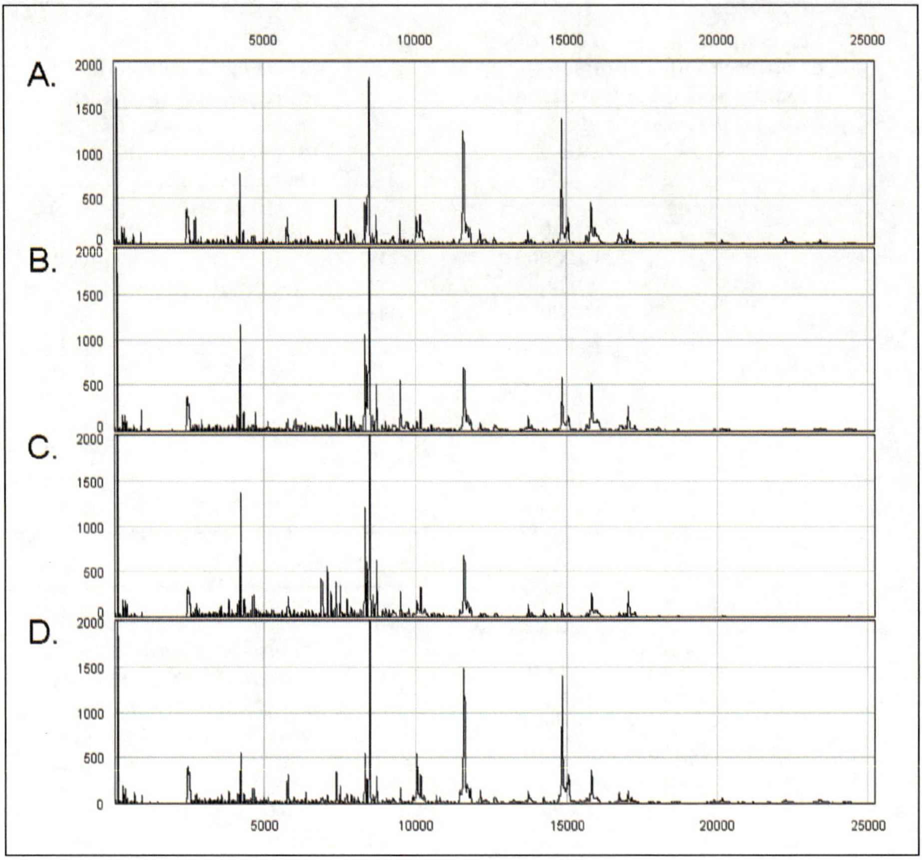

Figure 2. Representative spectra obtained from Ciphergen SELDI analysis. Panels A, B: T-cell clone 341-9 derived from a young donor at early (A) and late passage (B). Panels C, D: T-cell clone 435-7 derived from an elderly donor at early (C) and late (D) passage. Spectra were obtained using SELDI analysis conditions and software as described in Methods.

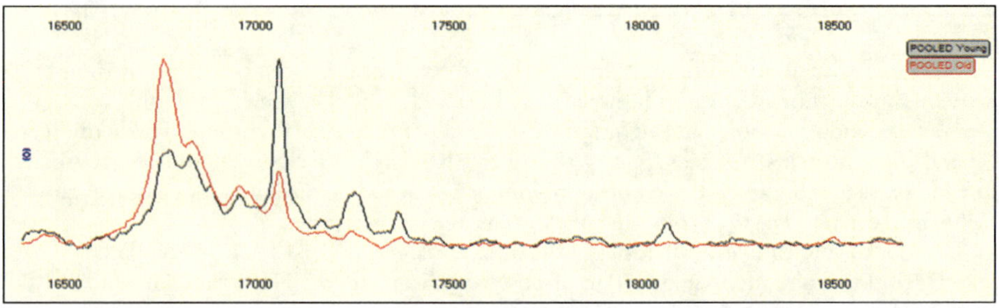

Figure 3. Spectra overlay using Ciphergen software. T-cell clones derived from young (black) and old (red) donors were pooled and spectra were overlaid to compare expression profiles between 16.5 kDa and 20 kDa.

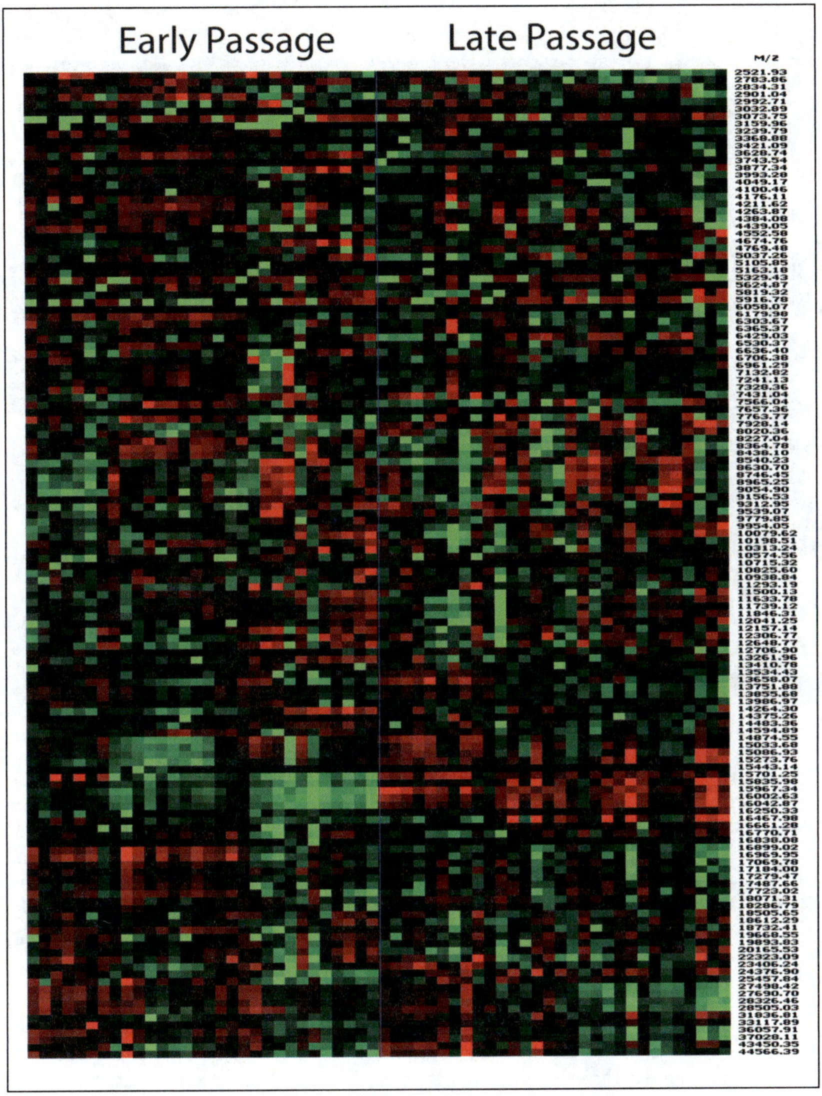

Figure 4. Heat map of early- and late-passage T-cell clones. Following SELDI/MS analysis using H50 ProteinChips, median intensity was determined for each protein/peptide peak. Peaks which showed heightened expression from the median are coloured red. Peaks with reduced expression compared to the median are coloured green. No change from the median expression is indicated by black colouring. Samples are grouped by time in passage (early, left; late, right).

SELDI were confirmed by MALDI analysis. Representative HPLC-purified fractions analysed by SELDI and MALDI are demonstrated in Figure 6 (Fig. 6A and 6B, respectively). MALDI-TOF was subsequently used for protein identification using the corresponding HPLC fractions.

Tables 5 and 6 show the protein identifications obtained from the respective HPLC fractions after tryptic digest and peptide mass fingerprinting by MALDI-TOF-MS/MS (Table 5) and SDS-PAGE prefractionation combined with peptide mass fingerprinting and ESI-Ion Trap

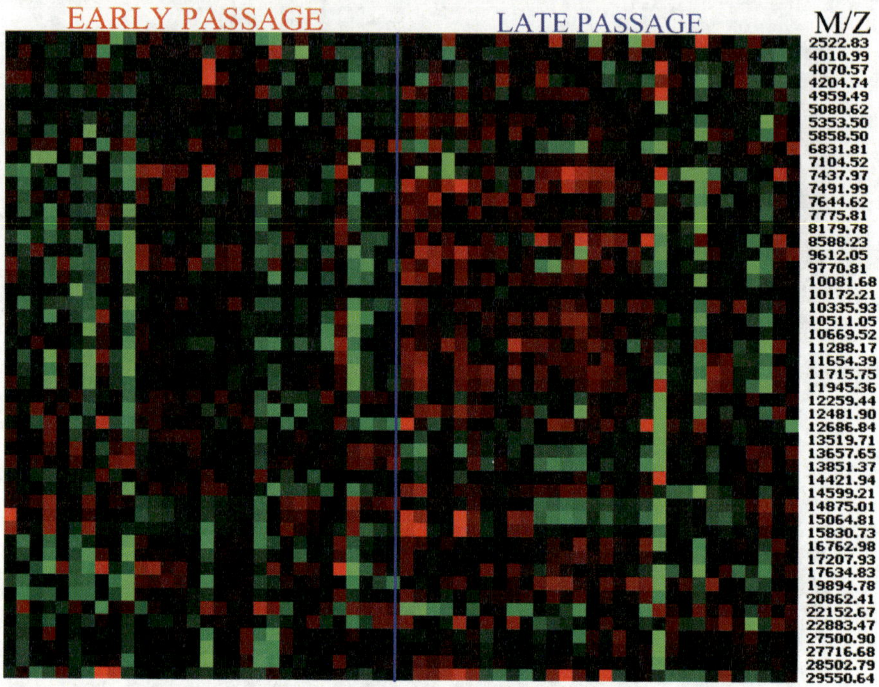

Figure 5: Heat map of early- and late-passage T-cell clones. Following SELDI/MS analysis using Q10 ProteinChips, median intensity was determined for each protein/peptide peak. Peaks which showed heightened expression from the median are coloured red. Peaks with reduced expression compared to the median are coloured green. No change from the median expression is indicated by black colouring. Samples are grouped by time in passage (early, left; late, right).

MS/MS (Table 6). Many of the identified high abundant proteins are bovine serum proteins (not shown) and likely derive from the fetal calf serum (FCS) used during the T-cell clone culturing. Because these proteins were more abundant than the human T-cell proteins, the analysis was negatively affected. A manual selection of particular tryptic peptides—which excluded high abundance bovine peptide—for Nanospray MS/MS identification revealed that these were only very low abundant peptides. Therefore, the obtained MS/MS spectra of the selected peptides did not give unambiguous sequence hits. The most prominent identified human proteins below 20 kDa belonged mainly to different histone proteins.

Discussion

In the current study, we used SELDI mass spectrometry to investigate differential expression of proteins during ageing and immunosenescence using H50 (reversed-phase chromatography) and Q10 (anion-exchange) ProteinChip arrays. With this approach, we first identified differentially expressed proteins by SELDI and confirmed the mass of these peaks and protein identities by HPLC combined with MALDI-TOF-MS/MS. We focussed our efforts in this initial study on investigating proteins altered in T-cell senescence identified by SELDI reversed-phase chromatography (H50) ProteinChip arrays. Advantages of the SELDI technique include (1) ease of analysis of protein or protein conjugates in serum and patient samples, (2) functional surface chemistries

Table 5. Peptide mass fingerprinting (PMF) identifications from reversed-phase HPLC fractions. Proteins from the indicated HPLC fractions (left column) were determined by tryptic digestion, PMF and MALDI-MS/MS followed by MASCOT database search. Protein identifier, name, confidence score and predicted m/z are shown, from left to right

RP-HPLC Fraction (min)	Accession Number	Protein ID	Mowse Score	Nominal Mass
63	S02061	Heterogeneous ribonuclear particle protein A1-beta (human)	82	34131
64	S02061	Heterogeneous ribonuclear particle protein A1-beta (human)	43	34131
65	B34504	Heterogeneous ribonuclear particle protein B1 (human)	110	37407
66	B34504	Heterogeneous ribonuclear particle protein B1 (human)	87	37407
67	1AK4B	Peptidylprolyl isomerase A, chain B (human)	54	17771
67	CAG32988	Cyclophilin a (human)	50	17987
72	H2BQ_HUMAN	Histone H2B.q (human)	56	13781
73	Q5JVTO_ HUMAN	Vimentin (human)	54	49623
73	HSHUA5	Histone H2A.5 (human)	46	14047
74	Q9BTM1_ HUMAN	H2A histone family member J(human)	43	14011
76	Q5JP8_ HUMAN	Serine V threonine kinase 19 (human)	57	40440
82	Q8WV32_ HUMAN	TALDO1 protein fragment (human)	77	37385

allow enrichment of proteins and peptides and (3) rapid scanning and high sample throughput. While MALDI-MS/MS provides higher information output than SELDI-MS, it is designed for a lower throughput of samples to be analysed. Therefore, by combining these two techniques, we maximise sample throughput while allowing heightened information output.

Protein identification has traditionally been performed through the use of two-dimensional (2-D) gels and the establishment of isoelectric point (pI). However, pI is often a poor indicator for identification purposes. Protein sequencing gives the best chance of identification but is time-consuming in the presence of multiple targets to identify. Peptide mass fingerprinting (PMF) is an alternative choice that utilises enzymatic digestion to create a "fingerprint" of smaller peptides that is unique to the starting protein. As long as the digestion is complete, cleavage of the molecule will produce a set of peptides, of varying masses, that are characteristic of that protein. The mass of each peptide determined by MS will be the sum of the amino acids present including any posttranslational modifications. This peptide information can subsequently be entered into a database which "predicts" the starting molecule. We initially utilised this technique in combination with MALDI-TOF-MS/MS to identify proteins which were differentially expressed in T-cell immunosenescence.

Although peptide mass fingerprinting combined with MALDI-TOF-MS/MS analysis identified several human proteins of interest, in these preliminary experiments, the majority were artefacts

Table 6. ESI-IonTrap MS/MS identifications from SDS-PAGE. Tryptic peptides were separated on an Agilent 1100 Series HPLC system by nano-RP-HPLC. Proteins were identified by ESI-IonTrap-MS/MS followed by MASCOT database search. Protein identifier, name, confidence score and predicted m/z are shown, from left to right

Accession Number	Protein Id	Mowse Score	Nominal Mass
ARF4_HUMAN	ADP-ribosylation factor 4 (human)	87	20481
COF1-HUMAN	Cofilin-1 (human)	47	18588
CSHUA	Peptidylprolyl-isomerase A (human)	83	18229
R3UH16	Ribosomal protein S16 (human)	87	16549
PROF1-HUMAN	Profilin-1 (human)	224	15085
H2AQ_HUMAN	Histone H2A.q (human)	63	13849
H2B_HUMAN	Histone HsB (human)	112	13752
HSHUB1	Histone H2B.1 (human)	71	13606
A48219	Calvasculin (human)	35	11721
LPHUA2	Apolipoprotein A-II precursor (human)	67	11168
LPHUC3	Apolipoprotein C-III precursor (human)	48	10846
BCHUY	Calcyclin (human)	36	10173
1C3TA	Ubiquitin mutant YES (human)	120	8560
1TBEB	Tetraubiquitin, chain (human)	140	8176

caused by the presence of larger, bovine serum proteins present in the extract. Several factors resulted in the comigration of bovine serum proteins with proteins of interest thereby hampering attempts to identify the proteins: (1) cells were cultured in the presence of bovine-derived serum and (2) proteins of interest were in low abundance, as evident by relatively low signal:noise ratios. In order to eliminate this problem, tryptic peptides were separated on HPLC by ESI-IonTrap-HPLC. The benefit of ESI-IonTrap is its MS capability, which allows improved determination of the sequence in cases of ambiguous spectral data or insufficient fragmentation ion series. This technique, however, has a lower sample throughput capability. After partial fractionation of proteins, which largely eliminated other proteins comigrating on the SDS-PAGE gel, bands were excised from the gel, subjected to tryptic digestion and the resulting peptide fingerprints allowed protein identification by ESI-IonTrap-MALDI-MS/MS analysis followed by a MASCOT database search, as shown in Table 6. From the combination of PMF-MALDI ESI-IonTrap approaches, we were able to identify eighteen human proteins under 20kDa, many of which may correspond to proteins that are differentially expressed in T-cell ageing and senescence and therefore represent potential bio-markers. The other proteins identified in Tables 5 and 6 had similar biochemical properties to the target proteins of interest (hydrophobicity) but did not represent proteins that were differentially expressed in T-cell clone senescence or aging, as listed in Tables 1-4 and were therefore not of interest.

The proteins which were shown to be differentially expressed in T-cell clone senescence following SELDI analysis and identified by MALDI/ESI-MS/MS were associated with SELDI peaks at 13-14 kDa (Histone H2B.q, H2A.5, H2A.q, H2B and H2B.1) and 8.3-8.5 kDa (Tetraubiquitin, chain B and Ubiquitin mutant YES). The proteins did not give unambiguous MOWSE scores

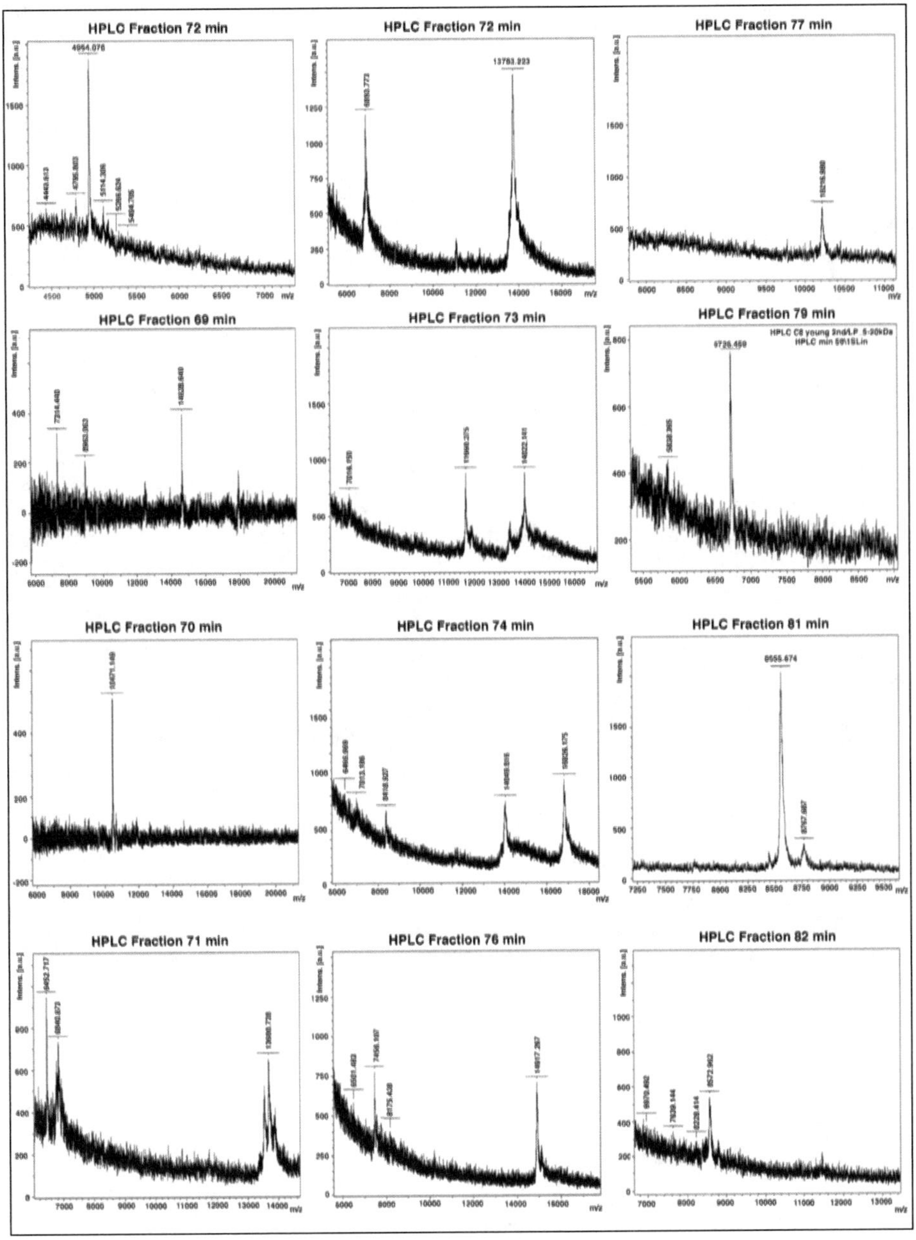

Figure 6A. Comparison of m/z obtained from SELDI-MS and MALDI-MS separated by reversed-phase high-performance liquid chromatography (HPLC). Young and old T-cell clones were pooled and protein was extracted using CHAPS/urea lysis buffer. Samples were subject to reversed-phase HPLC and fractions were isolated at one minute intervals for comparison of mass obtained by (A) MALDI-TOF-MS and (B) SELDI-MS. Representative fractions are shown, including m/z values on the X-axis, intensity on the y-axis and peak labels (in red, B, viewed on next page).

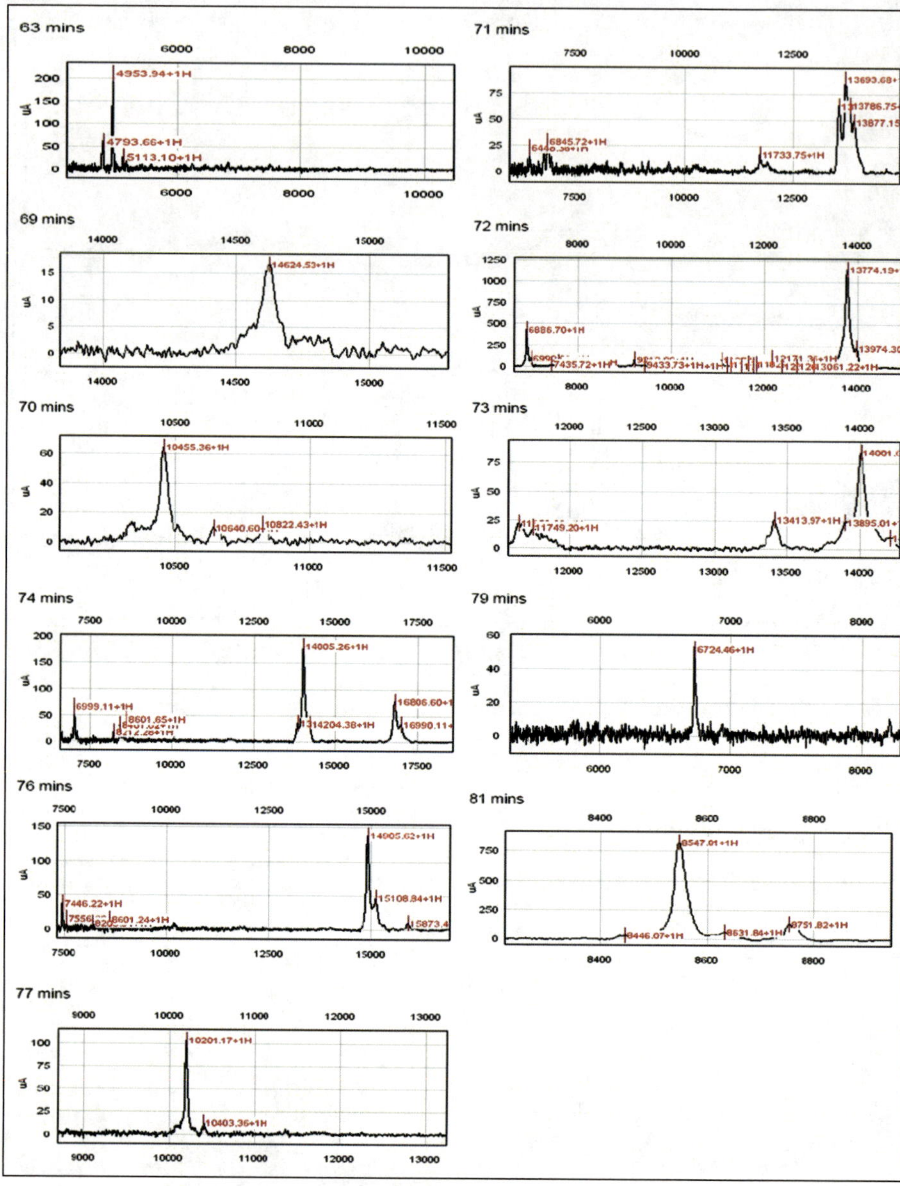

Figure 6B. See previous page for figure legend.

and ranged from 46 (Histone H2A.5) to 140 (Tetraubiquitin, chain B). A MASCOT database MOWSE score of 67 is associated with a p value $p < 0.05$. Therefore, while the expression profiles of each of these proteins needs to be confirmed by Western blot analysis or another protein expression technique, we remain confident that the majority of these proteins (with MASCOT scores >67) are indeed differentially expressed in senescent T-cell clones.

Recent evidence suggests that ubiquitin-mediated processing is important in cellular ageing and senescence. Ubiquitin-mediated proteolysis is critical for the removal of damaged proteins from the cell. Several groups have demonstrated that proteasome function is impaired in aging but is required to attain advanced age in centenarians.[26,27] In addition, alteration of the ubiquitin/proteasome system is often involved in neurodegenerative processes including Alzheimer's disease, Down syndrome, Huntington's disease and multi-system atrophy.[28,29] Taken together, these findings suggest that ubiquitin-mediated proteasome function is important in health and aging. In concordance with these findings, it is therefore unsurprising that expression of multiple ubiquitin-related proteins may be altered in aging and senescent T-cells.

Similarly, multiple studies have demonstrated that histone modifications, including deacetylation and gene silencing, occur as a function of aging and may impact on lifespan. Recently, SIR2p has been shown to posses an enzymatic function that allows it to remove acetyl groups from the N-terminal tails of the core histone H4 and thus may function in the repression of gene transcription.[30] Deletion of histone deacetylases reduces lifespan in yeast, suggesting that histone accessibility to allow gene transcription is critical for longevity.[31] In addition, ratios of H1 histone subfractions are altered in aging wheat and mice,[32,33] further demonstrating that relative expression of the histone variants and subunits may change during the aging process. However, it remains important to fully elucidate the effect of various histone-modifying activities on the lifespan in various organisms.

In this chapter, we have illustrated how combinations of newly-developed technologies can be begin to be applied to answer old questions in immunosenscence: here, SELDI and traditional MALDI-MS/MS are combined to identify proteins differentially expressed in T-cell aging and senescence. In this case, we enriched protein samples by reversed-phase chromatography prior to identification. In the current study, we only identified proteins which differed upon T-cell senescence and while proteins which are altered in aging would be an interesting investigation, identification of the protein/peptide peaks which differ between young and old donors has not been performed to date. We also enriched lysates on Q10 ProteinChips (anion-exchange chromatography) and identified several alternative protein/peptide peaks which differ between young and old donors and between early and late passage T-cell clones. The identity of many of these molecules has not been determined in this study and remains to be investigated further.

Here, we demonstrate the use of an advanced, high-throughput proteomics approach to investigate protein bio-markers of T-cell senescence. This approach has afforded many advantages over previously-used systems, including high-throughput capacity and applicability and translation of information obtained through SELDI analysis to traditional MALDI mass spectrometry. Through the combination of both approaches, we have identified several proteins that appear to be differentially expressed in a model of T-cell immunosenescence. The identity of these proteins remains to be confirmed by protein expression assays, including Western blot analysis. It is possible that several of these proteins may represent novel bio-markers of human ageing and immunosenescence and may help to define processes underlying T-cell dysfunction in ageing. An understanding of the molecular pathways that are affected during immunosenescence would allow potential interventions to be developed which target these functional changes and may ultimately impact on the prevention of the decline in immune health observed in the general ageing population.

Acknowledgements

The authors thank Ms Arnika Rehbein and Karin Hähnel (Tübingen, Germany) for project management and generating the T-cell clones for the current investigation. We also gratefully acknowledge Dr Thomas Flad, PANATecs GmbH, Tübingen, Germany for assistance with protein identification. We thank Birgitte Donaghy for technical expertise. This work was supported by the European Union Framework V RandD project T-Cell Immunity and Ageing (T-CIA) contract number QLK6-CT-2002-02283.

References

1. Pawelec G, Solana R. Immunosenescence. Immunology Today 1997; 18:514-516.
2. Franceschi C, Bonafe M, Valensin S et al. Inflamm-aging. An evolutionary perspective on immunosenescence. Ann N Y Acad Sci 2000; 908:244-254.
3. Pawelec G, Akbar A, Caruso C et al. Human immunosenescence: is it infectious? Immunol Rev 2005; 205:257-268.
4. Harris TB, Ferrucci L, Tracy RP et al. Associations of elevated interleukin-6 and C-reactive protein levels with mortality in the elderly. Am J Med 1999; 106:506-512.

5. Forsey RJ, Thompson JM, Ernedh J et al. Plasma cytokine profiles in elderly humans. Mech Aging Dev 2003; 124:487-493.
6. Bruunsgard H, Ladelund S, Pederson AN et al. Predicting death from tumour necrosis factor-alpha and interleukin-6 in 80-year-old people. Clin Exp Immunol 2003; 132:24-31.
7. Ferrucci L, Harris TB, Guralnik JM et al. Serum IL-6 level and the development of disability in older persons. J Am Geriatr Soc 1999; 47:639-646.
8. Harris TB, Ferrucci L, Tracy RP et al. Associations of elevated interleukin-6 and C-reactive protein levels with mortality in the elderly. Am J Med 1999; 106:506-512.
9. Hull M, Fiebich BL, Berger M et al. Interleukin-6 associated inflammatory process in Alzheimer's disease: new therapeutic options. Neurobiol Aging 1996:795-800.
10. Ershler WB, Keller ET. Age-associated increased interleukin-6 gene expression, late- life diseases and frailty. Ann Rev Med 2000; 51:245-270.
11. Wikby A, Johansson B, Ferguson F et al. Age-related changes in immune parameters in a very old population of Swedish people: A longitudinal study. Expt Geront 1994; 29:531-541.
12. Wikby A, Maxson P, Olsson J et al. Changes in CD8 and CD4 lymphocyte subsets, T-cell proliferation responses and nonsurvival in the very old: the Swedish longitudinal OCTO-immune study. Mech Aging Dev 1998; 102:187-198.
13. Huppert FA, Pinto EM, Morgan K et al. Survival in a population sample is predicted by proportions of lymphocyte subsets. Mech Aging Dev 2003; 124:449-541.
14. Wikby A, Johansson B, Olsson J et al. Expansions of peripheral blood CD8 T-lymphocyte subpopulations and an association with cytomegalovirus seropositivity in the elderly: the Swedish NONA immune study. Exp Gerontol 2002; 37:445-453.
15. Hadrup SR, Strindhall J, Kollgaard T et al. Longitudinal studies of clonally expanded CD8 T-cells reveal a repertoire shrinkage predicting mortality and an increased number of dysfunctional cytomegalovirus-specific T-cells in the very elderly. J Immunol 2006; 176:2645-2653.
16. Olsson J, Wikby A, Johansson B et al. Age-related change in peripheral blood T-lymphocyte subpopulations and cytomegalovirus infection in the very old: the Swedish longitudinal OCTO-immune study. Mech. Aging Dev 2002; 121:187-201.
17. O'Mahony L, Holland J, Jackson J et al. Quantitative intracellular cytokine measurement: age-related changes in proinflammatory cytokine production. Clin Exp Immunol 1998; 113:213-219.
18. Ouyang Q, Wagner WM, Wikby A et al. Compromised IFN-gamma production in the elderly leads to both acute and latent viral antigen stimulation: contribution to the immune risk phenotype? Eur Cytokine Netw 2002; 13:387-393.
19. Guidi L, Antico L, Bartoloni C et al. Changes in the amount and level of phosphorylation of p56lck in PBL from aging humans. Mech Aging Dev 1998; 102:177-186.
20. Pawelec G, Hirokawa K, Fulop T. Altered T-cell signalling in ageing. Mech Ageing Dev 2001; 122:1613-1637.
21. Hutchens TW, Yip TT. New desorption strategies for the mass spectrometric analysis of macromolecules. Rapid Commun Mass Spectrom 2003; 7:576-580
22. Merchant M, Weinberger SR. Recent advancements in surface-enhanced laser desorption/ionisation time-of-flight mass spectrometry. Electrophoresis 2000; 21:1164-1177.
23. Tang N, Tornatore P, Weinberger SR. Current developments in SELDI Affinity Technology. Mass Spectrometry Reviews 2004; 23:34-44.
24. Koopmann J, Zhang Z, White N et al. Serum diagnostics of pancreatic adenocarcinoma using surface-enhanced laser desorption and ionisation mass spectrometry. Clinical Cancer Research 2004; 10:860-868.
25. Pawelec G, Mariani E, Solana R et al. Human T-cell clones in long-term culture as models for the impact of chronic antigenic stress in aging. Handbook of Models for Human Aging. Academic Press, 2006:781-792.
26. Martinez-Vicente M, Sovak G, Cuervo AM. Protein degradation and aging. Exp Gerontol 2005; 40:622-633.
27. Mishto M, Santoro A, Bellavista E et al. Immunoproteasomes and immunosenescence. Aging Res Rev 2003; 2:419-432.
28. Lindsten K, de Vrij FM, Verhoef LG et al. Mutant ubiquitin found in neurodegenerative disorders is a ubiquitin fusion degradiation substrate that blocks proteasomal degradation. J Cell Biol 2002; 157:417-427.
29. van Leeuwen FW, de Kleijn DP, van den Hurk HH et al. Frameshift mutants of B-amyloid precursor protein and ubiquitin-B in Alzhemier's and Down patients. Science 1998; 279:242-247.
30. Imai S, Armstrong CM, Kaeberlein M et al. Transcriptional silencing and longevity protein Sir2 is an NAD-dependent histone deacetylase. Nature 2000; 403:795-800.
31. Kim S, Benguria A, Lai CY et al. Modulation of life-span by histone deacetylase genes in Saccharomyces cerevisiae. Mol Biol Cell 1999; 10:3125-3136.
32. Smirnova TA, Prusov AN, Kolomijtseva GY et al. H1 histone in developing and aging coleoptiles of etiolated wheat seedlings. Biochemistry 2004; 69:1128-1135.
33. Niedzweicki A, Lewis PN, Cinader B. Changes of histone H1 subtypes with aging in strains of mice that possess different immunological characteristics. J Gerontol 1985; 40:695-699.

Index

E

F

G

H

I

J

K

L